STATA SURVIVAL ANALYSIS AND EPIDEMIOLOGICAL TABLES
REFERENCE MANUAL
RELEASE 8

A Stata Press Publication
STATA CORPORATION
College Station, Texas

Stata Press, 4905 Lakeway Drive, College Station, Texas 77845

The suggested citation for this software is

StataCorp. 2003. *Stata Statistical Software: Release 8.0*. College Station, TX: Stata Corporation.

Table of Contents

Cross-Referencing the Documentation

When reading this manual, you will find references to other Stata manuals. For example,

[U] **29 Overview of Stata estimation commands**

[R] **regress**

[P] **matrix define**

The first is a reference to Chapter 29, *Overview of Stata estimation commands* in the *Stata User's Guide*, the second is a reference to the `regress` entry in the *Base Reference Manual*, and the third is a reference to the `matrix define` entry in the *Programming Reference Manual*.

All of the manuals in the Stata Documentation have a shorthand notation, such as [U] for the *User's Guide* and [R] for the *Base Reference Manual*.

The complete list of shorthand notations and manuals is as follows:

[GSM]	*Getting Started with Stata for Macintosh*
[GSU]	*Getting Started with Stata for Unix*
[GSW]	*Getting Started with Stata for Windows*
[U]	*Stata User's Guide*
[R]	*Stata Base Reference Manual*
[G]	*Stata Graphics Reference Manual*
[P]	*Stata Programming Reference Manual*
[CL]	*Stata Cluster Analysis Reference Manual*
[XT]	*Stata Cross-Sectional Time-Series Reference Manual*
[SVY]	*Stata Survey Data Reference Manual*
[ST]	*Stata Survival Analysis & Epidemiological Tables Reference Manual*
[TS]	*Stata Time-Series Reference Manual*

Detailed information about each of these manuals may be found online at

http://www.stata-press.com/manuals/

Title

---

intro — Introduction to survival analysis manual

Description

This entry describes the *Stata Survival Analysis & Epidemiological Tables Reference Manual.*

Remarks

This manual documents the survival analysis and epidemiological tables commands, and is referred to as [ST] in cross-references. Following this entry, [ST] **survival analysis** provides an overview of the commands.

This manual is arranged alphabetically. If you are new to Stata's survival analysis and epidemiological table commands, we recommend that you read the following sections first:

[ST] **survival analysis**	Introduction to survival analysis & epidemiological tables commands
[ST] **st**	Survival-time data
[ST] **stset**	Set variables for survival data

Stata is continually being updated. Stata users are always writing new commands, as well. To find out about the latest survival analysis features, type `search survival` after installing the latest official updates; see [R] **update**. To find out about the latest epidemiological features, type `search epi`.

What's new

This section is intended for previous Stata users. If you are new to Stata, you may as well skip it.

1. Existing command `stcox` has an important new feature and some minor improvements:

   a. `stcox` will now fit models with gamma-distributed frailty. In this model, frailty is assumed to be shared across groups of observations. Previously, if one wanted to use analyze multivariate survival data using the Cox model, one would fit a standard model and account for the correlation within groups by adjusting the standard errors for clustering. Now one may directly model the correlation by assuming a latent gamma-distributed random effect or frailty; observations within group are correlated because they share the same frailty. Estimation is via penalized likelihood. An estimate of the frailty variance is available and group-level frailty estimates can be retrieved.

   b. `fracpoly`, `mfp`, `sw`, and `linktest` now work after `stcox`.

   See [ST] **stcox**.

2. Existing command `streg` has an important new feature and some minor improvements:

   a. `streg`'s has new option `shared(varname)` for fitting parametric shared frailty models, analogous to random-effects models for panel data. `streg` could, and still can, fit frailty models where the frailties are assumed to be randomly distributed at the observation level.

   b. `fracpoly`, `mfp`, `sw`, and `linktest` now work after `streg`.

   c. `streg` has four other new options: `noconstant`, `offset()`, `noheader`, and `nolrtest`.

   See [ST] **streg**.

3. `predict` after `streg, frailty()` has two new options:

   a. `alpha1` generates predictions conditional on a frailty equal to 1.

   b. `unconditional` generates predictions that are "averaged" over the frailty distribution.

   These new options may also be used with `stcurve`. See [ST] **streg**.

4. `sts graph` and `stcurve` (after `stcox`) can now plot estimated hazard functions, which are calculated as weighted kernel smooths of the estimated hazard contributions; see [ST] **sts graph**.

5. `streg, dist(gamma)` is now faster and more accurate. In addition, you can now predict mean time after gamma; see [ST] **streg**.

6. Old commands `ereg`, `ereghet`, `llogistic`, `llogistichet`, `gamma`, `gammahet`, `weibull`, `weibullhet`, `lnormal`, `lnormalhet`, `gompertz`, `gompertzhet` are deprecated (they continue to work) in favor of `streg`. Old command `cox` is now removed (it continues to work) in favor of `stcox`.

For a complete list of all the new features in Stata 8, see [U] **1.3 What's new**.

References

Cleves, M. A., W. W. Gould, and R. G. Gutierrez. 2002. *An Introduction to Survival Analysis Using Stata*. College Station, TX: Stata Press.

Also See

Complementary: [U] **1.3 What's new**

Background: [R] **intro**

Title

> **survival analysis** — Introduction to survival analysis & epidemiological tables commands

Description

Stata's survival analysis routines are used to declare, convert, manipulate, summarize, and analyze survival data. Survival data is time-to-event data, and the field of survival analysis is full of jargon: truncation, censoring, hazard rates, etc. For a good Stata-specific introduction to survival analysis, see Cleves, Gould, & Gutierrez (2002).

Stata also has several commands for the analysis of contingency tables resulting from various forms of observational studies, such as cohort or matched case–control studies.

This manual contains documentation on the following commands, which are detailed in their respective manual entries.

Declaring and converting count data

ctset	[ST] **ctset**	Declare data to be count-time data
cttost	[ST] **cttost**	Convert count-time data to survival-time data

Converting snapshot data

snapspan	[ST] **snapspan**	Convert snapshot data to survival-time data

Declaring and summarizing survival-time data

stset	[ST] **stset**	Declare data to be survival-time data
stdes	[ST] **stdes**	Describing survival-time data
stsum	[ST] **stsum**	Summarizing survival-time data

Survival-time data manipulation

stvary	[ST] **stvary**	Report which variables vary over time
stfill	[ST] **stfill**	Fill in by carrying forward values of covariates
stgen	[ST] **stgen**	Generate variables reflecting entire histories
stsplit	[ST] **stsplit**	Split time-span records
stjoin	[ST] **stsplit**	Join time-span records
stbase	[ST] **stbase**	Form baseline dataset

Summary statistics, confidence intervals, tables, etc.

sts	[ST] **sts**	Generate, graph, list, and test the survivor and cumulative hazard functions
stir	[ST] **stir**	Report incidence-rate comparison
stci	[ST] **stci**	Confidence intervals for means and percentiles of survival time
strate	[ST] **strate**	Tabulate failure rate
stptime	[ST] **stptime**	Calculate person-time
stmh	[ST] **strate**	Calculate rate ratios using Mantel–Haenszel method
stmc	[ST] **strate**	Calculate rate ratios using Mantel–Cox method
ltable	[ST] **ltable**	Display and graph life tables

Regression models

stcox	[ST] **stcox**	Fit Cox proportional hazards model
stphtest	[ST] **stcox**	Test of Cox proportional hazards assumption
stphplot	[ST] **stphplot**	Graphical assessment of the Cox proportional hazards assumption
stcoxkm	[ST] **stphplot**	Graphical assessment of the Cox proportional hazards assumption
streg	[ST] **streg**	Fit parametric survival models
stcurve	[ST] **streg**	Plot fitted survival, cumulative hazard, and hazard functions

Converting survival-time data

sttocc	[ST] **sttocc**	Convert survival-time data to case–control data
sttoct	[ST] **sttoct**	Convert survival-time data to count data

Programmer's utilities

st_*	[ST] **st_is**	Survival analysis subroutines for programmers

Epidemiological tables

ir	[ST] **epitab**	Incidence rates for cohort studies
iri	[ST] **epitab**	Immediate form of ir
cs	[ST] **epitab**	Risk differences/ratios, odds ratios for cohort studies
csi	[ST] **epitab**	Immediate form of cs
cc	[ST] **epitab**	Odds ratios for case–control data
cci	[ST] **epitab**	Immediate form of cc
tabodds	[ST] **epitab**	Tests of log odds for case–control data
mhodds	[ST] **epitab**	Odds ratios controlled for confounding
mcc	[ST] **epitab**	Analysis of matched case–control data
mcci	[ST] **epitab**	Immediate form of mcc

Remarks

Remarks are presented under the headings

Introduction
Declaring and converting count data
Converting snapshot data
Declaring and summarizing survival-time data
Survival-time data manipulation
Summary statistics, confidence intervals, tables, etc.
Regression models
Converting survival-time data
Programmer's utilities
Epidemiological tables

Introduction

All but one of the entries in this manual deal with the analysis of survival data—data used to measure the time to an event of interest such as death or failure. Survival data can be organized in one of two ways. The first way is as *count data*. Count data refers to observations on populations, whether people or generators, with observations recording the number of units at a given time that failed or were lost due to censoring. The second way is as *survival-time*, or *time-span* data. In survival-time data, the observations represent time periods and contain three variables that record the start time of the period, the end time, and an indicator on whether failure or right-censoring occurred at the end of the period. It is the representation of the response as these three variables that makes survival data unique in terms of implementing the statistical methods in the software.

There is also a third way in which survival data may be organized, called *snapshot data*, but this is really just a small variation of the survival-time format. With snapshot data, observations depict an instance in time, rather than a time interval. When you have snapshot data, you simply use the `snapspan` command to convert it to survival-time data before proceeding.

Commands that begin with `ct` are for use with count data, and in fact, the only `ct` commands that exist in Stata are those which convert count data to survival-time data. Survival-time data is analyzed using Stata commands that begin with `st`, known in our terminology as `st` commands. You can express all the information contained in count data in an equivalent survival-time dataset, but the converse is not true. Thus, Stata commands are made to work with survival-time data, since it is the more general representation.

The one remaining entry is [ST] **epitab**, describing epidemiological tables. [ST] **epitab** covers many commands dealing with analyzing contingency tables arising from various observational studies, such as case–control or cohort studies. [ST] **epitab** is included in this manual because the concepts presented there are very much related to concepts of survival analysis, and both topics use the same terminology and are of equal interest to many researchers.

Declaring and converting count data

Count data must first be converted to survival-time data before Stata's `st` commands may be used. Count data can be thought of as aggregated survival-time data. Rather than having observations that are specific to a subject and to a time period, you have data that, at each recorded time, records the number lost due to failure, and optionally the number lost due to right-censoring.

`ctset` is used to tell Stata the names of the variables in your count data that record time, the number failed, and the number censored. You `ctset` your data for no reason other than to afterwards type `cttost` to convert it to survival-time data. Because you `ctset` your data, you can type `cttost` without any arguments in order to perform the conversion. Stata remembers how the data are `ctset`.

Converting snapshot data

Snapshot data are data such that each observation records the status of a given subject at a certain point in time. In most cases, you have multiple observations on each subject that chart the subject's progress through the study.

In order to use Stata's survival analysis commands with snapshot data, the data must first be converted to survival-time data, that is, the observations in the data should represent time-intervals. When you convert snapshot data, the existing time variable in your data is used to record the *end* of a time-span, and a new variable is created to record the beginning. Time-spans are created using the recorded snapshot times as break points where new intervals are to be created. The key issue with converting snapshot data to time-span data is the distinction between *enduring variables* and *instantaneous variables.* Enduring variables record characteristics of the subject that endure throughout the time-span, such as sex or smoking status. Instantaneous variables arc uscd to describe events that occur at the end of a time-span, such as failure or censoring. When converting snapshots to intervals, enduring variables obtain their values from the previous recorded snapshot, or are set to missing for the first interval. Instantaneous variables obtain their values from the current recorded snapshot, since the existing time variable is now understood to record the end of the span.

Stata's `snapspan` makes this whole process very easy. You specify an id variable identifying your subjects, the snapshot time variable, the name of the new variable to hold the begin times of the spans, and any variables you want treated as instantaneous variables. Stata does the rest for you.

Declaring and summarizing survival-time data

You declare your survival-time data to Stata using `stset`. `stset` is at the heart of the entire `st` suite of commands in Stata, as every other `st` command relies on the information provided when you `stset` your data. Survival-time data comes in all different kinds of forms. For example, your time variables may be dates, time measured from a fixed date, or time measured from some other point unique to each subject, such as enrollment in the study. What is the onset of risk for the subjects in your data? Is it time zero? Is it enrollment in the study or some other event such as a heart transplant? Do you have censoring, and if so, which variable records it? What values does this variable record for censoring/failure? Do you have delayed entry? That is, were some subjects at risk of failure before you actually observed them? Do you have simple data and wish to just treat everyone as entering and at risk at time zero?

Whatever the form of your data, you must first `stset` it before analyzing it, and so if you are new to Stata's `st` commands, we highly recommend you invest some time in learning about `stset`. It is really quite easy once you get the hang of it, and [ST] **stset** has lots of examples to help. For additional discussion of `stset`, see Chapter 6 of Cleves, Gould, & Gutierrez (2002).

Once you `stset`, you can use `stdes` to describe the aspects of your survival data. For example, you'll see how many subjects you were successful in declaring, the total number of records associated with these subjects, the total time at risk for these subjects, whether any of these subjects had any time gaps, whether you had any delayed entry, etc. You can use `stsum` to summarize your survival data, for example, to obtain the total time at risk and the quartiles of time-to-failure in analysis time units.

Survival-time data manipulation

Once your data have been `stset`, you may want to clean it up at bit before beginning your analysis. For example, suppose you had an enduring variable, and when you used `snapspan`, it recorded this variable as missing for the time interval leading up to the first recorded snapshot time. You can use `stfill` to fill in missing values of covariates, either by carrying forward the values from previous periods or by making the covariate equal to its earliest recorded (nonmissing) value for all time spans. You can use `stvary` to check for time-varying covariates or to confirm that certain variables, such as sex, are not time-varying. You can use `stgen` to generate new covariates based on functions of the time-spans for each given subject. For example, you can create a new variable called `eversmoked` that will equal one for all a subject's observations, if the variable `smoke` in your data is equal to one for *any* of the subject's time-spans. Think of `stgen` as just a convenient way to do things that could be done using by *subject_id*: with survival-time data.

stsplit is useful for creating multiple-record-per-subject data out of single-record-per-subject data. Suppose you have already stset your data and wish to introduce a time-varying covariate. You would first need to stsplit your data so that separate time-spans could be created for each subject, allowing the new covariate to assume different values over time within a subject. stjoin is the opposite of stsplit. Suppose you have multiple-record-per-subject data, but then realize that the data could be collapsed into single-subject records without any loss of information. Using stjoin would speed up any subsequent analysis using the st commands, with no change in the results.

stbase can be used in two ways. First, it can be used to set every variable in your multiple-record st data to the value at baseline, defined as the earliest time at which each subject was observed. Second, it can be used to convert st data to cross-sectional data.

Summary statistics, confidence intervals, tables, etc.

These commands are for nonparametric analysis of survival data and can produce a wide array of summary statistics, inference, tables, and graphs. sts is a truly powerful command. sts is used to obtain nonparametric estimates, inference, tests, and graphs of the survivor function, cumulative hazard function, and the hazard function. You can compare estimates across groups such as smoking vs. non-smoking, and you can adjust these estimates for the effects of other covariates in your data. sts can present these estimates as tables and graphically. sts can also be used to test the equality of survivor functions across groups.

stir is used to estimate incidence rates and to compare incidence rates across groups. stci is the survival-time data analog of ci, and is used to obtain confidence intervals for means and percentiles of time-to-failure. strate is used to tabulate failure rates. stptime is used to calculate person-time and standardized mortality/morbidity ratios (SMR). stmh calculates rate ratios using the Mantel—Haenszel method, and stmc calculates rate ratios using the Mantel—Cox method.

Regression models

Stata has routines for fitting both semiparametric and parametric regression models to survival data. stcox fits the Cox proportional hazards model, and can also be used to retrieve estimates of the baseline survival function, the baseline cumulative hazard function, and the baseline hazard contributions. stcox can also calculate a myriad of Cox regression diagnostic quantities, such as martingale residuals, efficient score residuals, and Schoenfeld residuals. stcox has four options for the treatment of tied failures. stcox can be used to fit stratified Cox models, where the baseline hazard is allowed to differ over the strata. stcox can also be used to model multivariate survival data using a *shared frailty* model, which can be thought of as a Cox model with random effects. After stcox, you can use stphtest to test the proportional hazards assumption, and stphplot and stcoxkm to graphically assess this assumption.

Stata offers six parametric regression models for survival data: exponential, Weibull, log-normal, log-logistic, Gompertz, and gamma. All six models are fitted using streg, and you specify which model you want with the distribution() option. All of these models, except for the exponential, have ancillary parameters that are estimated (along with the linear predictor) from the data. By default, these ancillary parameters are treating as constant, but you may optionally choose to model the ancillary parameters as functions of a linear predictor. Stratified models may also be fitted using streg. You can also fit frailty models with streg, and specify whether you want the frailties to be treated as spell-specific or shared across groups of observations.

stcurve is for use after stcox and streg, and will plot the estimated survivor, cumulative hazard, and hazard function for the fitted model. Covariates, by default, are held fixed at their mean values, but you can specify other values if you wish. `stcurve` is useful for comparing these functions across different levels of covariates.

Converting survival-time data

These are rarely used commands, since most of the analyses are performed using data in the survival-time format. `sttocc` is useful for converting survival data to case–control data suitable for estimation via `clogit`. `sttoct` is the opposite of `cttost`, and will convert survival-time data to count data.

Programmer's utilities

These are routines for programmers interested in writing their own `st` commands. They are basically utilities for setting, accessing, and verifying the information saved by `stset`. For example, `st_is` will verify that the data have in fact been `stset`, and will give the appropriate error if the data have not. `st_show` is used to preface the output of a program with key information on the `st` variables used in the analysis. Programmer's interested in writing `st` code should see [ST] **st_is**.

Epidemiological tables

See the *Description* section of [ST] **epitab** for an overview of all these commands.

References

Cleves, M. A., W. W. Gould, and R. G. Gutierrez. 2002. *An Introduction to Survival Analysis Using Stata.* College Station, TX: Stata Press.

Also See

Complementary:	[ST] **ct**, [ST] **ctset**, [ST] **cttost**, [ST] **discrete**, [ST] **epitab**, [ST] **ltable**,
	[ST] **snapspan**, [ST] **st**, [ST] **st_is**, [ST] **stbase**, [ST] **stci**, [ST] **stcox**, [ST] **stdes**,
	[ST] **stfill**, [ST] **stgen**, [ST] **stir**, [ST] **stphplot**, [ST] **stptime**, [ST] **strate**,
	[ST] **streg**, [ST] **sts**, [ST] **sts generate**, [ST] **sts graph**, [ST] **sts list**, [ST] **sts test**,
	[ST] **stset**, [ST] **stsplit**, [ST] **stsum**, [ST] **sttocc**, [ST] **sttoct**, [ST] **stvary**
Background:	[ST] **intro**

Title

> **ct** — Count-time data

Description

The term ct refers to count-time data and the commands—all of which begin with the letters ct—for analyzing them. If you have data on populations, whether people or generators, with observations recording the number of units under test at time t (subjects alive) and the number of subjects that failed or were lost due to censoring, you have what we call count-time data.

If, on the other hand, you have data on individual subjects with observations recording that this subject came under observation at time t_0 and then, later, at t_1, a failure or censoring was observed, you have what we call survival-time data. If you have survival-time data, you are in the wrong place; see [ST] **st**.

Do not confuse count-time data with counting-process data which can be analyzed using the st commands; see [ST] **st**.

There are two ct commands:

ctset	[ST] **ctset**	Declare data to be count-time data
cttost	[ST] **cttost**	Convert count-time data to survival-time data

The key is the cttost command. Once you have converted your count-time data to survival-time data, you can use the st commands to analyze the data. The entire process is

1. ctset your data so that Stata knows the data are count-time data; see [ST] **ctset**.

2. Type cttost to convert your data to survival-time data. You can see [ST] **cttost** but it is not necessary; you just have to type cttost.

3. Use the st commands; see [ST] **st**.

Also See

Complementary: [ST] **ctset**, [ST] **cttost**, [ST] **st**

Background: [ST] **survival analysis**

Title

ctset — Declare data to be count-time data

Syntax

ctset *timevar* *nfailvar* [*ncensvar* [*nentvar*]] [, by(*varlist*) no̲show]

ctset, { s̲how | no̲show }

ctset, clear

{ ctset | ct }

Description

ct refers to count-time data and is described in [ST] **ct** and below. Do not confuse count-time data with counting-process data, which can be analyzed using the st commands; see [ST] **st**.

In the first syntax, ctset declares the data in memory to be ct data, informing Stata of the key variables. When you ctset your data, ctset also runs various checks to ensure that what you have declared makes sense.

In the second syntax, ctset changes the value of show/noshow. In show mode—the default—the other ct commands display the identities of the key ct variables before their normal output. If you ctset, noshow, they will not do this. If you do that and then wish to restore their default behavior, type ctset, show.

In the third syntax, ctset, clear causes Stata to forget the ct markers; it makes the data no longer ct data to Stata. The dataset itself remains unchanged. It is not necessary to ctset, clear before doing another ctset. ctset, clear is used mostly by programmers.

In the fourth syntax, ctset—which can be abbreviated ct in this case—displays the identities of the key ct variables *and* it reruns the checks on your data. Thus, ct is useful to remind you of what you have ctset (especially if you have ctset, noshow) and to reverify your data if you make changes to the data.

In the above syntax diagrams, *timevar* refers to the time of failure, censoring, or entry. It should contain times ≥ 0.

nfailvar records the number failing at time *timevar*.

The optional *ncensvar* records the number censored at time *timevar*.

The optional *nentvar* records the number entering at time *timevar*.

Stata sequences events at the same time as

at *timevar*	*nfailvar* failures occurred,
then at *timevar* $+ 0$	*ncensvar* censorings occurred,
finally at *timevar* $+ 0 + 0$	*nentvar* subjects entered the data.

Options

by(*varlist*) indicates that counts are provided by group. For instance, consider data containing records such as

```
    t       fail      cens      sex     agecat
    5        10         2        0          1
    5         6         1        1          1
    5        12         0        0          2
```

These data indicate that, in the category sex = 0 and agecat = 1, 10 failed and 2 were censored at time 5; for sex = 1, 1 was censored and 6 failed; and so on.

The above data would be declared

```
. ctset t fail cens, by(sex agecat)
```

The order of the records is not important.

That there be a record at every time for every group is not important.

That there be only a single record for a time and group is not important.

All that is important is that the data contain the full table of events.

noshow and show specify whether the identities of the key ct variables be displayed at the start of every ct command. Some users find the report reassuring; others find it repetitive. In any case, you can set and unset show, and you can always type ct to see the summary.

clear makes Stata forget that this is ct data.

Remarks

About all you can do with ct data in Stata is convert it to survival-time (st) data, which is enough. All survival-analysis commands can be run on st data. To analyze count-time data with Stata,

```
. ctset ...
. cttost
. ( now use any of the st commands )
```

Example 1: Simple ct data

Generators are run until they fail. Here are your data:

```
. use http://www.stata-press.com/data/r8/ctset1
. list, sep(0)
```

```
        failtime    fail

  1.          22       1
  2.          30       1
  3.          40       2
  4.          52       1
  5.          54       4
  6.          55       2
  7.          85       7
  8.          97       1
  9.         100       3
 10.         122       2
 11.         140       1
```

For instance, at time 54, 4 generators failed. The `ctset` for these data is

```
. ctset failtime fail
    dataset name:  http://www.stata-press.com/data/r8/ctset1.dta
           time:  failtime
       no. fail:  fail
       no. lost:  --                    (meaning 0 lost)
      no. enter:  --                    (meaning all enter at time 0)
```

It is not important that there be only one observation per failure time. For instance, according to our data, at time 85 there were 7 failures. We could remove that observation and substitute two in its place—one stating that at time 85 there were 5 failures and another that at time 85 there were 2 more failures. `ctset` would interpret that data just as it did the previous data.

In more realistic examples, the generators might differ from one another. For instance, the following data shows the number failing with old-style (`bearings` = 0) and new-style (`bearings` = 1) bearings:

```
. use http://www.stata-press.com/data/r8/ctset2
. list, sepby(bearings)
```

	bearings	failtime	fail
1.	0	22	1
2.	0	40	2
3.	0	54	1
4.	0	84	2
5.	0	97	2
6.	0	100	1
7.	1	30	1
8.	1	52	1
9.	1	55	1
10.	1	100	3
11.	1	122	2
12.	1	140	1

That the data are sorted on bearings is not important. The `ctset` command for these data is

```
. ctset failtime fail, by(bearings)
    dataset name:  http://www.stata-press.com/data/r8/ctset2.dta
           time:  failtime
       no. fail:  fail
       no. lost:  --                    (meaning 0 lost)
      no. enter:  --                    (meaning all enter at time 0)
             by:  bearings
```

Example 2: ct data with censoring

In real data, not all units fail in the time alloted. Say the generator experiment was stopped after 150 days. The data might be

```
. use http://www.stata-press.com/data/r8/ctset3
. list
```

	bearings	failtime	fail	censored
1.	0	22	1	0
2.	0	40	2	0
3.	0	54	1	0
4.	0	84	2	0
5.	1	97	2	0
6.	0	100	1	0
7.	0	150	0	2
8.	1	30	1	0
9.	1	52	1	0
10.	1	55	1	0
11.	1	122	2	0
12.	1	140	1	0
13.	1	150	0	3

The `ctset` for this data is

```
. ctset failtime fail censored, by(bearings)
    dataset name:  http://www.stata-press.com/data/r8/ctset3.dta
            time:  failtime
        no. fail:  fail
        no. lost:  censored
       no. enter:  --                    (meaning all enter at time 0)
              by:  bearings
```

In some other data, observations might also be censored along the way; that is, the value of censored would not be 0 before time 150. For instance, a record might read

bearings	failtime	fail	censored
0	84	2	1

The would mean that at time 84, 2 failed and 1 was lost due to censoring. The failure and censoring occurred at the same time and, when you analyze these data, Stata will assume that the censored observation could have failed; that is, that the censoring occurred after the two failures.

Example 3: ct data with delayed entry

Data on survival time of patients with a particular kind of cancer are collected. Time is measured as time since diagnosis. After data collection started, the sample was enriched with some patients from hospital records who had been previously diagnosed. Some of the data are

(Continued on next page)

time	die	cens	ent	other variables
0	0	0	50	
1	0	0	5	...
:				
30	0	0	3	...
31	0	1	2	...
32	1	0	1	...
:				
100	1	1	0	...
:				

50 patients entered at time 0 (time of diagnosis); 5 patients entered 1 day after diagnosis; 3, 2, and 1 entered 30, 31, and 32 days after diagnosis; and, on the 32nd day, one of the previously entering patients died.

If the other variables are named sex and agecat, the ctset command for this data is

```
. ctset time die cens ent, by(sex agecat)
        time:  time
    no. fail:  die
    no. lost:  cens
   no. enter:  ent
          by:  sex agecat
```

The count-time format is an inferior way to record data like these—data in which every subject does not enter at time 0—because some information is already lost. When did the patient who died on the 32nd day enter? There is no way of telling.

For traditional survival-analysis calculations, it does not matter. More modern methods of estimating standard errors, however, seek to account for which patient is which, and these data will not support the use of such methods.

This issue concerns the robust estimates of variance and the robust options on some of the st analysis commands. After converting the data, you must not use the robust option even if an st command allows it because the identities of the subjects—tying together when a subject starts and ceases to be at risk—are assigned randomly by cttost when you convert your ct to st data. When did the patient who died on the 32nd day enter? For conventional calculations, it does not matter, and cttost chooses a time randomly from the available entry times.

Data errors flagged by ctset

ctset requires only two things of your data: that the counts all be positive or zero and, if you specify an entry variable, that the entering and exiting (failure + censored) balance.

If all subjects enter at time 0, we recommend you do not specify a number-that-enter variable. ctset can determine for itself the number who enter at time 0 by summing the failures and censorings.

Also See

Complementary:	[ST] **cttost**, [ST] **st**
Background:	[ST] **ct**, [ST] **survival analysis**

Title

> **cttost** — Convert count-time data to survival-time data

Syntax

cttost [, t0(*t0var*) w(*wvar*) clear no̲preserve]

cttost is for use with count-time data; see [ST] **ct**. You must have ctset your data before using this command.

Description

cttost converts count-time data to their survival-time format. This is the way count-time data are analyzed with Stata. Do not confuse count-time data with counting-process data, which can be analyzed with the st commands; see [ST] **ctset** for a definition and examples of count data.

Options

t0(*t0var*) specifies the name of the new variable to create that records entry time should such a variable be necessary. (For most ct data, no entry-time variable is necessary because everyone enters at time 0.)

Even if an entry-time variable is necessary, you need not specify this option. cttost will, by default, choose t0, time0, or etime according to which name does not already exist in the data.

w(*wvar*) specifies the name of the new variable to be created that records the frequency weights for the new pseudo-observations. Count-time data are actually converted to frequency-weighted st data and a variable is needed to record the weights. This sounds more complicated than it is. Understand that cttost needs a new variable name, which will become a permanent part of the st data.

If you do not specify w(), cttost will, by default, choose w, pop, weight, or wgt according to which name does not already exist in the data.

clear specifies that it is okay to proceed with the transformation even through the current dataset has not been saved on disk.

nopreserve speeds the transformation by not saving the original data, from which it can be restored should things go wrong or should the user press *Break*. nopreserve is intended for use by programmers using cttost as a subroutine. Programmers can specify this option if they have already preserved the original data. nopreserve changes nothing about the transformation that is made.

Remarks

There is nothing to converting ct to st data. We have some count-time data,

15

```
. webuse cttost

. ct
    dataset name:  http://www.stata-press.com/data/r8/cttost.dta
          time:  time
       no. fail:  ndead
       no. lost:  ncens
      no. enter:  --                         (meaning all enter at time 0)
            by:  agecat treat

. list in 1/5
```

	agecat	treat	time	ndead	ncens
1.	2	1	464	4	0
2.	3	0	268	3	1
3.	2	0	638	2	0
4.	1	0	803	1	4
5.	1	0	431	2	0

and, to convert it, we type `cttost`:

```
. cttost
(data are now st)
      failure event:  ndead != 0 & ndead < .
  obs. time interval:  (0, time]
  exit on or before:  failure
             weight:  [fweight=w]
```

```
       33  total obs.
        0  exclusions

       33  physical obs. remaining, equal to
       82  weighted obs., representing
       39  failures in single record/single failure data
    48726  total analysis time at risk, at risk from t =          0
                            earliest observed entry t =           0
                                 last observed exit t =        1227
```

Now that it is converted, we can use any of the st commands:

```
. sts test treat, logrank
        failure _d:  ndead
   analysis time _t:  time
            weight:  [fweight=w]
```

Log-rank test for equality of survivor functions

treat	Events observed	Events expected
0	22	17.05
1	17	21.95
Total	39	39.00

$$chi2(1) = 2.73$$
$$Pr>chi2 = 0.0986$$

Also See

Complementary: [ST] **ctset**, [ST] **st**

Background: [ST] **ct**

Title

As of the date that this manual was printed, Stata does not have a suite of built-in commands for discrete-time survival models matching the `st` suite for continuous-time models, but a good case could be made that it should. Instead, these models can be fitted easily using other existing estimation commands and data manipulation tools.

Discrete-time survival analysis concerns analysis of time-to-event data whenever survival times are either (a) intrinsically discrete (e.g. numbers of machine cycles), or (b) grouped into discrete intervals of time ("interval censoring"). If intervals are of equal length, then the same methods can be applied to both (a) and (b): survival times are positive integers.

The fitting of discrete-time survival models by the method of maximum likelihood may be accomplished straightforwardly. Data may contain completed or right-censored spells, and late entry (left truncation) can also be handled. So too can unobserved heterogeneity (also termed "frailty"). Estimation makes use of the property that the sample likelihood can be rewritten in a form identical to the likelihood for a binary dependent variable multiple regression model, applied to a specially organized dataset (Allison 1984, Jenkins 1995). For models without frailty, one may use, for example, `logistic` (or `logit`) to fit the discrete-time logistic hazard model, or `cloglog` to fit the discrete-time proportional hazards model (Prentice and Gloeckler 1978). Models incorporating normal frailty may be fitted using `xtlogit` and `xtcloglog`. A model with gamma frailty (Meyer 1990) may be fitted using `pgmhaz` (Jenkins 1997).

Estimation consists of three steps:

1. *Data organization step*: The dataset must be organized so that there is one observation for each period that a subject is at risk of experiencing the transition event. For example, if the original dataset contains one row for each subject, i, with information about their spell length T_i, the new dataset requires T_i rows for each subject, one row for each period at risk. This may be accomplished using `expand` or `stsplit`. (This step is episode splitting at each and every interval.) The result is data of the same form as a discrete panel (`xt`) dataset with repeated observations on each panel (subject).

2. *Variable creation step*: The analyst needs to create at least three types of variables. First, there is an interval identification variable. This is a sequence of positive integers $t = 1, \ldots, T_i$. For example,

   ```
   . sort subject_id
   . by subject_id: generate t = _n
   ```

   Second, there is the period-specific censoring indicator, d_i. If $d_i = 1$ if subject i's spell is complete, and $d_i = 0$ if the spell is right-censored, then the new indicator $d_{it}^* = 1$ if $d_i = 1$ and $t = T_i$, and $d_{it}^* = 0$ otherwise.

   Third, the analyst must define variables (as functions of t) to summarize the pattern of duration dependence. These variables are entered as covariates in the regression. For example, for a duration dependence pattern analogous to that in the continuous-time Weibull model, one would define a new variable $x_1 = \log t$. For a quadratic specification, define variables $x_1 = t$ and $x_2 = t^2$. A piece-wise constant specification is achieved by defining a set of dummy variables, with each group of periods sharing the same hazard rate; a semiparametric model (analogous to the Cox regression model for continuous survival-time data) is possible using separate dummy variables for each and every duration interval. No duration variable need be defined if the analyst is to fit a model with a constant hazard rate.

In addition to these three essentials, the analyst may define any time-varying covariates that are required.

3. *Estimation step*: Fit a binary dependent variable multiple regression model, with d_{it}^* as the dependent variable, and covariates including the duration variables and any other covariates.

Note that for estimation using spell data with late entry, the stages are exactly the same as outlined above, with one modification and one warning. To fit models without frailty, all intervals prior to each subject's entry to the study are simply dropped. For example, if entry is in period e_i, drop if $t < e_i$. If one wants to fit frailty models based on discrete-time data with late entry, then be aware that the estimation procedure outlined does not lead to correct estimates. (The sample likelihood in the reorganized data does not take appropriate account of conditioning for late entry in this case. In this case, users need to write their own likelihood function using ml; see [R] **maximize**.)

To derive predicted hazard rates, use the `predict` command. For example, after `logistic` or `cloglog`, use `predict, p`. After `xtlogit` or `xtcloglog`, use `predict, pu0` (which predicts the hazard assuming the individual effect is equal to the mean value). Estimates of the survivor function, S_{it}, can then be derived from the predicted hazard rates, p_{it}, since $S_{it} = (1 - p_{i1})(1 - p_{i2})(\cdots)(1 - p_{it})$.

References

Allison, P. D. 1984. *Event History Analysis. Regression for Longitudinal Event Data.* Newbury Park, CA: Sage Publications.

Jenkins, S. P. 1995. Easy estimation methods for discrete-time duration models. *Oxford Bulletin of Economics and Statistics* 57: 129–138.

———. 1997. sbe17: Discrete time proportional hazards regression. *Stata Technical Bulletin* 39: 22–32. Reprinted in *Stata Technical Bulletin Reprints*, vol. 7, pp. 109–121.

Meyer, B. D. 1990. Unemployment insurance and unemployment spells. *Econometrica* 58: 757–782.

Prentice, R. L. and L. Gloeckler. 1978. Regression analysis of grouped survival data with application to breast cancer data. *Biometrics* 34: 57–67.

Also See

Complementary:	[ST] **sts**, [ST] **stset**, [ST] **stsplit**,
	[R] **expand**, [R] **predict**
Related:	[ST] **stcox**, [ST] **streg**,
	[XT] **xtlogit**, [XT] **xtcloglog**
	[R] **logit**, [R] **logistic**, [R] **cloglog**
Background:	[U] **16.5 Accessing coefficients and standard errors**,
	[U] **23 Estimation and post-estimation commands**,
	[U] **23.14 Obtaining robust variance estimates**,
	[ST] **st**, [ST] **survival analysis**,
	[R] **maximize**

Title

> **epitab** — Tables for epidemiologists

Syntax

Cohort studies

    ir *var*~case~ *var*~exposed~ *var*~time~ $\big[$ *weight* $\big]$ $\big[$ if *exp* $\big]$ $\big[$ in *range* $\big]$ $\big[$, by(*varname*)

        tb <u>noc</u>rude <u>p</u>ool <u>noh</u>om <u>es</u>tandard <u>is</u>tandard <u>s</u>tandard(*varname*) ird

        <u>l</u>evel(*#*) $\big]$

    iri $\#_a$ $\#_b$ $\#_{N_1}$ $\#_{N_2}$ $\big[$, <u>l</u>evel(*#*) tb $\big]$

    cs *var*~case~ *var*~exposed~ $\big[$ *weight* $\big]$ $\big[$ if *exp* $\big]$ $\big[$ in *range* $\big]$ $\big[$, by(*varlist*) <u>e</u>xact or

        tb <u>w</u>oolf <u>noc</u>rude <u>p</u>ool <u>noh</u>om <u>es</u>tandard <u>is</u>tandard <u>s</u>tandard(*varname*) rd

        <u>b</u>inomial(*varname*) <u>l</u>evel(*#*) $\big]$

    csi $\#_a$ $\#_b$ $\#_c$ $\#_d$ $\big[$, <u>e</u>xact or tb <u>w</u>oolf <u>l</u>evel(*#*) $\big]$

Case–control studies

    cc *var*~case~ *var*~exposed~ $\big[$ *weight* $\big]$ $\big[$ if *exp* $\big]$ $\big[$ in *range* $\big]$ $\big[$, by(*varname*) <u>e</u>xact

        <u>cor</u>nfield tb <u>w</u>oolf <u>noc</u>rude <u>p</u>ool <u>noh</u>om bd

        <u>es</u>tandard <u>is</u>tandard <u>s</u>tandard(*varname*) <u>b</u>inomial(*varname*) <u>l</u>evel(*#*) $\big]$

    cci $\#_a$ $\#_b$ $\#_c$ $\#_d$ $\big[$, <u>e</u>xact <u>cor</u>nfield tb <u>w</u>oolf <u>l</u>evel(*#*) $\big]$

    tabodds *var*~case~ $\big[$ *expvar* $\big]$ $\big[$ *weight* $\big]$ $\big[$ if *exp* $\big]$ $\big[$ in *range* $\big]$ $\big[$, or <u>cor</u>nfield tb <u>w</u>oolf

        base(*#*) <u>adj</u>ust(*varlist*) <u>b</u>inomial(*varname*) <u>l</u>evel(*#*)

        <u>ci</u>plot graph plot(*plot*) <u>ci</u>opts(*rcap_options*) *connected_options twoway_options* $\big]$

    mhodds *var*~case~ *expvar* $\big[$ *vars*~adjust~ $\big]$ $\big[$ *weight* $\big]$ $\big[$ if *exp* $\big]$ $\big[$ in *range* $\big]$ $\big[$, by(*varlist*)

        <u>c</u>ompare(*level*~1~ , *level*~2~) <u>b</u>inomial(*varname*) <u>l</u>evel(*#*) $\big]$

Matched case–control studies

    mcc *var*~exposed_case~ *var*~exposed_control~ $\big[$ *weight* $\big]$ $\big[$ if *exp* $\big]$ $\big[$ in *range* $\big]$ $\big[$, tb <u>l</u>evel(*#*) $\big]$

    mcci $\#_a$ $\#_b$ $\#_c$ $\#_d$ $\big[$, tb <u>l</u>evel(*#*) $\big]$

fweights are allowed; see [U] **14.1.6 weight**.

Description

ir is used with incidence rate (incidence density or person-time) data. Point estimates and confidence intervals for the incidence rate ratio and difference are calculated along with attributable or prevented fractions for the exposed and total population. iri is the immediate form of ir; see [U] **22 Immediate commands**. Also see [R] **poisson** and [ST] **stcox** for related commands.

cs is used with cohort study data with equal follow-up time per subject and sometimes with cross-sectional data. Risk is then the proportion of subjects who become cases. Point estimates and confidence intervals for the risk difference, risk ratio, and (optionally) the odds ratio are calculated along with attributable or prevented fractions for the exposed and total population. csi is the immediate form of cs; see [U] **22 Immediate commands**. Also see [R] **logistic** and [R] **glogit** for related commands.

cc is used with case–control and cross-sectional data. Point estimates and confidence intervals for the odds ratio are calculated along with attributable or prevented fractions for the exposed and total population. cci is the immediate form of cc; see [U] **22 Immediate commands**. Also see [R] **logistic** and [R] **glogit** for related commands.

tabodds is used with case–control and cross-sectional data. It tabulates the odds of failure against a categorical explanatory variable *expvar*. If *expvar* is specified, tabodds will perform an approximate χ^2 test of homogeneity of odds and a test for linear trend of the log odds against the numerical code used for the categories of *expvar*. Both of these tests are based on the score statistic and its variance; see *Methods and Formulas* below. When *expvar* is absent, the overall odds are reported. The variable var_{case} is coded 0/1 for individual and simple frequency records, and equals the number of cases for binomial frequency records.

Optionally, tabodds will tabulate adjusted or unadjusted odds ratios using either the lowest levels of *expvar* or a user defined level as the reference group. If adjust(*varlist*) is specified, odds ratios adjusted for the variables in *varlist* will be produced along with a (score) test for trend.

mhodds is used with case–control and cross-sectional data. It estimates the ratio of the odds of failure for two categories of *expvar*, controlled for specified confounding variables, $vars_{\text{adjust}}$, and also tests whether this odds ratio is equal to one. When *expvar* has more than two categories but none are specified with the compare option, mhodds assumes that *expvar* is a quantitative variable and calculates a one-degree-of-freedom test for trend. It also calculates an approximate estimate of the rate ratio for a one unit increase in *expvar*. This is a one-step Newton–Raphson approximation to the maximum likelihood estimate calculated as the ratio of the score statistic, U, to its variance, V (Clayton and Hills 1993, 103).

mcc is used with matched case–control data. McNemar's chi-squared, point estimates and confidence intervals for the difference, ratio, and relative difference of the proportion with the factor, along with the odds ratio, are calculated. mcci is the immediate form of mcc; see [U] **22 Immediate commands**. Also see [R] **clogit** and [R] **symmetry** for related commands.

Options

Options are listed in alphabetical order.

adjust(*varlist*) is allowed only with tabodds. It specifies that odds ratios adjusted for the variables in *varlist* be calculated.

base(#) is allowed only with tabodds. It specifies the #th category of *expvar* to be used as the reference group for calculating odds ratios. If base() is not specified, the first category, corresponding to the minimum value of *expvar*, is used as the reference group.

bd specifies that Breslow and Day's χ^2 test of homogeneity be included in the output of a stratified analysis. This tests whether the exposure effect is the same across strata. bd is relevant only if by() is also specified.

binomial(*varname*) is allowed only with cs, cc, tabodds, and mhodds. It supplies the number of subjects (cases plus controls) for binomial frequency records. For individual and simple frequency records, this option is not used.

by(*varname*) specifies that the tables are stratified on *varname*. Within-stratum statistics are shown and then combined with Mantel–Haenszel weights. If estandard, istandard, or standard() is also specified (see below), the weights specified are used in place of Mantel–Haenszel weights. cs will accept a *varlist*.

ciopts(*rcap_options*) is allowed only with the ciplot option of tabodds. It affects the rendition of the confidence bands; see [G] **graph twoway rcap**.

ciplot is allowed only with tabodds. It produces the same plot as the graph option, except that it also includes the confidence intervals. This option is not allowed with either the or option or the adjust() option.

compare(v_1, v_2) is allowed only with mhodds. It gives the categories of *expvar* to be compared; v_1 defines the numerator and v_2 the denominator. When compare is absent and there are only two categories, the second is compared to the first; when there are more than two categories, an approximate estimate of the odds ratio for a unit increase in *expvar*, controlled for specified confounding variables, is given.

connected_options are allowed only with tabodds, and affect the rendition of the plotted points connected by lines; see [G] **graph twoway connected**.

cornfield requests that the Cornfield (1956) approximation be used for calculating the standard error of the odds ratio. Otherwise, standard errors are obtained as the square root of the variance of the score statistic or exactly in the case of cc and cci.

estandard, istandard, and standard(*varname*) request that within-stratum statistics be combined with external, internal, or user-specified weights to produce a standardized estimate. These options are mutually exclusive and can be used only when by() is also specified. (When by() is specified without one of these options, Mantel–Haenszel weights are used.)

estandard external weights are the person-time for the unexposed (ir), the total number of unexposed (cs), or the number of unexposed controls (cc).

istandard internal weights are person-time for the exposed (ir), the total number of exposed (cs), or the number of exposed controls (cc). istandard can be used for producing, among other things, standardized mortality ratios (SMRs).

standard(*varname*) allows user-specified weights. *varname* must contain a constant within stratum and be nonnegative. The scale of *varname* is irrelevant.

exact requests Fisher's exact p be calculated rather than the χ^2 and its significance level. We recommend specifying exact whenever samples are small. A conservative rule-of-thumb for 2×2 tables is to specify exact when the least-frequent cell contains fewer than 1,000 cases. When the least frequent cell contains 1,000 cases or more there will be no appreciable difference between the exact significance level and the significance level based on the χ^2, but the exact significance level will take considerably longer to calculate. Note that exact does *not* affect whether exact confidence intervals are calculated. Commands always calculate exact confidence intervals where they can unless tb or woolf is specified.

graph is allowed only with tabodds. It produces a graph of the odds against the numerical code used for the categories of *expvar*. Graph options other than connect() are allowed. This option is not allowed with the or option or the adjust() option.

ird may be used only with estandard, istandard, or standard(). It requests that ir calculate the standardized incidence rate difference rather than the default incidence rate ratio.

istandard; see estandard, above.

level(*#*) specifies the confidence level, in percent, for confidence intervals. The default is level(95) or as set by set level; see [R] **level**.

nocrude specifies that in a stratified analysis the crude estimate—the estimate one would obtain without regard to strata—not be displayed. nocrude is relevant only if by() is also specified.

nohom specifies that a χ^2 test of homogeneity not be included in the output of a stratified analysis. This tests whether the exposure effect is the same across strata and can be performed for any pooled estimate—directly pooled or Mantel–Haenszel. nohom is relevant only if by() is also specified.

or is allowed only with cs, csi, and tabodds. For cs and csi, or specified without by() reports the calculation of the odds ratio in addition to the risk ratio. With by(), or specifies that a Mantel–Haenszel estimate of the combined odds ratio be made rather than the Mantel–Haenszel estimate of the risk ratio. In either case, this is the same calculation as would be made by cc and cci. Typically, cc, cci, or tabodds is preferred for calculating odds ratios. For tabodds, or specifies that odds ratios be produced; see base() above for selection of reference category. By default, tabodds will calculate odds.

plot(*plot*) is allowed only with tabodds. It provides a way to add other plots to the generated graph; see [G] ***plot_option***.

pool specifies that in a stratified analysis the directly pooled estimate should also be displayed. The pooled estimate is a weighted average of the stratum-specific estimates using inverse-variance weights—weights that are the inverse of the variance of the stratum-specific estimate. pool is relevant only if by() is also specified.

rd may be used only with estandard, istandard, or standard(). It requests that cs calculate the standardized risk difference rather than the default risk ratio.

standard(*varname*); see estandard, above.

tb requests that test-based confidence intervals (Miettinen 1976) be calculated wherever appropriate in place of confidence intervals based on other approximations or exact confidence intervals. We recommend that test-based confidence intervals be used only for pedagogical purposes and never be used for research work.

twoway_options, allowed only with tabodds, are any of the options documented in [G] ***twoway_options*** excluding by(). These include options for titling the graph (see [G] ***title_options***) and options for saving the graph to disk (see [G] ***saving_option***).

woolf requests that the Woolf (1955) approximation, also known as the Taylor expansion, be used for calculating the standard error of the odds ratio. Otherwise, with the exception of tabodds and mhodds, the Cornfield (1956) approximation is used. The Cornfield approximation takes substantially longer (a few seconds) to calculate than the Woolf approximation. In the case of tabodds and mhodds standard errors of the odds ratios are obtained as the square root of the variance of the score statistic. This standard error is used in calculating a confidence interval for the odds ratio. (For matched case–control data, exact confidence intervals are always calculated.)

Remarks

Remarks are presented under the headings

> *Incidence rate data*
> *Stratified incidence rate data*
> *Standardized estimates with stratified incidence rate data*
> *Cumulative incidence data*
> *Stratified cumulative incidence data*
> *Standardized estimates with stratified cumulative incidence data*
> *Case–control data*
> *Stratified case–control*
> *Case–control data with multiple levels of exposure*
> *Case–control data with confounders and possibly multiple levels of exposure*
> *Standardized estimates with stratified case–control data*
> *Matched case–control data*

In order to calculate appropriate statistics and suppress inappropriate statistics, the ir, cs, cc, tabodds, mhodds, and mcc commands, along with their immediate counterparts, are organized in the way epidemiologists conceptualize data. ir processes incidence rate data from prospective studies; cs, cohort study data with equal follow-up time (cumulative incidence); cc, tabodds, and mhodds, case–control or cross-sectional (prevalence) data; and mcc, matched case–control data. With the exception of mcc, these commands work with both simple and stratified tables.

Epidemiological data are often summarized in a contingency table from which various statistics are calculated. The rows of the table reflect cases and noncases or cases and person-time, and the columns reflect exposure to a *risk factor*. To an epidemiologist, *cases* and *noncases* refer to the outcomes of the process being studied. For instance, a case might be a person with cancer and a noncase a person without cancer.

A *factor* is something that might affect the chances of being ultimately designated a case or a noncase. Thus, a case might be a cancer patient and the factor, smoking behavior. A person is said to be *exposed* or *unexposed* to the factor. Exposure can be classified as a dichotomy, smokes or does not smoke, or at multiple levels such as number of cigarettes smoked per week.

For an introduction to epidemiological methods, see Walker (1991). For an intermediate treatment, see Clayton and Hills (1993) and Lilienfeld and Stolley (1994). For a mathematically intermediate but otherwise advanced discussion, see Kelsey, Thompson, and Evans (1986). For other advanced discussions, see Fisher and van Belle (1993), Kleinbaum, Kupper, and Morgenstern (1982), and Rothman and Greenland (1998). For an anthology of writings on epidemiology since World War II, see Greenland (1987b).

❑ Technical Note

In many of the examples in this section, we provide output having specified the level(90) option, obtaining 90% confidence intervals. This was done in order that the results in our examples be the same as those given in various texts, especially Rothman and Greenland (1998). This should not be taken as an endorsement that a 90% confidence interval is somehow appropriate in professional work. ❑

Incidence rate data

In *incidence rate data* from a prospective study, you observe the transformation of noncases into cases. Starting with a group of noncase subjects, you follow them to determine if they become a case (e.g., struck by lightning or, more likely, stricken with cancer). You follow two populations—those exposed and those unexposed to the factor. A summary of the data is

	Exposed	Unexposed	Total
Cases	a	b	$a + b$
Person-time	N_1	N_0	$N_1 + N_0$

▷ Example

It will be easiest to understand these commands if we start with the immediate forms. Remember, in the immediate form, you specify the data on the command line rather than specify names of variables containing the data; see [U] **22 Immediate commands**. You have data (Boice and Monson 1977, reported in Rothman and Greenland 1998, 238) on breast cancer *cases* and person-years of observation for women with tuberculosis repeatedly *exposed* to multiple X-ray fluoroscopies, and those not so exposed:

| | X-ray fluoroscopy | |
	Exposed	Unexposed
Breast cancer cases	41	15
Person-years	28,010	19,017

Using the immediate form of `ir`, you specify the values in the table following the command:

```
. iri 41 15 28010 19017
```

	Exposed	Unexposed	Total
Cases	41	15	56
Person-time	28010	19017	47027

	Exposed	Unexposed	Total		
Incidence Rate	.0014638	.0007888	.0011908		

	Point estimate	[95% Conf. Interval]		
Inc. rate diff.	.000675	.0000749	.0012751	
Inc. rate ratio	1.855759	1.005722	3.60942	(exact)
Attr. frac. ex.	.4611368	.005689	.7229472	(exact)
Attr. frac. pop	.337618			

```
(midp)   Pr(k>=41) =                        0.0177  (exact)
(midp) 2*Pr(k>=41) =                        0.0355  (exact)
```

`iri` shows the table, reports the incidence rates for the exposed and unexposed populations, and then shows the point estimates of the difference and ratio of the two incidence rates along with their confidence intervals. The *incidence rate* is simply the frequency with which noncases are transformed into cases.

Next is reported the attributable fraction among the exposed population, an estimate of the proportion of exposed cases attributable to exposure. We estimate that 46.1% of the 41 breast cancer cases among the exposed were due to exposure. (Had the incidence rate ratio been less than 1, reported would have been the prevented fraction in the exposed population, an estimate of the net proportion of all potential cases in the exposed population that was prevented by exposure; see the following technical note.)

Following that is the attributable fraction in the total population, which is the net proportion of all cases attributable to exposure. This number, of course, depends on the proportion of cases that are exposed in the base population, which here is taken to be $41/56$ and so may not be relevant in all situations. We estimate that 33.8% of the 56 cases were due to exposure. Note that $.338 \times 56 = .461 \times 41 = 18.9$, that is, looked at either way, we estimate that 18.9 cases were caused by exposure.

At the bottom of the table are reported both one- and two-sided exact significance tests. For the one-sided test, the probability that the number of exposed cases is 41 or greater is .0177. This is a "midp" calculation; see *Methods and Formulas* below. The two-sided test is $2 \times .0177 = .0354$.

◁

❑ Technical Note

When the incidence rate is less than 1, `iri` (and `ir`, and `cs` and `csi`, and `cc` and `cci`) substitute the prevented fraction for the attributable fraction. Let us reverse the roles of exposure in the above data, treating as exposed a person who did not receive the X-ray fluoroscopy. You can think of this as a new treatment for preventing breast cancer—the suggested treatment being not to employ fluoroscopy.

```
. iri 15 41 19017 28010
```

	Exposed	Unexposed	Total
Cases	15	41	56
Person-time	19017	28010	47027
Incidence Rate	.0007888	.0014638	.0011908

	Point estimate	[95% Conf. Interval]		
Inc. rate diff.	−.000675	−.0012751	−.0000749	
Inc. rate ratio	.5388632	.2770528	.994311	(exact)
Prev. frac. ex.	.4611368	.005689	.7229472	(exact)
Prev. frac. pop	.1864767			

(midp) Pr(k<=15) =	0.0177	(exact)
(midp) 2*Pr(k<=15) =	0.0355	(exact)

The prevented fraction among the exposed is the net proportion of all potential cases in the exposed population that were prevented by exposure. We estimate that 46.1% of potential cases among the women receiving the new "treatment" were prevented by the treatment. (Previously, we estimated the same percent of actual cases among women receiving the X-rays were caused by the X-rays.)

The prevented fraction for the population, which is the net proportion of all potential cases in the total population that was prevented by exposure, as with the attributable fraction, depends on the proportion of cases that are exposed in the base population—here taken as 15/56—and so may not be relevant in all situations. We estimate that 18.6% of the potential cases were prevented by exposure.

Also see Greenland and Robins (1988) for a discussion of the interpretation of attributable and prevented fractions.

❑

▷ Example

`ir` works like `iri` except that it obtains the entries in the tables by summing data. You specify three variables—the first representing the number of cases represented by this observation, the second whether the observation is for subjects exposed to the factor, and the third recording the total time the subjects in this observation were observed. An observation may reflect a single subject or a group of subjects.

For instance, here is a two-observation dataset for the table in the previous example:

. list

	cases	exposed	time
1.	41	0	28010
2.	15	1	19017

If you typed `ir cases exposed time`, you would obtain the same output as we obtained above. Another way the data might be recorded is

. list

	cases	exposed	time
1.	20	0	14000
2.	21	0	14010
3.	15	1	19017

In this case the first two observations will be automatically summed by `ir` since both are exposed. Finally, the data might be individual-level data:

. list in 1/5

	cases	exposed	time
1.	1	1	10
2.	0	1	8
3.	0	0	9
4.	1	0	2
5.	0	1	1

The first observation represents a woman who got cancer, was exposed, and was observed for 10 years. The second is a woman who did not get cancer, was exposed, and was observed for 8 years, and so on.

◁

❑ Technical Note

`ir` (and all the other commands) assume a subject was exposed if the exposed variable is nonzero and not missing; assume the subject was not exposed if the variable is zero; and ignore the observation if the variable is missing. For `ir`, the case variable and the time variable are restricted to nonnegative integers and are summed within the exposed and unexposed groups to obtain the entries in the table.

❑

Stratified incidence rate data

▷ Example

`ir` can work with stratified as well as single tables. For instance, Rothman (1986, 185) discusses data from Rothman and Monson (1973) on the mortality by sex and age for patients with trigeminal neuralgia:

| | Age through 64 | | Age 65+ | |
	Males	Females	Males	Females
Deaths	14	10	76	121
Person-years	1516	1701	949	2245

Entering the data into Stata, we have the dataset:

```
. use http://www.stata-press.com/data/r8/rm
(Rothman and Monson 1973 data)

. list
```

	age	male	deaths	pyears
1.	<65	1	14	1516
2.	<65	0	10	1701
3.	65+	1	76	949
4.	65+	0	121	2245

And the stratified analysis of the incidence rate ratio is

```
. ir deaths male pyears, by(age) level(90)
```

Age category	IRR	[90% Conf. Interval]		M-H Weight	
<65	1.570844	.7380544	3.431878	4.712465	(exact)
65+	1.485862	1.153559	1.907441	35.95147	(exact)
Crude	1.099794	.8688097	1.388891		(exact)
M-H combined	1.49571	1.191914	1.876938		

```
Test of homogeneity (M-H)    chi2(1) =    0.02  Pr>chi2 = 0.8992
```

Here we also specified level(90) to obtain 90% confidence intervals. The row labeled M-H combined reflects the combined Mantel–Haenszel estimates.

As with the previous example, it was not important that each entry in the table correspond to a single observation in the data—ir sums the time (pyears) and case (deaths) variables within the exposure (male) category.

From the output, we learn that the Mantel–Haenszel estimate differs from the simple crude estimate—that incidence rates appear to vary by age category—and from the test of homogeneity, we find that the exposure effect (the effect of trigeminal neuralgia) is the same across age categories (i.e., we cannot reject the hypothesis that the effects do not differ). Thus, we are justified in our use of the Mantel–Haenszel estimate.

◁

❏ Technical Note

Stratification is one way to deal with confounding; that is, perhaps sex affects the incidence of trigeminal neuralgia and so does age, so the table was stratified by age in an attempt to uncover the sex effect. (You are concerned that age may confound the true association between sex and the incidence of trigeminal neuralgia because the less-than-65/65+ age distributions are so different for males and females. If age affects incidence, the difference in the age distributions would induce different incidences for males and females and thus confound the true effect of sex.)

One does not, however, have to use tables to uncover effects; the estimation alternative *when you have aggregate data* is Poisson regression and we can use the same data on which we ran `ir` with `poisson`. When you have individual-level data, Poisson regression is an alternative, but so is Cox regression; see below.

(Although `age` in the previous example appears to be a string, it is actually a numeric variable taking on values 0 and 1. We attached a value label to produce the labelings <65 and 65+ to make `ir`'s output look better; see [U] **15.6.3 Value labels**. Stata's estimation commands will ignore this labeling.)

```
. poisson deaths male age, exposure(pyears) level(90) irr

Iteration 0:    log likelihood = -10.836732
Iteration 1:    log likelihood = -10.734087
Iteration 2:    log likelihood = -10.733944
Iteration 3:    log likelihood = -10.733944

Poisson regression                          Number of obs   =          4
                                            LR chi2(2)      =     164.01
                                            Prob > chi2     =     0.0000
Log likelihood = -10.733944                 Pseudo R2       =     0.8843
```

deaths	IRR	Std. Err.	z	P>|z|	[90% Conf. Interval]	
male	1.495096	.2060997	2.92	0.004	1.191779	1.875611
age	8.888775	1.934943	10.04	0.000	6.213541	12.71583
pyears	(exposure)					

It is worth comparing these results with the Mantel–Haenszel estimates produced by `ir`:

Source	IR Ratio	90% Conf. Int.	
Mantel–Haenszel (ir)	1.50	1.19	1.88
poisson	1.50	1.19	1.88

Results are identical to two decimal places. Moreover, in addition to obtaining the incidence rate ratio for the `male` exposure variable, we also obtain an estimate and confidence interval for the incidence rate of being 65 and over relative to being less than age 65, although in this case the number is not of much interest since the outcome variable is total mortality and we already knew that older people have a higher mortality rate. In other contexts, however, the number may be of greater interest.

See [R] **poisson** for an explanation of the `poisson` command.

❏

❏ Technical Note

Both the model estimated above and the preceding table asserted that exposure effects are the same across age categories and, if they are not, then both of the previous results are equally inappropriate. The table presented a test of homogeneity, reassuring us that the exposure effects do indeed appear to be constant. The Poisson-regression alternative can be used to reproduce that test by including interactions between the groups and exposure:

```
. generate maleXage = male*age

. poisson deaths male age maleXage, exp(pyears) level(90) irr

Iteration 0:    log likelihood = -10.898799
Iteration 1:    log likelihood = -10.726225
Iteration 2:    log likelihood = -10.725904
Iteration 3:    log likelihood = -10.725904
```

```
Poisson regression                              Number of obs   =          4
                                                LR chi2(3)      =     164.03
                                                Prob > chi2     =     0.0000
    Log likelihood = -10.725904                 Pseudo R2       =     0.8843
```

deaths	IRR	Std. Err.	z	P>|z|	[90% Conf. Interval]	
male	1.660688	1.396496	0.60	0.546	.4164666	6.622099
age	9.167973	3.01659	6.73	0.000	5.33613	15.75144
maleXage	.9459	.41539	-0.13	0.899	.4593455	1.947829
pyears	(exposure)					

Note that the significance level of the `maleXage` effect is 0.899, the same as previously reported by `ir`.

In this case, forming the male-times-age interaction was easy because there were only two age groups. Had there been more groups, the test would have been slightly more difficult—see the next technical note.

❑

❑ Technical Note

A word of caution is in order when applying `poisson` (or any estimation technique) to more than two age categories. Let's pretend that in our data we had three age categories, which we will call categories 0, 1, and 2, and that they are stored in the variable `agecat`. You might think of the categories as corresponding to age less than 35, 35–64, and 65 and above.

With such data, you might type `ir deaths male pyears, by(agecat)`, but you would *not* type `poisson deaths male agecat, exposure(pyears)` to obtain the equivalent Poisson-regression estimated results. Such a model might be reasonable, but it is not equivalent because you would be constraining the age effect in category 2 to be (multiplicatively) twice the effect in category 1.

To `poisson` (and all of Stata's estimation commands other than `anova`), `agecat` is simply one variable and only one estimated coefficient is associated with it. Thus, the model is

$$\text{Poisson index} = P = \beta_0 + \beta_1 \text{male} + \beta_2 \text{agecat}$$

Without explanation, the expected number of deaths is then e^P and the incidence rate ratio associated with a variable is e^β; see [R] **poisson**. Thus, the value of the Poisson index when `male==0` and `agecat==1` is $\beta_0 + \beta_2$, and the possibilities are

	male==0	male==1
agecat==0	β_0	$\beta_0 + \beta_1$
agecat==1	$\beta_0 + \beta_2$	$\beta_0 + \beta_2 + \beta_1$
agecat==2	$\beta_0 + 2\beta_2$	$\beta_0 + 2\beta_2 + \beta_1$

What is important to note is that the age effect for `agecat==2` is constrained to be twice the age effect for `agecat==1`—the only difference between lines 3 and 2 of the table is that β_2 is replaced with $2\beta_2$. Under certain circumstances, such a constraint might be reasonable, but it does not correspond to the assumptions made in generating the Mantel–Haenszel combined results.

To obtain results equivalent to the Mantel–Haenszel, you must estimate a separate effect for each age group, meaning replace $2\beta_2$, the constrained effect, with β_3, a new coefficient that is free to take on any value. We can achieve this by creating two new variables and using them in place of `agecat`. `agecat1` will take on the value 1 when `agecat` is 1 and 0 otherwise; `agecat2` will take on the value 1 when `agecat` is 2 and 0 otherwise:

```
. generate agecat1 = (agecat==1)
. generate agecat2 = (agecat==2)
. poisson deaths male agecat1 agecat2 [freq=pop], exposure(pyears) irr
```

In Stata, we do not have to generate these variables for ourselves. We could use Stata's expand-interaction command xi:

```
. xi: poisson deaths male i.agecat [freq=pop], exposure(pyears) irr
```

See [R] **xi**. It is also worth reading [U] **28 Commands for dealing with categorical variables**. Although that discussion is in terms of linear regression, what is said there applies equally to Poisson regression.

To reproduce the homogeneity test with multiple age categories, we could type

```
. xi: poisson deaths i.agecat*male [freq=pop], exp(pyears) irr
. testparm _Ia*Xm*
```

Poisson regression combined with xi generalizes to multiway tables. For instance, perhaps there are three exposure categories. Assume exposure variable burn takes on the values 1, 2, and 3 for first-, second-, and third-degree burns. The table itself is estimated by typing

```
. xi: poisson deaths i.burn i.agecat [freq=pop], exp(pyears) irr
```

and the test of homogeneity by typing

```
. xi: poisson deaths i.burn*i.agecat [freq=pop], exp(pyears) irr
. testparm _Ib*Xa*
```

❑

❑ Technical Note

When you have individual-level data, Cox regression and Poisson regression are alternatives to stratified tables. For instance, if we had the individual-level data for deaths by sex and age for the patients with trigeminal neuralgia, the first few observations might be

```
. list in 1/3
```

	patient	male	age	died	studyt~e
1.	1	0	64	0	2
2.	2	1	58	1	3.5
3.	3	1	44	0	3

Patient 1 is a female, age 64, who did not die and was followed for 2 years. Patient 2 is a male, age 58, who did die after being followed for 3.5 years. With these data, we could obtain the same incidence-rate table we obtained previously by typing

```
. generate agecat = age>=65
. ir died male studytime, by(agecat)
```

but there would be a problem. We would be assuming that patient 1 contributes 2 person-years to the age 64 group, which is not true. Patient 1 contributes 1 person-year to the age 64 group and 1 more person-year to the age 65 group. We would not have this problem if age did not vary over time, so for the moment, pretend that age is really something else—say race—that does not vary over time. In that case, the Cox regression alternative could be obtained by typing

```
. stset studytime, failure(died)
. stcox male agecat
```

Now let's deal with the fact that age does vary over time. The following commands will transform our data:

```
. expand studytime+cond(int(studytime)==studytime,0,1)
. by patient, sort: generate time=cond(_n!=_N,1,studytime-(_N-1))
. by patient: generate cumtime=sum(time)
. by patient: replace died=0 if _n!=_N
. by patient: replace age=age+_n-1
```

See [R] **expand** and [U] **16 Functions and expressions**, and for an explanation of cond() and sum(), see [U] **16.3 Functions**. After issuing these commands, our data are

```
. list in 1/10
```

	patient	male	age	died	studyt~e	time	cumtime
1.	1	0	64	0	2	1	1
2.	1	0	65	0	2	1	2
3.	2	1	58	0	3.5	1	1
4.	2	1	59	0	3.5	1	2
5.	2	1	60	0	3.5	1	3
6.	2	1	61	0	3.5	1	4
7.	2	1	62	1	3.5	-.5	3.5
8.	3	1	44	0	3	1	1
9.	3	1	45	0	3	1	2
10.	3	1	46	0	3	1	3

That is, there are now two observations for patient 1, one at age 64 and another at age 65. In neither case did the patient die. For patient 2, there are now four observations, three corresponding to ages 58–60, during which the patient did not die, and one more (contributing half a year) corresponding to age 61, at which time the patient did die. The variable time records the contribution of the observation to exposure; cumtime records the total time from entry into the study (diagnosis of trigeminal neuralgia) to the end of the observation.

We can obtain our original table from these data by typing

```
. generate agecat = age>=65
. ir died sex time, by(agecat)
```

The corresponding Cox regression model could be estimated by typing

```
. stset cumtime, failure(died) id(patient)
. stcox sex agecat
```

Note that it is with the stset command that we specify that this is a dataset with repeated observations on patients; see [ST] **stset**. Once the dataset is stset, stcox will know to estimate a time-varying regression. As an aside, it might be better to replace agecat with age since, with Cox regression, one can estimate continuous effects so there is no reason to categorize the data.

So what would happen if we did all of this? We do not have the original data, so results must be speculative. We can make up various datasets that correspond to the aggregate data in the table. One possibility, for instance, is that the 6,411 person-years reported in the aggregate table correspond to 6,411 patients, each observed for 1 year. In that case:

```
. use http://www.stata-press.com/data/r8/rm2
(Rothman and Monson 1973 data)

. stcox male agecat, level(90)

        failure _d:  dead
   analysis time _t:  cumtime

Iteration 0:   log likelihood = -1937.2353
Iteration 1:   log likelihood = -1858.8456
Iteration 2:   log likelihood = -1855.3196
Iteration 3:   log likelihood = -1855.2292
Iteration 4:   log likelihood = -1855.2291
Refining estimates:
Iteration 0:   log likelihood = -1855.2291

Cox regression -- Breslow method for ties

No. of subjects =          6411                 Number of obs   =       6411
No. of failures =           221
Time at risk    =          6411
                                               LR chi2(2)      =     164.01
Log likelihood  =    -1855.2291                Prob > chi2     =     0.0000
```

_t _d	Haz. Ratio	Std. Err.	z	P>|z|	[90% Conf. Interval]	
male	1.495096	.2060997	2.92	0.004	1.191779	1.875611
agecat	8.888775	1.934943	10.04	0.000	6.213541	12.71583

Thus, comparing these results with those previously obtained:

Source	IR Ratio	90% Conf. Int.	
Mantel–Haenszel (ir)	1.50	1.19	1.88
poisson	1.50	1.19	1.88
stcox	1.50	1.19	1.88

Remember, we obtained the Cox regression estimates under the assumption that each patient was observed for only one year (that is, outside of this technical note, we concocted a dataset that would yield the same results as the aggregate data if we ran the ir command on them, and we concocted that data assuming that each patient was observed for a single year). If we had the real underlying data, and so knew exactly how long each patient was observed, the stcox estimates would be preferable to those of ir or poisson.   ❑

Standardized estimates with stratified incidence rate data

The by() option specifies that the data are stratified and, by default, produce a Mantel–Haenszel combined estimate of the incidence rate ratio. With the estandard, istandard, or standard(*varname*) options, you can specify your own weights and obtain standardized estimates of the incidence rate ratio or difference.

▷ Example

Rothman and Greenland (1998, 259) report results from Doll and Hill (1966) on age-specific coronary disease deaths among British male doctors by cigarette smoking:

		Smokers		Nonsmokers
Age	Deaths	Person-years	Deaths	Person-years
35–44	32	52,407	2	18,790
45–54	104	43,248	12	10,673
55–64	206	28,612	28	5,710
65–74	186	12,663	28	2,585
75–84	102	5,317	31	1,462

We have entered these data into Stata:

```
. use http://www.stata-press.com/data/r8/dollhill2

. list
```

	age	smokes	deaths	pyears
1.	35-44	1	32	52407
2.	35-44	0	2	18790
3.	45-54	1	104	43248
4.	45-54	0	12	10673
5.	55-64	1	206	28612
6.	55-64	0	28	5710
7.	65-74	1	186	12663
8.	65-74	0	28	2585
9.	75-84	1	102	5317
10.	75-84	0	31	1462

We can obtain the Mantel–Haenszel combined estimate along with the crude estimate for ignoring stratification of the incidence rate ratio and 90% confidence intervals by typing

```
. ir deaths smokes pyears, by(age) level(90)
```

age	IRR	[90% Conf. Interval]		M-H Weight	
35-44	5.736638	1.704242	33.62016	1.472169	(exact)
45-54	2.138812	1.274529	3.813215	9.624747	(exact)
55-64	1.46824	1.044925	2.110463	23.34176	(exact)
65-74	1.35606	.9625995	1.953472	23.25315	(exact)
75-84	.9047304	.6375086	1.305422	24.31435	(exact)
Crude	1.719823	1.437554	2.068803		(exact)
M-H combined	1.424682	1.194375	1.699399		

Test of homogeneity (M-H)   chi2(4) =   10.41   Pr>chi2 = 0.0340

Note the presence of heterogeneity revealed by the test; the effect of smoking is not the same across age categories. Moreover, the listed stratum-specific estimates show an effect that appears to be declining with age. (Even if the test of homogeneity was not significant, one should always examine estimates carefully when stratum-specific effects occur on both sides of 1 for ratios and 0 for differences.)

Rothman and Greenland (1998, 264) obtain the standardized incidence rate ratio and 90% confidence intervals weighting each age category by the population of the exposed group, thus producing the standardized mortality ratio (SMR). This calculation can be reproduced by specifying by(age) to indicate the table is stratified, and istandard to specify we want the internally standardized rate. We may also specify that we would like to see the pooled estimate (weighted average where the weights are based on the variance of the strata calculations):

```
. ir deaths smokes pyears, by(age) level(90) istandard pool
```

age	IRR	[90% Conf. Interval]		Weight	
35-44	5.736638	1.704242	33.62016	52407	(exact)
45-54	2.138812	1.274529	3.813215	43248	(exact)
55-64	1.46824	1.044925	2.110463	28612	(exact)
65-74	1.35606	.9625995	1.953472	12663	(exact)
75-84	.9047304	.6375086	1.305422	5317	(exact)
Crude	1.719823	1.437554	2.068803		(exact)
Pooled (direct)	1.355343	1.134356	1.619382		
I. Standardized	1.417609	1.186641	1.693676		

```
Test of homogeneity (direct) chi2(4) =     10.20  Pr>chi2 = 0.0372
```

We obtained the simple pooled results because we specified the `pool` option. Note the significance of the homogeneity test; it provides the motivation for standardizing the rate ratios.

If we wanted the externally standardized ratio (weights proportional to the population of the unexposed group), we would substitute `estandard` for `istandard` in the above command.

Not shown by Rothman and Greenland, but as easily obtained, are the internally and externally weighted standardized incidence rate difference. We will obtain the internally weighted difference:

```
. ir deaths smokes pyears, by(age) level(90) istandard ird
```

age	IRD	[90% Conf. Interval]		Weight
35-44	.0005042	.0002877	.0007206	52407
45-54	.0012804	.0006205	.0019403	43248
55-64	.0022961	.0005628	.0040294	28612
65-74	.0038567	.0000521	.0076614	12663
75-84	-.0020201	-.0090201	.00498	5317
Crude	.0018537	.001342	.0023654	
I. Standardized	.0013047	.000712	.0018974	

◁

▷ Example

In addition to calculating results using internal or external weights, `ir` (and `cs` and `cc`) can calculate results for arbitrary weights. If we wanted to obtain the incidence rate ratio weighting each age category equally:

```
. generate conswgt=1
. ir deaths smokes pyears, by(age) level(90) standard(conswgt)
```

age	IRR	[90% Conf. Interval]		Weight	
35-44	5.736638	1.704242	33.62016	1	(exact)
45-54	2.138812	1.274529	3.813215	1	(exact)
55-64	1.46824	1.044925	2.110463	1	(exact)
65-74	1.35606	.9625995	1.953472	1	(exact)
75-84	.9047304	.6375086	1.305422	1	(exact)
Crude	1.719823	1.437554	2.0688		(exact)
Standardized	1.155026	.9373745	1.423214		

◁

❑ Technical Note

estandard and istandard are convenience features; they do nothing different from what you could accomplish by creating the appropriate weights and using the standard() option. For instance, we could duplicate the previously shown results of istandard (example before last) by typing

```
. sort age smokes
. by age: generate wgt=pyears[_N]
. list in 1/4
```

	age	smokes	deaths	pyears	conswgt	wgt
1.	35-44	0	2	18790	1	52407
2.	35-44	1	32	52407	1	52407
3.	45-54	0	12	10673	1	43248
4.	45-54	1	104	43248	1	43248

```
. ir deaths smokes pyears, by(age) level(90) standard(wgt) ird
  (output omitted)
```

sort age smokes made the exposed group (smokes = 1) the last observation within each age category. by age: gen wgt=pyears[_N] created wgt equal to the last observation in each age category.

❑

Cumulative incidence data

In cumulative incidence (follow-up or longitudinal) data, rather than using the time a subject was at risk to normalize ratios, you use the number of subjects. A group of noncases is followed for some period of time and during that time some become cases. Each subject is also known to be exposed or unexposed. A summary of the data is

	Exposed	Unexposed	Total
Cases	a	b	$a + b$
Noncases	c	d	$c + d$
Total	$a + c$	$b + d$	$a + b + c + d$

Data of this type are generally summarized using the risk ratio. A ratio of 2 means that an exposed subject is twice as likely to become a case as is an unexposed subject, a ratio of one-half means half as likely, and so on. The "null" value—the number corresponding to no effect—is a ratio of 1. It should be noted that if cross-sectional data are analyzed in this format, a prevalence ratio can be obtained.

▷ Example

You have data on diarrhea during a 10-day follow-up period among 30 breast-fed infants colonized with *Vibrio cholerae* 01 according to antilipopolysaccharide antibody titers in the mother's breast milk (Glass et al. 1983, reported in Rothman and Greenland 1998, 243):

	Antibody Level	
	High	Low
Diarrhea	7	12
No Diarrhea	9	2

The csi command works much like the iri command. We recommend specifying the exact option, however, whenever the least frequent cell contains fewer than 1,000 observations. We have very few observations here:

```
. csi 7 12 9 2, exact
```

	Exposed	Unexposed	Total
Cases	7	12	19
Noncases	9	2	11
Total	16	14	30
Risk	.4375	.8571429	.6333333

	Point estimate	[95% Conf. Interval]	
Risk difference	-.4196429	-.7240828	-.1152029
Risk ratio	.5104167	.2814332	.9257086
Prev. frac. ex.	.4895833	.0742914	.7185668
Prev. frac. pop	.2611111		

```
                 1-sided Fisher's exact P = 0.0212
                 2-sided Fisher's exact P = 0.0259
```

We find that high antibody levels reduce the risk of diarrhea (the risk falls from .86 to .44). The difference is just significant at the 2.59% two-sided level. (Had we not specified the exact option, a χ^2 value and its significance level would have been reported in place of Fisher's exact p. The calculated χ^2 two-sided significance level would have been .0173, but this calculation is inferior for small samples.)

◁

❑ Technical Note

By default, cs and csi do not report the odds ratio, but they will if you specify the or option. If you want odds ratios, however, you should use the cc or cci commands—the commands appropriate for case–control data—because cs and csi calculate the attributable (prevented) fraction using the risk ratio even if you specify or:

```
. csi 7 12 9 2, or exact
```

	Exposed	Unexposed	Total
Cases	7	12	19
Noncases	9	2	11
Total	16	14	30
Risk	.4375	.8571429	.6333333

	Point estimate	[95% Conf. Interval]		
Risk difference	-.4196429	-.7240828	-.1152029	
Risk ratio	.5104167	.2814332	.9257086	
Prev. frac. ex.	.4895833	.0742914	.7185668	
Prev. frac. pop	.2611111			
Odds ratio	.1296296	.0246233	.7180882	(Cornfield)

```
                 1-sided Fisher's exact P = 0.0212
                 2-sided Fisher's exact P = 0.0259
```

Sometimes, the lower or upper confidence bound for the odds ratio will be missing, meaning the Cornfield approximation did not converge. This can occur when the number of observations is quite small. Two other approximations, however, are available—the Woolf confidence intervals (obtained by specifying woolf) and the test-based intervals (obtained by specifying tb).

❏

❏ Technical Note

As with iri and ir, csi and cs report either the attributable or the prevented fraction for the exposed and total populations; see the discussion under *Incidence rate data* above. In the previous example, we estimated that 49% of potential cases in the exposed population were prevented by exposure. We also estimated that exposure accounted for a 26% reduction in cases over the entire population, but that is based on the exposure distribution of the (small) population (16/30) and probably is of little interest.

Fleiss (1981, 77) reports infant mortality by birth weight for 72,730 live white births in 1974 in New York City:

```
. csi 618 422 4597 67093
```

	Exposed	Unexposed	Total
Cases	618	422	1040
Noncases	4597	67093	71690
Total	5215	67515	72730
Risk	.1185043	.0062505	.0142995

	Point estimate	[95% Conf. Interval]	
Risk difference	.1122539	.1034617	.121046
Risk ratio	18.95929	16.80661	21.38769
Attr. frac. ex.	.9472554	.9404996	.9532441
Attr. frac. pop	.5628883		

```
                    chi2(1) =  4327.92  Pr>chi2 = 0.0000
```

In these data, exposed means a premature baby (birth weight 2,500 grams or less) and a case is a dead baby at the end of one year. We find that being premature accounts for 94.7% of deaths among the premature population. We also estimate, paraphrasing from Fleiss (1981, 77), that 56.3% of all white infant deaths in New York City in 1974 could have been prevented if prematurity had been eliminated. (Moreover, Fleiss puts a standard error on the attributable fraction for the population. The formula is given in *Methods and Formulas* but is appropriate only for the population on which the estimates are based.)

❏

▷ Example

cs works like csi except that it obtains its information from the data. The data equivalent to typing csi 7 12 9 2 are

```
. list
```

	case	exp	pop
1.	1	1	7
2.	1	0	12
3.	0	1	9
4.	0	0	2

We could then type cs case exp [freq=pop]. If we had individual-level data, so that each observation reflected a patient and we had 30 observations, we would type cs case exp.

◁

Stratified cumulative incidence data

▷ Example

Rothman and Greenland (1998, 255) reprint the following age-specific information for deaths from all causes for tolbutamide and placebo treatment groups (University Group Diabetes Program 1970):

	Age through 54		Age 55 and above	
	Tolbutamide	Placebo	Tolbutamide	Placebo
Dead	8	5	22	16
Surviving	98	115	76	79

The data corresponding to these results are

```
. use http://www.stata-press.com/data/r8/ugdp
. list
```

	age	case	exposed	pop
1.	<55	0	0	115
2.	<55	0	1	98
3.	<55	1	0	5
4.	<55	1	1	8
5.	55+	0	0	69
6.	55+	0	1	76
7.	55+	1	0	16
8.	55+	1	1	22

The order of the observations is unimportant. If we were now to type cs case exposed [freq=pop], we would obtain a summary for all the data, ignoring the stratification by age. To incorporate the stratification, we type

(Continued on next page)

```
. cs case exposed [freq=pop], by(age)
    Age category |        RR      [95% Conf. Interval]    M-H Weight
    -------------+------------------------------------------------
             <55 |   1.811321     .6112044    5.367898      2.345133
             55+ |   1.192602     .6712664     2.11883      8.568306
    -------------+------------------------------------------------
           Crude |   1.435574     .8510221    2.421645
    M-H combined |   1.325555      .797907    2.202132
    -------------+------------------------------------------------
Test of homogeneity (M-H)        chi2(1) =     0.447  Pr>chi2 = 0.5037
```

Mantel–Haenszel weights are appropriate when the risks may differ according to the strata but the risk ratio is believed to be the same (homogeneous across strata). Under these assumptions, Mantel–Haenszel weights are designed to use the information efficiently. They are not intended to measure a composite risk ratio when the within-strata risk ratios differ. In that case, one wants a standardized ratio (see below).

The risk ratios above appear to differ markedly, but the confidence intervals are also broad due to the small sample sizes. The test of homogeneity shows that the differences can be attributed to chance; the use of the Mantel–Haenszel combined is probably justified.

◁

❑ Technical Note

Stratified cumulative incidence tables, as with stratified incidence rate tables, are only one way to control for the effect of a confounding factor. The estimation alternative is logistic regression, although this requires using the odds ratio rather than the risk ratio as the measure of risk. For the above data, the odds ratio is

```
. cs case exposed [freq=pop], by(age) or
    Age category |        OR      [95% Conf. Interval]    M-H Weight
    -------------+------------------------------------------------
             <55 |   1.877551     .6238165    5.637046      2.168142 (Cornfield)
             55+ |   1.248355     .6112772    2.547411      6.644809 (Cornfield)
    -------------+------------------------------------------------
           Crude |   1.510673     .8381198    2.722012
    M-H combined |   1.403149     .7625152    2.582015
    -------------+------------------------------------------------
Test of homogeneity (M-H)        chi2(1) =     0.347  Pr>chi2 = 0.5556
             Test that combined OR = 1:
                         Mantel-Haenszel chi2(1) =       1.19
                                       Pr>chi2 =       0.2750
```

In this case, the event is sufficiently unlikely to proceed; the risk ratio is 1.33 and the odds ratio is 1.40. We also present the crude estimate obtained for analyzing the data without stratifying.

Recasting the problem in the language of modeling, the outcome variable is case and it is to be explained by age and exposed. (As in the incidence rate example, age may appear to be a string variable in our data—we listed the data in the previous example—but it is actually a numeric variable taking on values 0 and 1 with value labels disguising that fact; see [U] **15.6.3 Value labels**.)

```
. logistic case exposed age [freq=pop]
Logit estimates                              Number of obs  =        409
                                             LR chi2(2)     =      22.47
                                             Prob > chi2    =     0.0000
Log likelihood =  -142.6212                  Pseudo R2      =     0.0730
```

case	Odds Ratio	Std. Err.	z	P>|z|	[95% Conf. Interval]
exposed	1.404674	.4374454	1.09	0.275	.7629451 2.586175
age	4.216299	1.431519	4.24	0.000	2.167361 8.202223

It is worth comparing these results with the Mantel–Haenszel estimates obtained with cs:

Source	Odds Ratio	95% Conf. Int.	
Mantel–Haenszel (cs)	1.40	0.76	2.58
logistic	1.40	0.76	2.59

They are virtually identical.

Logistic regression has advantages over the stratified-table approach. First, we obtained an estimate of the age effect: Being 55 years or over significantly increases the odds of death. In addition to the point estimate, 4.22, we have a confidence interval for the effect: 2.17 to 8.20.

Given the measured age effect, one would expect that it does not apply only to persons on either side of the 55-year cutoff. It would be more reasonable to assume that a 54 year-old patient has a higher probability of death, due merely to age, than a 53 year-old patient; a 53 year-old, a higher probability than a 52; and so on. If we had the underlying data, where each patient's age is presumably known, we could include the actual age in the model and so better control for the age effect. This would improve our estimate of the effect of being exposed to tolbutamide.

See [R] **logistic** for an explanation of the logistic command. Also see the technical note in *Incidence rate data* above; what is said there concerning categorical variables applies to logistic regression as well as Poisson regression.

❏

Standardized estimates with stratified cumulative incidence data

As with ir, cs can produce standardized estimates and the method is basically the same, although the options for which estimates are to be combined or standardized make it confusing. We showed above that cs can produce Mantel–Haenszel weighted estimates of the risk ratio (the default) or the odds ratio (obtained by specifying or). cs can also produce standardized estimates of the risk ratio (the default) or the risk difference (obtained by specifying rd).

▷ Example

To produce an estimate of the internally standardized risk ratio using our age-specific data on deaths from all causes for tolbutamide and placebo treatment groups (example above), we type

```
. cs case exposed [freq=pop], by(age) istandard
```

Age category	RR	[95% Conf. Interval]		Weight
<55	1.811321	.6112044	5.367898	106
55+	1.192602	.6712664	2.11883	98
Crude	1.435574	.8510221	2.421645	
I. Standardized	1.312122	.7889772	2.182147	

We could obtain externally standardized estimates by substituting `estandard` for `istandard`.

If we wish to produce an estimate of the risk ratio weighting each age category equally:

```
. generate wgt=1
. cs case exposed [freq=pop], by(age) standard(wgt)
```

Age category	RR	[95% Conf. Interval]		Weight
<55	1.811321	.6112044	5.367898	1
55+	1.192602	.6712664	2.11883	1
Crude	1.435574	.8510221	2.421645	
Standardized	1.304737	.7844994	2.169967	

If we instead wanted the rate difference:

```
. cs case exposed [freq=pop], by(age) standard(wgt) rd
```

Age category	RD	[95% Conf. Interval]		Weight
<55	.033805	-.0278954	.0955055	1
55+	.0362545	-.0809204	.1534294	1
Crude	.0446198	-.0192936	.1085332	
Standardized	.0350298	-.0311837	.1012432	

If we wanted to weight the less-than-55 age group five times as heavily as the 55-and-over group, we would have created `wgt` to contain 5 for the first age group and 1 for the second (or 10 for the first group and 2 for the second—the scale of the weights does not matter).

◁

Case–control data

In case–control data, one selects a sample on the basis of the outcome under study or, said differently, cases and noncases are sampled at different rates. If one were examining the link between coffee consumption and heart attacks, for instance, one selects a sample of subjects with and without the heart problem and then examines their coffee-drinking behavior. A subject who has suffered a heart attack is called a *case* just as with cohort study data. A subject who has never suffered a heart attack, however, is called a *control* rather than merely a noncase, emphasizing that the sampling was performed with respect to the outcome.

In case–control data, all hope of identifying the risk (i.e., incidence) of the outcome (heart attacks) associated with the factor (coffee drinking) vanishes, at least without information on the underlying sampling fractions, but one can examine the proportion of coffee drinkers among the two populations and reason that, if there is a difference, that coffee drinking may be associated with the risk of heart attacks. Remarkably, even without the underlying sampling fractions, one can also measure the ratio of the odds of heart attacks if one does drink coffee to the odds if one does not—the so-called odds ratio.

What is lost is the ability to compare absolute rates, which is not always the same as comparing relative rates; see Fleiss (1981, 91).

▷ Example

`cci` calculates the odds ratio and the attributable risk associated with a 2 × 2 table. Rothman et al. (1979) (reprinted in Rothman 1986, 161 and reprinted in Rothman and Greenland 1998, 245) presents case–control data on the history of chlordiazopoxide use in early pregnancy for mothers of children born with and without congenital heart defects:

```
                       Chlordiazopoxide use
                             Yes    No
                   ─────────────────────────
                   Case mothers      4    386
                   Control mothers   4   1250
```

We will use the `level(90)` option to obtain 90% confidence intervals as reported by Rothman and Greenland:

```
. cci 4 386 4 1250, level(90)
                                                            Proportion
                   │   Exposed   Unexposed  │    Total      Exposed
         ──────────┼────────────────────────┼─────────────────────────
            Cases  │      4          386     │     390       0.0103
         Controls  │      4         1250     │    1254       0.0032
         ──────────┼────────────────────────┼─────────────────────────
            Total  │      8         1636     │    1644       0.0049

                   │  Point estimate         │  [90% Conf. Interval]
         ──────────┼────────────────────────┼─────────────────────────
       Odds ratio  │      3.238342           │   .7698467    13.59663  (exact)
    Attr. frac. ex.│          .6912          │  -.2989599     .9264524  (exact)
   Attr. frac. pop │       .0070892          │
                   ─────────────────────────────────────────────────────
                       chi2(1) =      3.07  Pr>chi2 = 0.0799
```

We obtain a point estimate of the odds ratio as 3.24 and a χ^2 value, which is a test that the odds ratio is 1, significant at the 10% level.

◁

❑ Technical Note

Although one tends to think of statistical tests and confidence intervals as being different ways of presenting the same information, that is not exactly true. The statistical test is derived under the assumption of the null hypothesis that the odds ratio is 1. A "test-based" confidence interval is a direct transformation of the statistical test and there can be no disagreement between them. If you specify the `tb` option, `csi` and the other epitab commands will produce test-based confidence intervals. When you do not specify this option, however, either exact or approximate confidence intervals are calculated. Exact confidence intervals are unbeatable—if there is a disagreement between the exact interval and the test statistic, believe the interval. (Within exact results, midp intervals more closely approximate asymptotic test statistics.) All the classical test statistics are derived under various simplifying assumptions, assumptions not present in exact calculations. (The epitab commands always place the word "exact" next to exact calculation results.)

The Cornfield confidence intervals, on the other hand, are an example of an approximation. The Cornfield approximation, along with the other approximations, attempts to produce confidence intervals around a point central to the confidence interval, producing intervals that should more accurately reflect the exact confidence interval when the parameter deviates from the null hypothesis. This serves the purpose of making the confidence interval focus on estimation rather than statistical testing.

In this case, here is the result of calculating the confidence interval for the odds ratio using various methods:

Method	90% Conf. Int.		Source
exact	0.77	13.60	Rothman 1986, 174 or cci
Woolf	1.01	10.40	Rothman 1986, 175 or cci, woolf
test-based	1.07	9.77	Rothman 1986, 175 or cci, tb
Cornfield	1.07	9.83	Rothman 1986, 175 or cci

There are only 8 exposed patients in this case and the differences shown above are larger than typically observed. The fact that they are also on both sides of the knife-edge of 1 is (pedagogically speaking) fortuitous. The exact answer, which is just that, includes 1 in the interval and so one cannot reject (at the 10% level) the hypothesis that the odds ratio is 1. That is the "right" answer (but see the next technical note). The Woolf, test-based, and Cornfield intervals exclude 1 and so reassuringly conform to the χ^2 value. In any case, all results are so close to 1 that one should still think about chance as an explanation.

❏

❏ Technical Note

If you specify the `exact` option, `cc` and `cci` will report the exact significance of the table. We could have, and should have, done that above because of the small number of cases. We did not for expositional purposes, but we will do so now:

```
. cci 4 386 4 1250, level(90) exact
                                                       Proportion
                        Exposed   Unexposed   |   Total    Exposed
            Cases          4         386      |    390     0.0103
          Controls         4        1250      |   1254     0.0032

            Total          8        1636      |   1644     0.0049

                        Point estimate        |  [90% Conf. Interval]

       Odds ratio        3.238342             |  .7698467   13.59663   (exact)
    Attr. frac. ex.        .6912              | -.2989599    .9264524  (exact)
    Attr. frac. pop      .0070892             |

                    1-sided Fisher's exact P = 0.0964
                    2-sided Fisher's exact P = 0.0964
```

The first thing to note is that in this table the 1- and 2-sided significance values are equal. This is not a mistake, but it does not happen often. Exact significance values are calculated by summing the probabilities for tables that have the same marginals (row and column sums) but that are less likely (given an odds ratio of 1) than the observed table. When considering each possible table, one asks if the table is in the same or opposite tail as the observed table. If it is in the same tail, one counts the table under consideration in the 1-sided test and, either way, one counts it in the 2-sided test. In this case, it just turns out that all the tables more extreme than this table are in the same tail, so the 1- and 2-sided tests are the same.

Note that the exact p is .096 < .100, that is, the table is significant at the 10% level. In the previous example, we said "exact confidence intervals are unbeatable—if there is a disagreement between the exact interval and the test statistic, believe the interval". We also reported (Rothman 1986, 174) that the "exact" confidence interval for the odds ratio is 0.77 to 13.6, which includes 1. How can both statements be true and yet both statistics be "exact"?

Confidence intervals and test statistics measure different things. The .0964 is the fraction of all tables that are less likely than the observed table *conditional on the odds ratio being 1*. The lower bound of the confidence interval 0.77 corresponds to the statement that the probability of observing tables with 4 or more exposed cases *conditional on the odds ratio being .77* is 0.05. The upper bound of the confidence interval 13.6 corresponds to the statement that the probability of observing tables with 4 or less exposed cases *conditional on the odds ratio being 13.6* is 0.05. No two of these are the same question, and there is no reason why the answers to these different questions should somehow agree.

You expect them to agree because you are used to thinking in terms of large-sample approximations. For infinite-sized samples, the answers to the questions above will agree, and for large but finite samples, the answers will tend to agree. If we had a large number of exposed individuals (as we do not), we would not have had this "problem".

❏

❏ Technical Note

The reported value of the attributable or prevented fraction among the exposed is based on using the odds ratio as a proxy for the risk ratio. This can only be justified if the outcome is exceedingly unlikely in the population. The extrapolation to the attributable or prevented fraction for the population assumes the control group is a random sample of the corresponding group in the underlying population.

❏

▷ Example

Equivalent to typing `cci 4 386 4 1250` would be typing `cc case exposed [freq=pop]` with the data:

```
. list
```

	case	exposed	pop
1.	1	1	4
2.	1	0	386
3.	0	1	4
4.	0	0	1250

◁

Stratified case–control

▷ Example

`cc` has the ability to work with stratified tables. Rothman and Greenland (1998, 273) reprint and discuss data from a case–control study on infants with congenital heart disease and Down's syndrome and healthy controls, according to maternal spermicide use before conception and maternal age at delivery (Rothman 1982):

	Maternal age to 34		Maternal age 35+	
	Spermicide used	not used	Spermicide used	not used
Down's syndrome	3	9	1	3
Controls	104	1059	5	86

The data corresponding to these tables are

```
. use http://www.stata-press.com/data/r8/downs
```

```
. list
```

	case	exposed	pop	age
1.	1	1	3	<35
2.	1	0	9	<35
3.	0	1	104	<35
4.	0	0	1059	<35
5.	1	1	1	35+
6.	1	0	3	35+
7.	0	1	5	35+
8.	0	0	86	35+

The stratified results for the odds ratio are

```
. cc case exposed [freq=pop], by(age) woolf
```

Maternal age	OR	[95% Conf. Interval]		M-H Weight	
<35	3.394231	.9048403	12.73242	.7965957	(Woolf)
35+	5.733333	.5016418	65.52706	.1578947	(Woolf)
Crude	3.501529	1.110362	11.04208		(Woolf)
M-H combined	3.781172	1.18734	12.04142		

```
Test of homogeneity (M-H)      chi2(1) =     0.14  Pr>chi2 = 0.7105
                 Test that combined OR = 1:
                       Mantel-Haenszel chi2(1) =      5.81
                                     Pr>chi2 =      0.0159
```

For no particular reason, we also specified the woolf option to obtain Woolf approximations to the within-strata confidence intervals rather than Cornfield approximations. Had we wanted test-based confidence intervals and the crude estimate, we would have used

```
. cc case exposed [freq=pop], by(age) tb bd
```

Maternal age	OR	[95% Conf. Interval]		M-H Weight	
<35	3.394231	.976611	11.79672	.7965957	(tb)
35+	5.733333	.6402941	51.33752	.1578947	(tb)
Crude	3.501529	1.189946	10.30358		(tb)
M-H combined	3.781172	1.282056	11.15183		(tb)

```
Test of homogeneity (M-H)      chi2(1) =     0.14  Pr>chi2 = 0.7105
Test of homogeneity (B-D)      chi2(1) =     0.14  Pr>chi2 = 0.7092
                 Test that combined OR = 1:
                       Mantel-Haenszel chi2(1) =      5.81
                                     Pr>chi2 =      0.0159
```

We recommend that test-based confidence intervals only be used for pedagogical reasons and never for research work.

Whatever method is chosen for calculating confidence intervals, also reported is a test of homogeneity, which in our case is $\chi^2(1) = .14$ and "insignificant". That is, the odds of Down's syndrome might vary with maternal age, but we cannot reject the hypothesis that the odds ratios of those exposed to spermicide and those not exposed are the same in the two maternal age strata. This is thus a test to reject the appropriateness of the single, Mantel–Haenszel combined odds ratio—a rejection not justified by these data.

◁

❑ Technical Note

Note that in the last example we requested that Breslow and Day's test of homogeneity be reported by specifying the bd option. For this data, both the Mantel–Haenszel and the Breslow and Day tests produced virtually identical results, but this is not always the case. When data are sparse, you may have strata with zero count in one or more cells. The Mantel–Haenszel test will not include these strata in the homogeneity test because the odds ratios for these strata cannot be estimated. On the other hand, the Breslow and Day homogeneity test includes all strata in its calculation. Note that although Breslow and Day's test of homogeneity is more appropriate when data are sparse, if the number of strata is large and there are few observations per strata, the distribution of the test statistic may not approximate the nominal chi-squared even if the null hypothesis is true.

❑

❑ Technical Note

As with cohort study data, an alternative to stratified tables for uncovering effects is logistic regression. From the logistic point of view, case–control data is no different from cohort study data—one must merely ignore the estimated intercept, which is not reported by logistic in any case. (logit, on the other hand, makes the same estimates as logistic but displays the coefficients rather than transforming them to odds ratios and so does display the estimated intercept. The intercept is meaningless in case–control data because it reflects the baseline prevalence of the outcome which, by sampling, you controlled.)

The data we used with cs can be used directly by logistic. (The age variable, which appears to be a string, is really numeric with an associated value label; see [U] **15.6.3 Value labels**. age takes on the value 0 for the age-less-than-35 group and 1 for the 35+ group.)

```
. logistic case exposed age [freq=pop]
Logit estimates                                Number of obs   =       1270
                                               LR chi2(2)      =       8.74
                                               Prob > chi2     =     0.0127
Log likelihood = -81.517532                    Pseudo R2       =     0.0509
```

case	Odds Ratio	Std. Err.	z	P>|z|	[95% Conf. Interval]	
exposed	3.787779	2.241922	2.25	0.024	1.187334	12.0836
age	4.582857	2.717351	2.57	0.010	1.433594	14.65029

Comparing the results with those presented by cc in the previous example:

Source	Odds Ratio	95% Conf. Int.	
Mantel–Haenszel (cc)	3.78	1.19	12.04
logistic	3.79	1.19	12.08

As with the cohort study data, results are virtually identical and all the same comments we made previously apply once again.

To demonstrate, let us now ask a question that would be difficult to answer on the basis of a stratified table analysis. We now know that spermicide use appears to increase the risk of having a baby with Down's syndrome and we also know that the mother's age also increases the risk. Is the effect of spermicide use statistically different for mothers in the two age groups?

```
. generate ageXex = age*exposed
. logistic case exposed age ageXex [freq=pop]
Logit estimates                                    Number of obs   =        1270
                                                   LR chi2(3)      =        8.87
                                                   Prob > chi2     =      0.0311
Log likelihood = -81.451332                        Pseudo R2       =      0.0516
```

case	Odds Ratio	Std. Err.	z	P>|z|	[95% Conf. Interval]	
exposed	3.394231	2.289544	1.81	0.070	.9048403	12.73242
age	4.104651	2.774868	2.09	0.037	1.091034	15.44237
ageXex	1.689141	2.388785	0.37	0.711	.1056563	27.0045

The answer is that the effect is not statistically different. The odds ratio and confidence interval reported for exposed now measure the spermicide effect for an age==0 (age less than 35) mother. The odds ratio and confidence interval reported for ageXex are the (multiplicative) difference in the odds ratio for an age==1 (age 35+) mother relative to a young mother. The point estimate is that the effect is larger for older mothers, suggesting grounds for future research, but the difference is not significant.

See [R] **logistic** for an explanation of the logistic command. Also see the technical note under *Incidence rate data* above. What was said there concerning Poisson regression applies equally to logistic regression.

❑

Case–control data with multiple levels of exposure

As previously noted, in a case–control study, subjects with the disease of interest (cases) are compared to disease-free individuals (controls) to assess the relationship between exposure to one or more risk factors and disease incidence. Often, exposure is measured qualitatively at several discrete levels, or measured on a continuous scale and then grouped into 3 or more levels. The data can be summarized as

	Exposure level				
	1	2	...	k	Total
Cases	a_1	a_2	...	a_k	M_1
Controls	c_1	c_2	...	c_k	M_0
Total	N_1	N_2	...	N_k	T

An advantage afforded by having multiple levels of exposure is the ability to examine dose–response relationships. If the association between a risk factor and a disease or outcome is real, we expect the strength of that association to increase with the level and duration of exposure. Demonstrating the existence of a dose–response relationship provides strong support for a direct or even causal relationship between the risk factor and the outcome. On the other hand, the lack of a dose–response is usually seen as an argument against causality.

We can use the tabodds command to tabulate and examine the odds of "failure" or odds ratios against a categorical exposure variable. The test for trend calculated by tabodds can serve as a test for dose–response if the exposure variable is at least ordinal. Note that if the exposure variable has no natural ordering the trend test is meaningless and should be ignored. See the technical note at the end of this section for more information regarding the test for trend.

Before looking at an example, consider three possible data arrangements for case–control and prevalence studies. The most common data arrangement is individual records, where each subject in the study has his or her own record. Closely related are frequency records where identical individual records are included only once, but with a variable giving the frequency with which the record occurs. The *weight* option is used for this data to specify the frequency variable. Data can also be arranged as binomial frequency records where each record contains a variable D, the number of cases, another variable N, the number of total subject (cases plus controls), and other variables. An advantage of binomial frequency records is that otherwise large datasets can be entered succinctly into a Stata database.

▷ Example

Consider the following data from the Ille-et-Villaine study of esophageal cancer discussed in Breslow and Day (1980, chapter 4):

| | Alcohol consumption (g/day) | | | | |
	0–39	40–79	80–119	120+	Total
Cases	2	9	9	5	25
Controls	47	31	9	5	92
Total	49	40	18	10	117

corresponding to subjects age 55 to 64 that use from 0 to 9 grams of tobacco per day. There are 24 such tables, each representing one of four levels of tobacco use and one of six age categories. The data can be used to create a binomial frequency record dataset by simply entering each table's data by typing

```
. input alcohol D N agegrp tobacco
           alcohol         D          N       agegrp       tobacco
  1.             1         2         49            4             1
  2.             2         9         40            4             1
  3.             3         9         18            4             1
  4.             4         5         10            4             1
  5. end
  .
```

where, D is the number of esophageal cancer cases and N is the number of total subjects (cases plus controls) for each combination of six age-groups (agegrp), four levels of alcohol consumption in g/day (alcohol), and four levels of tobacco use in g/day (tobacco).

Both the tabodds and mhodds commands can correctly handle all three data arrangements. Binomial frequency records require that the number of total subjects (cases plus controls) represented by each record N be specified with the binomial() option.

We could also enter the data as frequency-weighted data:

```
. input alcohol case freq agegrp tobacco
           alcohol        case        freq       agegrp       tobacco
  1.             1         1           2            4             1
  2.             1         0          47            4             1
  3.             2         1           9            4             1
  4.             2         0          31            4             1
  5.             3         1           9            4             1
  6.             3         0           9            4             1
  7.             4         1           5            4             1
  8.             4         0           5            4             1
  9. end
  .
```

If you are planning on using any of the other estimation commands, such as `poisson` or `logistic`, we recommend that you enter your data either as individual records or as frequency weighted records and not as binomial frequency records because the estimation commands currently do not recognize the `binomial` option.

We have entered all the esophageal cancer data into Stata as a frequency-weighted record dataset as previously described. In our data, `case` indicates the esophageal cancer cases and controls and `freq` is the number of subjects represented by each record (the weight).

We added value labels to the variables `agegrp`, `alcohol`, and `tobacco` in our dataset to ease interpretation in outputs but note that these variables are numeric.

We are interested in the association between alcohol consumption and esophageal cancer. We first use `tabodds` to tabulate the odds of esophageal cancer against alcohol consumption:

```
. use http://www.stata-press.com/data/r8/bdesop
. tabodds case alcohol [fweight=freq]
```

alcohol	cases	controls	odds	[95% Conf. Interval]	
0-39	29	386	0.07513	0.05151	0.10957
40-79	75	280	0.26786	0.20760	0.34560
80-119	51	87	0.58621	0.41489	0.82826
120+	45	22	2.04545	1.22843	3.40587

```
Test of homogeneity (equal odds): chi2(3)  =    158.79
                                   Pr>chi2  =    0.0000

Score test for trend of odds:      chi2(1)  =    152.97
                                   Pr>chi2  =    0.0000
```

The test of homogeneity clearly indicates that the odds of esophageal cancer differ by level of alcohol consumption and the test for trend indicates a significant increase in odds with increasing alcohol use. This is suggestive of a strong dose–response relation. The `graph` option can be used to study the shape of the relationship of the odds with alcohol consumption. Note that most of the heterogeneity in these data can be "explained" by the linear increase in risk of esophageal cancer with increased dosage (alcohol consumption).

We could also have requested that the odds ratios at each level of alcohol consumption be calculated by specifying the `or` option. For example, `tabodds case alcohol [fweight=freq], or` would produce odds ratios using the minimum value of `alcohol`, i.e., `alcohol = 1` (0–39) as the reference group, and the command `tabodds case alcohol [fweight=freq], or base(2)` would use `alcohol = 2` (40–79) as the reference group.

Although our results appear to provide strong evidence in support of an association between alcohol consumption and esophageal cancer, we need to be concerned with the possible existence of confounders, specifically age and tobacco use, in our data. We can again use `tabodds` to tabulate and examine the odds of esophageal cancer against age and against tobacco use, independently:

```
. tabodds case agegrp [fweight=freq]
```

agegrp	cases	controls	odds	[95% Conf. Interval]	
25-34	1	115	0.00870	0.00121	0.06226
35-44	9	190	0.04737	0.02427	0.09244
45-54	46	167	0.27545	0.19875	0.38175
55-64	76	166	0.45783	0.34899	0.60061
65-74	55	106	0.51887	0.37463	0.71864
75+	13	31	0.41935	0.21944	0.80138

```
Test of homogeneity (equal odds): chi2(5)  =      96.94
                                   Pr>chi2  =     0.0000

Score test for trend of odds:     chi2(1)  =      83.37
                                   Pr>chi2  =     0.0000

. tabodds case tobacco [fweight=freq]
```

tobacco	cases	controls	odds	[95% Conf.	Interval]
0-9	78	447	0.17450	0.13719	0.22194
10-19	58	178	0.32584	0.24228	0.43823
20-29	33	99	0.33333	0.22479	0.49428
30+	31	51	0.60784	0.38899	0.94983

```
Test of homogeneity (equal odds): chi2(3)  =      29.33
                                   Pr>chi2  =     0.0000

Score test for trend of odds:     chi2(1)  =      26.93
                                   Pr>chi2  =     0.0000
```

We can see that there is evidence to support our concern that both age and tobacco use are potentially important confounders. Clearly, before we can make any statements regarding the association between esophageal cancer and alcohol use, we must examine and, if necessary, adjust for the effect of any confounder. We will return to this example in the following section.

◁

❏ Technical Note

The score test for trend performs a test for linear trend of the log odds against the numerical code used for the exposure variable. The test depends not only on the relationship between dose level and the outcome, but also on the numeric values assigned to each level, or to be more accurate, to the distance between the numeric values assigned. For example, the trend test on a dataset with four exposure levels coded 1, 2, 3, and 4 gives the same results as coding the levels 10, 20, 30, and 40 because the distance between the levels in each case is constant. In the first case, the distance is one unit and in the second case, it is 10 units. However, if we code the exposure levels as 1, 10, 100, and 1000, we would obtain different results because the distance between exposure levels is not constant. Thus, care must be taken when assigning values to exposure levels. You must determine if equally spaced numbers make sense for your data, or if other more meaningful values should be used.

One last comment about the trend test: remember that we are testing whether a log-linear relationship exists between the odds and the outcome variable. For your particular problem, this relationship may not be correct or even make sense, so you must be careful in interpreting the output of this trend test.

❏

Case–control data with confounders and possibly multiple levels of exposure

In the esophageal cancer data example introduced in the previous section, we determined that the apparent association between alcohol consumption and esophageal cancer could be confounded by age and/or tobacco use. The effect of possible confounding factors can be adjusted for by stratifying on these factors. This is the method used by both tabodds and mhodds to adjust for other variables in the dataset. We will compare and contrast these two commands in the following example.

▷ Example

We begin by using `tabodds` to tabulate unadjusted odd ratios. We do this to compare the adjusted and unadjusted odds ratios.

```
. tabodds case alcohol [fweight=freq], or
```

alcohol	Odds Ratio	chi2	P>chi2	[95% Conf. Interval]	
0-39	1.000000	.	.	.	.
40-79	3.565271	32.70	0.0000	2.237981	5.679744
80-119	7.802616	75.03	0.0000	4.497054	13.537932
120+	27.225705	160.41	0.0000	12.507808	59.262107

```
Test of homogeneity (equal odds): chi2(3)  =    158.79
                                   Pr>chi2  =    0.0000
Score test for trend of odds:      chi2(1)  =    152.97
                                   Pr>chi2  =    0.0000
```

The `alcohol = 1` group (0–39) was used by `tabodds` as the reference category for calculating the odds ratios. We could have selected a different group by specifying the `base()` option, however, because the lowest dosage level is most often the adequate reference group, as it is in these data, the `base()` option is seldom used.

We use `tabodds` with the `adjust()` option to tabulate Mantel–Haenszel age-adjusted odds ratios:

```
. tabodds case alcohol [fweight=freq], adjust(age)
Mantel-Haenszel odds ratios adjusted for age
```

alcohol	Odds Ratio	chi2	P>chi2	[95% Conf. Interval]	
0-39	1.000000	.	.	.	.
40-79	4.268155	37.36	0.0000	2.570025	7.088314
80-119	8.018305	59.30	0.0000	4.266893	15.067922
120+	28.570426	139.70	0.0000	12.146409	67.202514

```
Score test for trend of odds: chi2(1)  =  135.09
                              Pr>chi2  =  0.0000
```

We observe that the age-adjusted odds ratios are just slightly higher than the unadjusted ones, thus it appears that age is not as strong a confounder as it first appeared. Note that even after adjusting for age, the dose–response relationship, as measured by the trend test, remains strong.

We now perform the same analysis but this time adjust for tobacco use instead of age.

```
. tabodds case alcohol [fweight=freq], adjust(tobacco)
Mantel-Haenszel odds ratios adjusted for tobacco
```

alcohol	Odds Ratio	chi2	P>chi2	[95% Conf. Interval]	
0-39	1.000000	.	.	.	.
40-79	3.261178	28.53	0.0000	2.059764	5.163349
80-119	6.771638	62.54	0.0000	3.908113	11.733306
120+	19.919526	123.93	0.0000	9.443830	42.015528

```
Score test for trend of odds: chi2(1)  =  135.04
                              Pr>chi2  =  0.0000
```

Again we observe a significant dose–response relationship and not much difference between the adjusted and unadjusted odds ratios. We could also adjust for the joint effect of both age and tobacco use by specifying `adjust(tobacco age)`, but we will not bother in this case.

◁

A different approach to the analysis of these data may be performed using the mhodds command. As previously mentioned, mhodds estimates the ratio of the odds of failure for two categories of an exposure variable, controlling for any specified confounding variables, and also tests whether this odds ratio is equal to one. In the case of multiple exposures, if two exposure levels are not specified with compare(), mhodds assumes that exposure is quantitative and calculates a one-degree-of-freedom test for trend. This test for trend is the same as tabodds reports.

▷ Example

We first use mhodds to estimate the effect of alcohol controlled for age:

```
. mhodds case alcohol agegrp [fweight=freq]
Score test for trend of odds with alcohol
controlling for agegrp
(The Odds Ratio estimate is an approximation to the odds ratio
for a one unit increase in alcohol)
```

Odds Ratio	chi2(1)	P>chi2	[95% Conf. Interval]	
2.845895	135.09	0.0000	2.385749	3.394792

Because alcohol has more than two levels, mhodds estimated and reported an approximate age adjusted odds ratio for a one unit increase in alcohol consumption. Note that the χ^2 value reported is identical to that reported by tabodds for the score test for trend, on the previous page.

We now use mhodds to estimate the effect of alcohol controlled for age, and while we are at it, we may as well do this by levels of tobacco consumption:

```
. mhodds case alcohol agegrp [fweight=freq], by(tobacco)
Score test for trend of odds with alcohol
controlling for agegrp
by tobacco
note: only 19 of the 24 strata formed in this analysis contribute
      information about the effect of the explanatory variable
(The Odds Ratio estimate is an approximation to the odds ratio
for a one unit increase in alcohol)
```

tobacco	Odds Ratio	chi2(1)	P>chi2	[95% Conf. Interval]	
0-9	3.579667	75.95	0.0000	2.687104	4.768710
10-19	2.303580	25.77	0.0000	1.669126	3.179196
20-29	2.364135	13.27	0.0003	1.488098	3.755890
30+	2.217946	8.84	0.0029	1.311837	3.749921

Mantel-Haenszel estimate controlling for agegrp and tobacco

Odds Ratio	chi2(1)	P>chi2	[95% Conf. Interval]	
2.751236	118.37	0.0000	2.292705	3.301471

```
Test of homogeneity of ORs (approx): chi2(3)  =     5.46
                              Pr>chi2  =   0.1409
```

Again, because `alcohol` has more than two levels, `mhodds` estimated and reported an approximate Mantel–Haenszel age and tobacco-use adjusted odds ratio for a one unit increase in alcohol consumption. The χ^2 test for trend reported with the Mantel–Haenszel estimate is again the same as `tabodds` produces if `adjust(agegrp tobacco)` is specified.

The results from this analysis also show an effect of alcohol, controlled for age, of about $\times 2.7$, which is consistent across different levels of tobacco consumption. Similarly

```
. mhodds case tobacco agegrp [fweight=freq], by(alcohol)
```

```
Score test for trend of odds with tobacco
controlling for agegrp
by alcohol
```

```
note: only 18 of the 24 strata formed in this analysis contribute
      information about the effect of the explanatory variable
```

```
(The Odds Ratio estimate is an approximation to the odds ratio
for a one unit increase in tobacco)
```

alcohol	Odds Ratio	chi2(1)	P>chi2	[95% Conf. Interval]	
0-39	2.420650	15.61	0.0001	1.561214	3.753197
40-79	1.427713	5.75	0.0165	1.067168	1.910070
80-119	1.472218	3.38	0.0659	0.974830	2.223387
120+	1.214815	0.59	0.4432	0.738764	2.997628

```
Mantel-Haenszel estimate controlling for agegrp and alcohol
```

Odds Ratio	chi2(1)	P>chi2	[95% Conf. Interval]	
1.553437	20.07	0.0000	1.281160	1.883580

```
Test of homogeneity of ORs (approx): chi2(3)  =    5.26
                                      Pr>chi2  =  0.1540
```

shows an effect of tobacco, controlled for age, of about $\times 1.5$, which is consistent across different levels of alcohol consumption.

Comparisons between particular levels of alcohol and tobacco consumption can be made by generating a new variable with levels corresponding to all combinations of alcohol and tobacco, as in

```
. egen alctob = group(alcohol tobacco)
. mhodds case alctob [fweight=freq], compare(16,1)
Maximum likelihood estimate of the odds ratio
Comparing alctob==16 vs. alctob==1
```

Odds Ratio	chi2(1)	P>chi2	[95% Conf. Interval]	
93.333333	103.21	0.0000	14.766136	589.938431

which yields an odds ratio of 93 between subjects with the highest levels of alcohol and tobacco, and those with the lowest levels. Similar results can be obtained simultaneously for all levels of `alctob` using `alctob = 1` as the comparison group by specifying `tabodds D alctob , bin(N) or`.

◁

Standardized estimates with stratified case–control data

▷ Example

You obtain standardized estimates (in this case, for the odds ratio) using cc just as you obtain standardized estimates using ir or cs. Along with the by() option, you specify one of estandard, istandard, or standard(*varname*).

Rothman and Greenland (1998, 272) report the standardized mortality rate (SMR) along with a 90% confidence interval for the case–control study on infants with congenital heart disease and Down's syndrome. We can reproduce their estimates along with the pooled estimates by typing

```
. use http://www.stata-press.com/data/r8/downs
. cc case exposed [freq=pop], by(age) istandard level(90) pool
```

Maternal age	OR	[90% Conf. Interval]		Weight	
<35	3.394231	.7761115	11.5405	104	(exact)
35+	5.733333	.1856098	61.20938	5	(exact)
Crude	3.501529	1.024302	10.04068		(exact)
Pooled (direct)	3.824166	1.442194	10.14028		
I. Standardized	3.779749	1.423445	10.03657		

Test of homogeneity (direct)   chi2(1) =   0.14   Pr>chi2 = 0.7109

Using the distribution of the nonexposed subjects in the source population as the standard, we can obtain an estimate of the standardized risk ratio (SRR):

```
. cc case exposed [freq=pop], by(age) estan level(90)
```

Maternal age	OR	[90% Conf. Interval]		Weight	
<35	3.394231	.7761115	11.5405	1059	(exact)
35+	5.733333	.1856098	61.20938	86	(exact)
Crude	3.501529	1.024302	10.04068		(exact)
E. Standardized	3.979006	1.430689	11.06634		

Finally, if we wanted to weight the two age groups equally, we could type

```
. generate wgt=1
. cc case exposed [freq=pop], by(age) level(90) standard(wgt)
```

Maternal age	OR	[90% Conf. Interval]		Weight	
<35	3.394231	.7761115	11.5405	1	(exact)
35+	5.733333	.1856098	61.20938	1	(exact)
Crude	3.501529	1.024302	10.04068		(exact)
Standardized	5.275104	.8787523	31.66617		

◁

Matched case–control data

Matched case–control studies are performed to gain sample size efficiency and to control for important confounding factors. In a matched case–control design, each case is matched with a control on the basis of demographic characteristics, clinical characteristics, etc. Thus, their difference with respect to the outcome must be due to something other than the matching variables. If the only difference between them was exposure to the factor, one could attribute any difference in outcome to the factor.

A summary of the data is

	Controls		
Cases	Exposed	Unexposed	Total
Exposed	a	b	$a + b$
Unexposed	c	d	$c + d$
Total	$a + c$	$b + d$	$n = a + b + c + d$

Each entry in the table represents the number of case–control pairs. For instance, in a of the pairs both members were exposed; in b of the pairs the case was exposed but the control was not, and so on. In total, n pairs were observed.

▷ Example

Rothman (1986, 257) discusses data from Jick et al. (1973) on a matched case–control study of myocardial infarction and drinking six or more cups of coffee per day (persons drinking from 1 to 5 cups per day were excluded):

	Controls	
Cases	6+ cups	0 cups
6+ cups	8	8
0 cups	3	8

mcci analyzes matched case–control data:

```
. mcci 8 8 3 8
                     | Controls
        Cases        | Exposed    Unexposed  |        Total
---------------------+-----------------------+--------------
            Exposed  |       8            8  |           16
          Unexposed  |       3            8  |           11
---------------------+-----------------------+--------------
              Total  |      11           16  |           27

McNemar's chi2(1) =       2.27     Prob > chi2 = 0.1317
Exact McNemar significance probability       = 0.2266

Proportion with factor
        Cases        .5925926
        Controls     .4074074        [95% Conf. Interval]
                     ---------
        difference   .1851852        -.0822542     .4526246
        ratio        1.454545         .891101      2.374257
        rel. diff.   .3125           -.0243688     .6493688

        odds ratio   2.666667         .6400699     15.60439   (exact)
```

The relationship is not significant at better than the 13.17% level, but if one justifies a one-sided test, the table is significant at the $13.17/2 = 6.59\%$ level. The point estimate is that drinkers of 6+ cups of coffee per day are 2.67 times more likely to suffer myocardial infarction. The interpretation of the relative difference is that for every 100 controls who fail to have heart attacks, 31.25 might be expected to get heart attacks if they became heavy coffee drinkers.

◁

mcc works like the other nonimmediate commands but does not handle stratified data. If you have stratified matched case–control data, you can use conditional logistic regression to estimate odds ratios; see [R] **clogit**.

Matched case–control studies can also be analyzed using mhodds by controlling on the variable used to identify the matched sets. For example, if the variable set is used to identify the matched set for each subject,

```
. mhodds fail xvar set
```

will do the job. Note that any attempt to control for further variables will restrict the analysis to the comparison of cases and matched controls that share the same values of these variables. In general, this would lead to the omission of many records from the analysis. Similar considerations usually apply when investigating effect modification using the by() option. An important exception to this general rule is that a variable used in matching cases to controls may appear in the by() option without loss of data.

▷ Example

Let us use mhodds to analyze matched case–control studies using the study of endometrial cancer and exposure to estrogen described in Breslow and Day (1980, chapter 4). In this study, there are four controls matched to each case. Cases and controls were matched on age, marital status, and time living in the community. The data collected included information on the daily dose of conjugated estrogen therapy. Breslow and Day created four levels of the dose variable and began by analyzing the 1:1 study formed by using the first control in each set. We examine the effect of exposure to estrogen:

```
. use http://www.stata-press.com/data/r8/bdendo11, clear
. describe
Contains data from http://www.stata-press.com/data/r8/bdendo11.dta
  obs:           126
  vars:           13                          3 Sep 2002 12:25
  size:         2,898 (99.9% of memory free)
```

variable name	storage type	display format	value label	variable label
set	int	%8.0g		Set number
fail	byte	%8.0g		Case=1/Control=0
gall	byte	%8.0g		Gallbladder dis
hyp	byte	%8.0g		Hypertension
ob	byte	%8.0g		Obesity
est	byte	%8.0g		Estrogen
dos	byte	%8.0g		Ordinal dose
dur	byte	%8.0g		Ordinal duration
non	byte	%8.0g		Non-estrogen drug
duration	int	%8.0g		months
age	int	%8.0g		years
cest	byte	%8.0g		Conjugated est dose
agegrp	float	%9.0g		age group of set

```
Sorted by:  set
. mhodds fail est set
Mantel-Haenszel estimate of the odds ratio
Comparing est==1 vs. est==0, controlling for set
note: only 32 of the 63 strata formed in this analysis contribute
      information about the effect of the explanatory variable
```

Odds Ratio	chi2(1)	P>chi2	[95% Conf. Interval]	
9.666667	21.12	0.0000	2.944702	31.733072

In the case of the 1:1 matched study, the Mantel–Haenszel methods are equivalent to conditional likelihood methods. The maximum conditional likelihood estimate of the odds ratio is given by the ratio of the off-diagonal frequencies in the following table:

```
. tabulate case control [fweight=freq]

           |      control
      case |      0          1  |   Total
-----------+--------------------+--------
         0 |      4          3  |      7
         1 |     29         27  |     56
-----------+--------------------+--------
     Total |     33         30  |     63
```

This is $29/3 = 9.67$, which agrees exactly with the value obtained from mhodds and from mcci. In the more general $1:m$ matched study, however, the Mantel–Haenszel methods are no longer equal to the maximum conditional likelihood, although they are usually quite close.

To illustrate the use of the by() option in matched case–control studies, we look at the effect of exposure to estrogen, stratified by age3, which codes the sets into three age groups (55–64, 65–74, and 75+) as follows:

```
. use http://www.stata-press.com/data/r8/bdendo11, clear

. generate age3 =agegrp

. recode age3 1/2=1 3/4=2 5/6=3
(age3: 124 changes made)

. mhodds fail est set, by(age3)

Mantel-Haenszel estimate of the odds ratio
Comparing est==1 vs. est==0, controlling for set
by age3

note: only 32 of the 63 strata formed in this analysis contribute
      information about the effect of the explanatory variable
```

age3	Odds Ratio	chi2(1)	P>chi2	[95% Conf. Interval]	
1	6.000000	3.57	0.0588	0.722351	49.83724
2	15.000000	12.25	0.0005	1.981409	113.5556
3	8.000000	5.44	0.0196	1.000586	63.96252

```
Mantel-Haenszel estimate controlling for set and age3
```

Odds Ratio	chi2(1)	P>chi2	[95% Conf. Interval]	
9.666667	21.12	0.0000	2.944702	31.733072

```
Test of homogeneity of ORs (approx): chi2(2)  =    0.41
                               Pr>chi2  =  0.8128
```

Note that there is no further loss of information when we stratify by age3 because age was one of the matching variables.

The full set of matched controls can be used in the same way. For example, the effect of exposure to estrogen is obtained (using the full dataset) by

```
. use http://www.stata-press.com/data/r8/bdendo, clear

. mhodds fail est set

Mantel-Haenszel estimate of the odds ratio
Comparing est==1 vs. est==0, controlling for set
note: only 58 of the 63 strata formed in this analysis contribute
      information about the effect of the explanatory variable
```

Odds Ratio	chi2(1)	P>chi2	[95% Conf. Interval]	
8.461538	31.16	0.0000	3.437773	20.826746

The effect of exposure to estrogen, stratified by age3, is obtained by

```
. generate age3 = agegrp

. recode age3 1/2=1 3/4=2 5/6=3
(310 changes made)

. mhodds fail est set, by(age3)

Mantel-Haenszel estimate of the odds ratio
Comparing est==1 vs. est==0, controlling for set
by age3
note: only 58 of the 63 strata formed in this analysis contribute
      information about the effect of the explanatory variable
```

age3	Odds Ratio	chi2(1)	P>chi2	[95% Conf. Interval]	
1	3.800000	3.38	0.0660	0.821651	17.57438
2	10.666667	18.69	0.0000	2.787731	40.81376
3	13.500000	9.77	0.0018	1.598317	114.0262

Mantel-Haenszel estimate controlling for set and age3

Odds Ratio	chi2(1)	P>chi2	[95% Conf. Interval]	
8.461538	31.16	0.0000	3.437773	20.826746

```
Test of homogeneity of ORs (approx): chi2(2)   =    1.41
                                      Pr>chi2  =  0.4943
```

◁

Saved Results

ir and iri save in r():

Scalars

r(p)	one-sided p-value	r(afe)	attributable (prev.) fraction among exposed
r(ird)	incidence rate difference	r(lb_afe)	lower bound of CI for afe
r(lb_ird)	lower bound of CI for ird	r(ub_afe)	upper bound of CI for afe
r(ub_ird)	upper bound of CI for ird	r(afp)	attributable fraction for the population
r(irr)	incidence rate ratio	r(chi2_mh)	Mantel–Haenszel heterogeneity χ^2 (ir only)
r(lb_irr)	lower bound of CI for irr	r(chi2_p)	pooled heterogeneity χ^2 (pool only)
r(ub_irr)	upper bound of CI for irr	r(df)	degrees of freedom (ir only)

cs and csi save in r():

Scalars

r(p)	two-sided p-value	r(ub_or)	upper bound of CI for or
r(rd)	risk difference	r(afe)	attributable (prev.) fraction among exposed
r(lb_rd)	lower bound of CI for rd	r(lb_afe)	lower bound of CI for afe
r(ub_rd)	upper bound of CI for rd	r(ub_afe)	upper bound of CI for afe
r(rr)	risk ratio	r(afp)	attributable fraction for the population
r(lb_rr)	lower bound of CI for rr	r(chi2_mh)	Mantel–Haenszel heterogeneity χ^2 (cs only)
r(ub_rr)	upper bound of CI for rr	r(chi2_p)	pooled heterogeneity χ^2 (pool only)
r(or)	odds ratio	r(df)	degrees of freedom
r(lb_or)	lower bound of CI for or	r(chi2)	χ^2

cc and cci save in r():

Scalars

r(p)	two-sided p-value	r(lb_afe)	lower bound of CI for afe
r(p1_exact)	χ^2 or one-sided exact significance	r(ub_afe)	upper bound of CI for afe
r(p_exact)	two-sided significance (χ^2 or exact)	r(afp)	attributable fraction for the population
		r(chi2_p)	pooled heterogeneity χ^2
r(or)	odds ratio	r(chi2_bd)	Breslow–Day χ^2
r(lb_or)	lower bound of CI for or	r(df_bd)	degrees of freedom for Breslow–Day χ^2
r(ub_or)	upper bound of CI for or	r(df)	degrees of freedom
r(afe)	attributable (prev.) fraction among exposed	r(chi2)	χ^2

tabodds saves in r():

Scalars

r(odds)	odds	r(p_hom)	p-value for test of homogeneity
r(lb_odds)	lower bound for odds	r(df_hom)	degrees of freedom for χ^2 test of homogeneity
r(ub_odds)	upper bound for odds	r(chi2_tr)	χ^2 for score test for trend
r(chi2_hom)	χ^2 test of homogeneity	r(p_trend)	p-value for score test for trend

mhodds saves in r():

Scalars

r(p)	two-sided p-value	r(chi2_hom)	χ^2 test of homogeneity
r(or)	odds ratio	r(df_hom)	degrees of freedom for χ^2 test of homogeneity
r(lb_or)	lower bound of CI for or	r(chi2)	χ^2
r(ub_or)	upper bound of CI for or		

mcc and mcci save in r():

Scalars

r(p_exact)	two-sided significance (χ^2 or exact)	r(R_f)	ratio of proportion with factor
		r(lb_R_f)	lower bound of CI for R_f
r(or)	odds ratio	r(ub_R_f)	upper bound of CI for R_f
r(lb_or)	lower bound of CI for or	r(RD_f)	relative difference in proportion with factor
r(ub_or)	upper bound of CI for or		
r(D_f)	difference in proportion with factor	r(lb_RD_f)	lower bound of CI for RD_f
r(lb_D_f)	lower bound of CI for D_f	r(ub_RD_f)	upper bound of CI for RD_f
r(ub_D_f)	upper bound of CI for D_f	r(chi2)	χ^2

Methods and Formulas

All of the epitab commands are implemented as ado-files.

The notation for incidence-rate data is

	Exposed	Unexposed	Total
Cases	a	b	M_1
Person-time	N_1	N_0	T

The notation for $2 \times k$ tables is

	\multicolumn{4}{c}{Exposure level}				Total
	1	2	...	k	Total
Cases	a_1	a_2	...	a_k	M_1
Controls	c_1	c_2	...	c_k	M_0
Total	N_1	N_2	...	N_k	T

If tables are stratified, all quantities are indexed by i, the stratum number.

We will refer to Fleiss (1981), Kleinbaum, Kupper, and Morgenstern (1982), and Rothman (1986) so often that we will adopt the notation F-23 to mean Fleiss (1981) page 23, KKM-52 to mean Kleinbaum et al. (1982) page 52, and R-164 to mean Rothman (1986) page 164.

It is also worth noting that, in all cases, we have avoided making the continuity corrections to χ^2 statistics, following the advice of KKM-292: "[...] the use of a continuity correction has been the subject of considerable debate in the statistical literature [...] On the basis of our evaluation of this debate and other evidence, we do *not* recommend the use of the continuity correction". Breslow and Day (1980, 133), on the other hand, argue for inclusion of the correction, but not strongly. Their summary is that for very small datasets, one should use exact statistics. In practice, we believe the adjustment makes little difference for reasonably sized datasets.

Unstratified incidence rate data

The incidence rate difference is defined $I_d = a/N_1 - b/N_0$ (R-164). The standard error of the incidence rate is $s_{I_d} \approx \sqrt{a/N_1^2 + b/N_0^2}$ (R-170), from which confidence intervals are calculated. For test-based confidence intervals, define

$$\chi = \frac{a - N_1 M_1 / T}{\sqrt{M_1 N_1 N_0 / T^2}}$$

(R-155). Test-based confidence intervals are $I_d(1 \pm z/\chi)$ (R-171), where z is obtained from the normal distribution.

The incidence rate ratio is defined $I_r = (a/N_1)/(b/N_0)$ (R-164). Let p_l and p_h be the exact confidence interval of the binomial probability for observing a successes out of M_1 trials (obtained from cii, see [R] ci). The exact confidence interval for the incidence ratio is then $(p_l N_0)/\{(1-p_l)N_1\}$ to $(p_u N_0)/\{(1-p_u)N_1\}$ (R-166). Test-based confidence intervals are $I_r^{1 \pm z/\chi}$ (R-172).

The attributable fraction among exposed is defined AFE $= (I_r - 1)/I_r$ for $I_r \geq 1$ (KKM-164, R-38); the confidence interval is obtained by similarly transforming the interval values of I_r. The attributable fraction for the population is AF $=$ AFE $\cdot a/M_1$ (KKM-161); no confidence interval is reported. For $I_r < 1$, the prevented fraction among exposed is defined PFE $= 1 - I_r$ (KKM-166, R-39); the confidence interval is obtained by similarly transforming the interval values of I_r. The prevented fraction for the population is PF $=$ PFE $\cdot N_1/T$ (KKM-165); no confidence interval is reported.

The "midp" one-sided exact significance (R-155) is calculated as the binomial probability (with $n = M_1$ and $p = N_1/T$) $\Pr(k = a)/2 + \Pr(k > a)$ if $I_r \geq 1$ and $\Pr(k = a)/2 + \Pr(k < a)$ otherwise. The two-sided significance is twice the one-sided significance (R-155). If preferred, one can obtain non-midp exact probabilities (and, to some ways of thinking, a more reasonable definition of two-sided significance) using bitest; see [R] **bitest**.

Unstratified cumulative incidence data

The risk difference is defined $R_d = a/N_1 - b/N_0$ (R-164). Its standard error is

$$s_{R_d} \approx \left\{ \frac{a(N_1 - a)}{N_1^3} + \frac{b(N_0 - b)}{N_0^3} \right\}^{1/2}$$

(R-172), from which confidence intervals are calculated. For test-based confidence intervals, define

$$\chi = \frac{a - N_1 M_1/T}{\sqrt{(M_1 M_0 N_1 N_0)/\{T^2(T - 1)\}}}$$

(R-163). Test-based confidence intervals are $R_d(1 \pm z/\chi)$ (R-172).

The risk ratio is defined $R_r = (a/N_1)/(b/N_0)$ (R-165). The standard error of $\ln R_r$ is

$$s_{\ln R_r} \approx \left(\frac{c}{aN_1} + \frac{d}{bN_0} \right)^{1/2}$$

(R-173), from which confidence intervals are calculated. Test-based confidence intervals are $R_r^{1 \pm z/\chi}$ (R-173).

For $R_r \geq 1$, the attributable fraction among the exposed is calculated as AFE $= (R_r - 1)/R_r$ (KKM-164, R-38); the confidence interval is obtained by similarly transforming the interval values for R_r. The attributable fraction for the population is calculated as AF $=$ AFE $\cdot\, a/M_1$ (KKM-161); no confidence interval is reported, but F-76 provides

$$\left\{ \frac{c + (a + d)\text{AFE}}{bT} \right\}^{1/2}$$

as the approximate standard error of $\ln(1 - \text{AF})$.

For $R_r < 1$, the prevented fraction among the exposed is calculated as PFE $= 1 - R_r$ (KKM-166, R-39); the confidence interval is obtained by similarly transforming the interval values for R_r. The prevented fraction for the population is calculated as PF $=$ PFE $\cdot\, N_1/T$; no confidence interval is reported.

The odds ratio is defined $\psi = (ad)/(bc)$ (R-165). The Woolf estimate (Woolf 1955) of the standard error of $\ln \psi$ is

$$s_{\ln \psi} = \left(\frac{1}{a} + \frac{1}{b} + \frac{1}{c} + \frac{1}{d} \right)^{1/2}$$

(R-173; Schlesselman 1982, 176), from which confidence intervals are calculated. Test-based confidence intervals are $\psi^{1 \pm z/\chi}$ (R-174). Alternatively, the Cornfield (1956) calculation is

$$\psi_l = a_l(M_0 - N_1 + a_l) \Big/ \big\{ (N_1 - a_l)(M_1 - a_l) \big\}$$

$$\psi_u = a_u(M_0 - N_1 + a_u) \Big/ \big\{ (N_1 - a_u)(M_1 - a_u) \big\}$$

where a_u and a_l are determined iteratively from

$$a_{i+1} = a \pm z_\alpha \left(\frac{1}{a_i} + \frac{1}{N_1 - a_i} + \frac{1}{M_1 - a_i} + \frac{1}{M_0 - N_1 + a_i} \right)^{-1/2}$$

where z_α is the index from the normal distribution for an α significance level (Schlesselman 1982, 177, but without the continuity correction). a_{i+1} converges to a_u using the plus signs and a_l using the minus signs. a_0 is taken as a. With small numbers, the iterative technique may fail. It is then restarted by decrementing (a_l) or incrementing (a_u) a_0. If that fails, a_0 is again decremented or incremented and iterations restarted, and so on, until a terminal condition is met ($a_0 < 0$ or $a_0 > M_1$), at which point the value is not calculated.

The χ^2 is defined

$$\chi^2 = \sum_{i=1}^{2} \sum_{j=1}^{2} \frac{(p_{ij} - p_{i\cdot}p_{\cdot j})^2}{p_{i\cdot}p_{\cdot j}}$$

(F-22, but without the continuity correction) where $p_{11} = a/T$, $p_{12} = b/T$, etc.

Fisher's exact p is calculated as described in [R] **tabulate**.

Unstratified case–control data

Calculation of the odds ratio ψ and χ^2 is as described for unstratified cumulative incidence data. The other calculations described there are inappropriate.

The odds ratio ψ is used as an estimate of the risk ratio in calculating attributable or prevented fractions. For $\psi \geq 1$, the attributable fraction among the exposed is calculated as AFE $= (\psi - 1)/\psi$ (KKM-164); the confidence interval is obtained by similarly transforming the interval values for ψ. The attributable fraction for the population is calculated as AF $=$ AFE $\cdot a/M_1$ (KKM-161). No confidence interval is reported; however, F-94 provides

$$\left(\frac{a}{M_1 b} + \frac{c}{M_0 d} \right)^{1/2}$$

as the standard error of $\ln(1 - \text{AF})$.

For $\psi < 1$, the prevented fraction among the exposed is calculated as PFE $= 1 - \psi$ (KKM-166); the confidence interval is obtained by similarly transforming the interval values for ψ. The prevented fraction for the population is calculated as PF $= \{(a/M_1)\text{PFE}\}/\{(a/M_1)\text{PFE} + \psi\}$ (KKM-164); no confidence interval is reported.

Unstratified matched case–control data

The columns of the 2×2 table reflect controls; the rows, cases. Each entry in the table reflects a pair of a matched case and control.

McNemar's χ^2 (McNemar 1947) is defined as

$$\chi^2 = \frac{(b - c)^2}{b + c}$$

(R-259).

The proportion of controls with the factor is $p_1 = N_1/T$, and the proportion of cases with the factor is $p_2 = M_1/T$.

The difference in the proportions is $P_d - p_2 - p_1$. An estimate of its standard error when the two underlying proportions are *not* hypothesized to be equal is

$$s_{P_d} \approx \frac{\{(a+d)(b+c) + 4bc\}^{1/2}}{T^{3/2}}$$

(F-117), from which confidence intervals are calculated.

The ratio of the proportions is $P_r = p_2/p_1$ (R-276, R-278). The standard error of $\ln P_r$ is

$$s_{\ln P_r} \approx \left(\frac{b+c}{M_1 N_1}\right)^{1/2}$$

(R-276), from which confidence intervals are calculated.

The relative difference in the proportions $P_e = (b-c)/(b+d)$ (F-118) is a measure of the relative value of the factor under the assumption that the factor can affect only those patients who are unexposed controls. Its standard error is

$$s_{P_e} \approx (b+d)^{-2} \{(b+c+d)(bc + bd + cd) - bcd\}^{-1/2}$$

(F-118), from which confidence intervals are calculated.

The odds ratio is $\psi = b/c$ (F-115), and the exact Fisher confidence interval is obtained by transforming into odds ratios the exact binomial confidence interval for the binomial parameter from observing b successes in $b+c$ trials (R-264). Binomial confidence limits are obtained from `cii` (see [R] **ci**) and are transformed by $p/(1-p)$. Test-based confidence intervals are $\psi^{1 \pm z/\chi}$ (R-267) where χ is the square root of McNemar's χ^2, $(b-c)/\sqrt{b+c}$.

Stratified incidence-rate data

Statistics presented for each stratum are calculated independently according to the formulas in *Unstratified incidence-rate data* above. Within strata, the Mantel–Haenszel style weight is $W_i = b_i N_{1i}/T_i$ and the combined incidence rate ratio (Rothman and Boice 1982) is

$$I_{\text{mh}} = \frac{\sum_i a_i N_{0i}/T_i}{\sum_i W_i}$$

(R-196). The standard error is obtained by considering each a_i to be an independent binomial variate conditional on N_{1i} (Greenland and Robins 1985)

$$s_{\ln I_{\text{mh}}} \approx \left\{ \frac{\sum_i M_{1i} N_{1i} N_{0i}/T_i^2}{\left(\sum_i a_i N_{0i}/T_i\right)\left(\sum_i b_i N_{1i}/T_i\right)} \right\}^{1/2}$$

(R-213), from which confidence intervals are calculated.

For standardized rates, let w_i be the user-specified weight within category i. The standardized rate difference and rate ratio are defined as

$$\mathrm{SRD} = \frac{\sum_i w_i (R_{1i} - R_{0i})}{\sum_i w_i}$$

$$\mathrm{SRR} = \frac{\sum_i w_i R_{1i}}{\sum_i w_i R_{0i}}$$

(R-229). The standard error of SRD is

$$s_{\mathrm{SRD}} \approx \left\{ \frac{1}{(\sum_i w_i)^2} \sum_i w_i^2 \left(\frac{a_i}{N_{1i}^2} + \frac{b_i}{N_{0i}^2} \right) \right\}^{1/2}$$

(R-231), from which confidence intervals are calculated. The standard error of $\ln(\mathrm{SRR})$ is

$$s_{\ln(\mathrm{SRR})} \approx \left\{ \frac{\sum_i w_i^2 a_i / N_{1i}^2}{(\sum_i w_i R_{1i})^2} + \frac{\sum_i w_i^2 b_i / N_{0i}^2}{(\sum_i w_i R_{0i})^2} \right\}^{1/2}$$

(R-231) from which confidence intervals are calculated.

Internally and externally standardized measures are calculated using $w_i = N_{1i}$ and $w_i = N_{0i}$, respectively.

For directly pooled estimates of risk differences and risk ratios, stratum-specific point estimates and variances are calculated as given in *Unstratified incidence rate data*. (Note that for the risk ratio, this calculation is performed in the logs.) Inverse variance weights $w_i = 1/s^2$ are then used to sum the stratum-specific values (which are then exponentiated in the case of the risk ratio). The overall variance of the pooled estimate is calculated as $1/\sum w_i$, from which confidence intervals are calculated (R-183–188).

For risk differences, the χ^2 test of homogeneity is calculated as $(\sum R_{di} - \widehat{R}_d)^2 / \mathrm{var}(R_{di})$ where R_{di} are the stratum-specific rate differences and $\widehat{R}_d$ is the pooled estimate (Mantel–Haenszel or directly pooled). Degrees of freedom are one less than the number of strata (R-222).

For risk ratios, the same calculation is made except that it is made on a logarithmic scale using $\ln(R_{ri})$ (R-222).

Stratified cumulative incidence data

Statistics presented for each stratum are calculated independently according to the formulas in *Unstratified cumulative incidence data* above. The Mantel–Haenszel χ^2 test (Mantel and Haenszel 1959) is

$$\chi_{\mathrm{mh}}^2 = \frac{\left(\left| \sum_i a_i - N_{1i} M_{1i} / T_i \right| \right)^2}{\sum_i (N_{1i} N_{0i} M_{1i} M_{0i}) / \left\{ T_i^2 (T_i - 1) \right\}}$$

(R-206).

For the odds ratio, the Mantel–Haenszel weight is $W_i = b_i c_i / T_i$ and the combined odds ratio (Mantel and Haenszel 1959) is

$$\psi_{\mathrm{mh}} = \frac{\sum_i a_i d_i / T_i}{\sum_i W_i}$$

(R-195). The standard error (Robins, Breslow, and Greenland 1986) is

$$s_{\ln \psi_{\text{mh}}} \approx \left\{ \frac{\sum_i P_i R_i}{2\left(\sum_i R_i\right)^2} + \frac{\sum_i P_i S_i + Q_i R_i}{2 \sum_i R_i \sum_i S_i} + \frac{\sum_i Q_i S_i}{2\left(\sum_i S_i\right)^2} \right\}^{1/2}$$

where

$$P_i = (a_i + d_i)/T_i$$
$$Q_i = (b_i + c_i)/T_i$$
$$R_i = a_i d_i/T_i$$
$$S_i = b_i c_i/T_i$$

(R-220). Alternatively, test-based confidence intervals are calculated as $\psi_{\text{mh}}^{1 \pm z/\chi}$ (R-220).

For the risk ratio the Mantel–Haenszel style weight is $W_i = b_i N_{1i}/T_i$ and the combined risk ratio (Rothman and Boice 1982) is

$$R_{\text{mh}} = \frac{\sum_i a_i N_{0i}/T_i}{\sum_i W_i}$$

(R-196). The standard error (Greenland and Robins 1985) is

$$s_{\ln R_{\text{mh}}} \approx \frac{\sum_i (M_{1i} N_{1i} N_{0i} - a_i b_i T_i)/T_i^2}{\left(\sum_i a_i N_{0i}/T_i\right)\left(\sum_i b_i N_{1i}/T_i\right)}$$

(R-216) from which confidence intervals are calculated.

For standardized rates, let w_i be the user-specified weight within category. The standardized rate difference (SRD) and rate ratios (SRR) are defined as above (*Stratified incidence rate data*), where the individual risks are defined $R_{1i} = a_i/N_{1i}$ and $R_{0i} = b_i/N_{0i}$. The standard error of SRD is

$$s_{\text{SRD}} \approx \left[\frac{1}{(\sum_i w_i)^2} \sum_i w_i^2 \left\{ \frac{a_i(N_{1i} - a_i)}{N_{1i}^3} + \frac{b_i(N_{0i} - b_i)}{N_{0i}^3} \right\} \right]^{1/2}$$

(R-231), from which confidence intervals are calculated. The standard error of ln(SRR) is

$$s_{\ln(\text{SRR})} \approx \left\{ \frac{\sum_i w_i^2 a_i(N_{1i} - a_i)/N_{1i}^3}{(\sum_i w_i R_{1i})^2} + \frac{\sum_i w_i^2 b_i(N_{0i} - b_i)/N_{0i}^3}{(\sum_i w_i R_{0i})^2} \right\}^{1/2}$$

(R-231), from which confidence intervals are calculated.

Internally and externally standardized measures are calculated using $w_i = N_{1i}$ and $w_i = N_{0i}$, respectively.

For directly pooled estimates of the odds ratio, stratum-specific odds ratios are summed with stratum-specific inverse variance weights with the calculation performed in the logs and the point estimate and variances being as given in *Unstratified cumulative incidence data*. The overall variance of the pooled estimate is calculated as the sum of the inverse weights, from which confidence intervals are calculated (R-189–190).

For pooled estimates of risk differences and risk ratios, stratum-specific point estimates and variances are calculated as given in *Unstratified cumulative incidence data*. (Note that for the risk ratio, this calculation is performed in the logs.) Inverse variance weights $w_i = 1/s^2$ are then used to sum the stratum-specific values (which are then exponentiated in the case of the risk ratio). The overall variance of the pooled estimate is calculated as $1/\sum w_i$, from which confidence intervals are calculated (R-183–188).

For risk differences, the χ^2 test of homogeneity is calculated as $(\sum R_{di} - \widehat{R}_d)^2/\mathrm{var}(R_{di})$ where R_{di} are the stratum-specific rate differences and $\widehat{R}_d$ is the pooled estimate (Mantel–Haenszel or directly pooled). Degrees of freedom are one less than the number of strata (R-222).

For risk and odds ratios, the same calculation is made except that it is made in the logs using $\ln(R_{ri})$ or $\ln(\psi_i)$ (R-222).

Stratified case–control data

Statistics presented for each stratum are calculated independently according to the formulas in *Unstratified cumulative incidence data* above. The combined odds ratio ψ_{mh} and the test that $\psi_{\mathrm{mh}} = 1$ (χ^2_{mh}) are calculated as described in *Stratified cumulative incidence data* above.

For standardized weights, let w_i be the user-specified weight within category. The standardized odds ratio is calculated as

$$\mathrm{SOR} = \frac{\sum_i w_i a_i/c_i}{\sum_i w_i b_i/d_i}$$

(Greenland 1987a). The standard error of $\ln(\mathrm{SOR})$ is

$$s_{\ln(\mathrm{SOR})} = \frac{\sum_i (w_j a_j/c_j)^2 (\frac{1}{a_i} + \frac{1}{b_i} + \frac{1}{c_i} + \frac{1}{d_i})}{\left(\sum_i w_i a_i/c_i\right)^2}$$

(Greenland 1987a), from which confidence intervals are calculated. The internally and externally standardized odds ratios are calculated using $w_i = c_i$ and $w_i = d_i$, respectively.

The directly pooled estimate of the odds ratio is calculated as described in *Stratified cumulative incidence data* above.

The direct and Mantel–Haenszel χ^2 tests of homogeneity are calculated as $\sum \{\ln(R_{ri}) - \ln(\widehat{R}_r)\}^2/\mathrm{var}\{\ln(R_{ri})\}$ where R_{ri} are the stratum-specific odds ratios and $\widehat{R}_r$ is the pooled estimate (Mantel–Haenszel or directly pooled). The number of degrees of freedom is one less than the number of strata (R-222).

The Breslow–Day χ^2 test of homogeneity is calculated as the sum over all strata of the stratum specific squared deviations of the observed and fitted values based on the overall odds ratio, divided by the variance of the fitted values:

$$\sum_i \frac{\{a_i - A_i(\widehat{\psi})\}^2}{\mathrm{Var}(a_i;\widehat{\psi})}$$

where $A_i(\widehat{\psi})$ is the fitted count for cell a determined as the root of the quadratic equation

$$A(M_0 - N_1 + A) = (\widehat{\psi})(M_1 - A)(N_1 - A)$$

which makes all cell values of the 2×2 table for the ith stratum positive, $\mathrm{Var}(a_i;\widehat{\psi})$ is the variance of the fitted table for the ith stratum, and $\widehat{\psi}$ is the Mantel–Haenszel estimate of the common odds ratio.

By default, both `tabodds` and `mhodds` produce test statistics and confidence intervals based on score statistics (Clayton and Hills, 1993). Using the notation for $2 \times k$ tables, the confidence interval for the odds of the ith exposure level, odds_i, $i = 1, \ldots, k$, is given by

$$\mathrm{odds}_i \cdot \exp\left(\pm z\sqrt{1/a_i + 1/c_i}\right)$$

The score χ^2 test of homogeneity of odds is calculated as

$$\chi^2_{k-1} = \frac{T(T-1)}{M_1 M_0} \sum_{i=1}^{k} \frac{(a_i - E_i)^2}{N_i}$$

where $E_i = (M_1 N_i)/T$.

Let l_i denote the value of the exposure at the ith level. The score χ^2 test for trend of odds is calculated as

$$\chi^2_1 = \frac{U^2}{V}$$

where

$$U = \frac{M_1 M_0}{T} \left(\sum_{i=1}^{k} \frac{a_i l_i}{M_1} - \sum_{i=1}^{k} \frac{c_i l_i}{M_0} \right)$$

and

$$V = \frac{M_1 M_0}{T} \left\{ \frac{\sum_{i=1}^{k} N_i l_i^2 - (\sum_{i=1}^{k} N_i l_i)^2/T}{T-1} \right\}$$

Acknowledgments

We would like to thank Hal Morgenstern, Department of Epidemiology, UCLA School of Public Health; Ardythe Morrow, Center for Pediatric Research, Norfolk, Virginia; and the late Stewart West, Baylor College of Medicine, for their assistance in designing these commands. We also extend our appreciation to Jonathan Freeman, Department of Epidemiology, Harvard School of Public Health, for encouraging us to extend these commands to include tests for homogeneity, for helpful comments on the default behavior of the commands, and for his comments on an early draft of this section. We would also like to thank David Clayton, MRC Biostatistical Research Unit, Cambridge, and Michael Hills, London School of Hygiene and Tropical Medicine (retired); the original versions of mhodds and tabodds were written by them. Finally, we would like to thank William Dupont and Dale Plummer for their contribution to the implementation of exact confidence intervals for the odds ratios for cc and cci.

References

Abramson, J. H. and Z. H. Abramson. 2001. *Making Sense of Data: A Self-Instruction Manual on the Interpretation of Epidemiological Data*. 3d ed. New York: Oxford University Press.

Boice, J. D. and R. R. Monson. 1977. Breast cancer in women after repeated fluoroscopic examinations of the chest. *Journal of the National Cancer Institute* 59: 823–832.

Breslow, N. E. and N. E. Day. 1980. *Statistical Methods in Cancer Research*, vol. 1. Lyon: International Agency for Research on Cancer.

Carlin, J. and S. Vidmar. 2000. sbe35: Menus for epidemiological statistics. *Stata Technical Bulletin* 56: 15–16.

Clayton, D. and M. Hills. 1993. *Statistical Models in Epidemiology*. Oxford: Oxford University Press.

———. 1995. ssa8: Analysis of case–control and prevalence studies. *Stata Technical Bulletin* 27: 26–31. Reprinted in *Stata Technical Bulletin Reprints*, vol. 5, pp. 227–233.

Cornfield, J. 1956. A statistical problem arising from retrospective studies. In *Proceedings of the Third Berkeley Symposium* vol. 4, ed. J. Neyman, 135–148. Berkeley, CA: University of California Press.

Doll, R. and A. B. Hill. 1966. Mortality of British doctors in relation to smoking: observations on coronary thrombosis. In *Epidemiological Approaches to the Study of Cancer and Other Chronic Diseases*, ed. W. Haenszel. *National Cancer Institute Monograph* 19: 205–268.

Dupont, W. D. 2003. *Statistical Modeling for Biomedical Researchers*. Cambridge: Cambridge University Press. (*Forthcoming in first quarter 2003*.)

Dupont, W. D. and D. Plummer. 1999. sbe31: Exact confidence intervals for odds ratios from case–control studies. *Stata Technical Bulletin* 52: 12–16. Reprinted in *Stata Technical Bulletin Reprints*, vol. 9, pp. 150–154.

Fisher, L. D. and G. van Belle. 1993. *Biostatistics: A Methodology for the Health Sciences*. New York: John Wiley & Sons.

Fleiss, J. L. 1981. *Statistical Methods for Rates and Proportions*. 2d ed. New York: John Wiley & Sons.

Glass, R. I., A. M. Svennerholm, B. J. Stoll, M. R. Khan, K. M. B. Hossain, M. I. Huq, and J. Holmgren. 1983. Protection against cholera in breast-fed children by antibodies in breast milk. *New England Journal of Medicine* 308: 1389–1392.

Gleason, J. R. 1999. sbe30: Improved confidence intervals for odds ratios. *Stata Technical Bulletin* 51: 24–27. Reprinted in *Stata Technical Bulletin Reprints*, vol. 9, pp. 146–150.

Greenland, S. 1987a. Interpretation and choice of effect measures in epidemiologic analysis. *American Journal of Epidemiology* 125: 761–768.

——, ed. 1987b. *Evolution of Epidemiologic Ideas: Annotated Readings on Concepts and Methods*. Newton Lower Falls, MA: Epidemiology Resources.

Greenland, S. and J. M. Robins. 1985. Estimation of a common effect parameter from sparse follow-up data. *Biometrics* 41: 55–68.

——. 1988. Conceptual problems in the definition and interpretation of attributable fractions. *American Journal of Epidemiology* 128: 1185–1197.

Jick, H., O. S. Miettinen, R. K. Neff, S. Shapiro, O. P. Heinonen, and D. Slone. 1973. Coffee and myocardial infarction. *New England Journal of Medicine* 289: 63–67.

Kelsey, J. L., W. D. Thompson, and A. S. Evans. 1986. *Methods in Observational Epidemiology*. New York: Oxford University Press.

Kleinbaum, D. G. 2003. *ActivEpi multimedia CD*. New York: Springer.

Kleinbaum, D. G., L. L. Kupper, and H. Morgenstern. 1982. *Epidemiologic Research*. New York: Van Nostrand Reinhold.

Lilienfeld, D. E. and P. D. Stolley. 1994. *Foundations of Epidemiology*. 3d ed. New York: Oxford University Press.

López-Vizcaíno, M. E., S. Pérez-Hoyos, and L. Abraira-García. 2000. sbe32: Automated outbreak detection from public health surveillance data. *Stata Technical Bulletin* 54: 23–25. Reprinted in *Stata Technical Bulletin Reprints*, vol. 9, pp. 154–157.

——. 2000. sbe32.1: Automated outbreak detection from public health surveillance data: errata. *Stata Technical Bulletin* 55: 2.

MacMahon B., S. Yen, D. Trichopoulos, K. Warren and G. Nardi. 1981. Coffee and cancer of the pancreas. *New England Journal of Medicine* 11: 630–633.

Mantel, N. and W. Haenszel. 1959. Statistical aspects of the analysis of data from retrospective studies of disease. *Journal of the National Cancer Institute* 22: 719–748. Reprinted in *Evolution of Epidemiologic Ideas*, ed. S. Greenland, 112–141. Newton Lower Falls, MA: Epidemiology Resources.

McNemar, Q. 1947. Note on the sampling error of the difference between correlated proportions or percentages. *Psychometrika* 12: 153–157.

Miettinen, O. S. 1976. Estimability and estimation in case-referent studies. *American Journal of Epidemiology* 103: 226–235. Reprinted in *Evolution of Epidemiologic Ideas*, ed. S. Greenland, 181–190. Newton Lower Falls, MA: Epidemiology Resources.

Pearce, M. S. and R. Feltbower. 2000. sg149: Tests for seasonal data via the Edwards and Walter & Elwood tests. *Stata Technical Bulletin* 56: 47–49.

Reilly, M. and A. Salim. 2000. sxd2: Computing optimal sampling designs for two-stage studies. *Stata Technical Bulletin* 58: 37–41. Reprinted in *Stata Technical Bulletin Reprints*, vol. 10, pp. 376–382.

Robins, J. M., N. E. Breslow, and S. Greenland. 1986. Estimators of the Mantel–Haenszel variance consistent in both sparse data and large-strata limiting models. *Biometrics* 42: 311–323.

Rothman, K. J. 1982. Spermicide use and Down's syndrome. *American Journal of Public Health* 72: 399–401.

——. 1986. *Modern Epidemiology*. Boston: Little, Brown and Company.

——. 2002. *Epidemiology: An Introduction*. New York: Oxford University Press.

Rothman, K. J. and J. D. Boice. 1982. *Epidemiologic Analysis with a Programmable Calculator*. Brookline, MA: Epidemiology Resources. (First edition published June 1979 by U.S. Government Printing Office, Washington D.C., NIH Publication No. 79-1649.)

Rothman, K. J., D. C. Fyler, A. Goldblatt, and M. B. Kreidberg. 1979. Exogenous hormones and other drug exposures of children with congenital heart disease. *American Journal of Epidemiology* 109: 433–439.

Rothman, K. J. and S. Greenland. 1998. *Modern Epidemiology*. 2d ed. Philadelphia: Lippincott–Raven.

Rothman, K. J. and R. R. Monson. 1973. Survival in trigeminal neuralgia. *Journal of Chronic Diseases* 26: 303–309.

Royston, P. and A. Babiker. 2002. A menu-drive facility for complex sample size calculation in randomized controlled trials with a survival or a binary outcome. *The Stata Journal* 2: 151–163.

Schlesselman, J. J. 1982. *Case–Control Studies: Design, Conduct, Analysis*. New York: Oxford University Press.

University Group Diabetes Program. 1970. A study of the effects of hypoglycemic agents on vascular complications in patients with adult onset diabetes. *Diabetes* 19, supplement 2: 747–830.

Walker, A. M. 1991. *Observation and Inference: An Introduction to the Methods of Epidemiology*. Newton Lower Falls, MA: Epidemiology Resources.

Wang, Z. 1999. sbe27: Assessing confounding effects in epidemiological studies. *Stata Technical Bulletin* 49: 12–15. Reprinted in *Stata Technical Bulletin Reprints*, vol. 9, pp. 134–138.

Woolf, B. 1955. On estimating the relation between blood group and disease. *Annals of Human Genetics* 19: 251–253. Reprinted in *Evolution of Epidemiologic Ideas*, ed. S. Greenland, 108–110. Newton Lower Falls, MA: Epidemiology Resources.

Also See

Related: [ST] **st**, [ST] **stcox**,

[R] **bitest**, [R] **ci**, [R] **clogit**, [R] **dstdize**, [R] **logistic**, [R] **poisson**, [R] **tabulate**

Background: [U] **22 Immediate commands**,

[ST] **survival analysis**

Title

> **ltable** — Life tables for survival data

Syntax

ltable *timevar* [*deadvar*] [*weight*] [if *exp*] [in *range*] [, by(*groupvar*) level(*#*)

    <u>surv</u>ival <u>f</u>ailure <u>h</u>azard <u>i</u>ntervals(w | *numlist*) <u>t</u>est <u>tv</u>id(*varname*) <u>noa</u>djust

    <u>nota</u>b graph noconf plot(*plot*) ciopts(*rspike_options*) *connected_options*

    twoway_options]

fweights are allowed; see [U] **14.1.6 weight**.

Description

ltable displays and graphs life tables for individual-level or aggregate data and optionally presents the likelihood-ratio and log-rank tests for equivalence of groups. ltable also allows examining the empirical hazard function through aggregation. Also see [ST] **sts** for alternative commands.

timevar specifies the time of failure or censoring. If *deadvar* is not specified, all values of *timevar* are interpreted as failure times; otherwise, *timevar* is interpreted as a failure time where *deadvar* $\neq 0$ and as a censoring time otherwise. Observations with *timevar* or *deadvar* equal to missing are ignored.

Note carefully that *deadvar* does *not* specify the *number* of failures. An observation with *deadvar* equal to 1 or 50 has the same interpretation—the observation records one failure. Specify frequency weights for aggregated data (e.g., ltable timc [freq=number]).

Options

by(*groupvar*) creates separate tables (or graphs within the same image) for each value of *groupvar*. *groupvar* may be string or numeric.

level(*#*) specifies the confidence level, in percent, for confidence intervals. The default is level(95) or as set by set level; see [R] **level**.

survival, failure, and hazard indicate the table to be displayed. If not specified, the default is the survival table. Specifying failure would display the cumulative failure table. Specifying survival failure would display both the survival and the cumulative failure table. If graph is specified, multiple tables may not be requested.

(Continued on next page)

71

intervals(w | *numlist*) specifies the time intervals into which the data are to be aggregated for tabular presentation. A single numeric argument is interpreted as the width of the interval. For instance, interval(2) aggregates data into the time intervals $0 \leq t < 2$, $2 \leq t < 4$, and so on. Not specifying interval() is equivalent to specifying interval(1). Since in most data, failure times are recorded as integers, this amounts to no aggregation except that implied by the recording of the time variable and so produces Kaplan–Meier product-limit estimates of the survival curve (with an actuarial adjustment; see the noadjust option below). Also see [ST] **sts list**. Although it is possible to examine survival and failure without aggregation, some form of aggregation is almost always required for examining the hazard.

When more than one argument is specified, time intervals are aggregated as specified. For instance, interval(0,2,8,16) aggregates data into the intervals $0 \leq t < 2$, $2 \leq t < 8$, $8 \leq t < 16$, and (if necessary) the open-ended interval $t \geq 16$.

interval(w) is equivalent to interval(0,7,15,30,60,90,180,360,540,720), corresponding to one week, (roughly) two weeks, one month, two months, three months, six months, 1 year, 1.5 years, and 2 years when failure times are recorded in days. The w suggests widening intervals.

test presents two χ^2 measures of the differences between groups when by() is specified. test does nothing if by() is not specified.

tvid(*varname*) is for use with longitudinal data with time-varying parameters. Each subject appears in the data more than once, and equal values of *varname* identify observations referring to the same subject. When tvid() is specified, only the last observation on each subject is used in making the table. The order of the data does not matter, and "last" here means the last observation chronologically.

noadjust suppresses the actuarial adjustment for deaths and censored observations. The default is to consider the adjusted number at risk at the start of the interval as the total at the start minus (the number dead or censored)/2. If noadjust is specified, the number at risk is simply the total at the start, corresponding to the standard Kaplan and Meier assumption. noadjust should be specified when using ltable to list results corresponding to those produced by sts list; see [ST] **sts list**.

notab suppresses displaying the table. This option is often used with graph.

graph requests that the table be presented graphically as well as in tabular form; when notab is also specified, only the graph is presented. When specifying graph, only one table can be calculated and graphed at a time; see survival, failure, and hazard above.

graph may not be specified with hazard. Use sts graph to graph estimates of the hazard function.

noconf suppresses graphing the confidence intervals around survival, failure, or hazard.

plot(*plot*) provides a way to add other plots to the generated graph; see [G] *plot_option*.

ciopts(*rspike_options*) affect the rendition of the confidence intervals for the graphed survival, failure, or hazard. See [G] **graph twoway rspike**. This option may not be combined with the noconf option.

connected_options affect the rendition of the plotted points connected by lines. See [G] **graph twoway connected**.

twoway_options are any of the options documented in [G] *twoway_options* excluding by(). These include options for titling the graph (see [G] *title_options*) and options for saving the graph to disk (see [G] *saving_option*).

Remarks

Life tables date back to the seventeenth century; Edmund Halley (1693) is often credited with their development. ltable is for use with "cohort" data and, although one often thinks of such tables as following a population from the "birth" of the first member to the "death" of the last, more generally, such tables can be thought of as a reasonable way to list any kind of survival data. For an introductory discussion of life tables, see Pagano and Gauvreau (2000, 489–495); for an intermediate discussion, see, for example, Armitage and Berry (1994, 470–477) or Selvin (1996, 311–355); and for a more complete discussion, see Chiang (1984).

▷ Example

In Pike (1966), two groups of rats were exposed to a carcinogen and the number of days to death from vaginal cancer was recorded (reprinted in Kalbfleisch and Prentice 2002, 2):

Group 1	143	164	188	188	190	192	206	209	213	216
	220	227	230	234	246	265	304	216*	244*	
Group 2	142	156	163	198	205	232	232	233	233	233
	233	239	240	261	280	280	296	296	323	204*
	344*									

The '*' on a few of the entries indicates that the observation was censored—as of the recorded day, the rat had still not died due to vaginal cancer but was withdrawn from the experiment for other reasons.

Having entered these data into Stata, the first few observations are

```
. use http://www.stata-press.com/data/r8/rat
. list in 1/5
```

	group	t	died
1.	1	143	1
2.	1	164	1
3.	1	188	1
4.	1	188	1
5.	1	190	1

That is, the first observation records a rat from group 1 that died on the 143rd day. The variable died records whether that rat died or was withdrawn (censored):

```
. list if died==0
```

	group	t	died
18.	1	216	0
19.	1	244	0
39.	2	204	0
40.	2	344	0

Four rats, two from each group, did not die but were withdrawn.

The survival table for group 1 is

. ltable t died if group==1

Interval		Beg. Total	Deaths	Lost	Survival	Std. Error	[95% Conf. Int.]	
143	144	19	1	0	0.9474	0.0512	0.6812	0.9924
164	165	18	1	0	0.8947	0.0704	0.6408	0.9726
188	189	17	2	0	0.7895	0.0935	0.5319	0.9153
190	191	15	1	0	0.7368	0.1010	0.4789	0.8810
192	193	14	1	0	0.6842	0.1066	0.4279	0.8439
206	207	13	1	0	0.6316	0.1107	0.3790	0.8044
209	210	12	1	0	0.5789	0.1133	0.3321	0.7626
213	214	11	1	0	0.5263	0.1145	0.2872	0.7188
216	217	10	1	1	0.4709	0.1151	0.2410	0.6713
220	221	8	1	0	0.4120	0.1148	0.1937	0.6194
227	228	7	1	0	0.3532	0.1125	0.1502	0.5648
230	231	6	1	0	0.2943	0.1080	0.1105	0.5070
234	235	5	1	0	0.2355	0.1012	0.0751	0.4459
244	245	4	0	1	0.2355	0.1012	0.0751	0.4459
246	247	3	1	0	0.1570	0.0931	0.0312	0.3721
265	266	2	1	0	0.0785	0.0724	0.0056	0.2864
304	305	1	1	0	0.0000	.	.	.

The reported survival rates are the survival rates at the end of the interval. Thus, 94.7% of rats survived 144 days or more.

◁

❏ Technical Note

If you compare the table just printed with the corresponding table in Kalbfleisch and Prentice (2002, 16), you will notice that the survival estimates differ beginning with the interval 216–217, the first interval containing a censored observation. ltable treats censored observations as if they were withdrawn half-way through the interval. The table printed in Kalbfleisch and Prentice treated censored observations as if they were withdrawn at the end of the interval even through Kalbfleisch and Prentice (2002, 19) mention how results could be adjusted for censoring.

In this case, the same results as printed in Kalbfleisch and Prentice could be obtained by incrementing the time of withdrawal by 1 for the four censored observations. We say "in this case" because there were no deaths on the incremented dates. For instance, one of the rats was withdrawn on the 216th day, a day on which there was also a real death. There were no deaths on day 217, however, so moving the withdrawal forward one day is equivalent to assuming the withdrawal occurred at the end of the day 216–217 interval. If the adjustments are made and ltable is used to calculate survival in both groups, the results are as printed in Kalbfleisch and Prentice except that for group 2 in the interval 240–241, they report the survival as .345 when they mean .354.

In any case, the one-half adjustment for withdrawals is generally accepted, but it is important to remember that it is only a crude adjustment that becomes cruder the wider the intervals.

❏

▷ Example

When you do not specify the intervals, ltable uses unit intervals. The only aggregation performed on the data was aggregation due to deaths or withdrawals occurring on the same "day". If we wanted to see the table aggregated into 30-day intervals, we would type

```
. ltable t died if group==1, interval(30)
```

Interval		Beg. Total	Deaths	Lost	Survival	Std. Error	[95% Conf. Int.]	
120	150	19	1	0	0.9474	0.0512	0.6812	0.9924
150	180	18	1	0	0.8947	0.0704	0.6408	0.9726
180	210	17	6	0	0.5789	0.1133	0.3321	0.7626
210	240	11	6	1	0.2481	0.1009	0.0847	0.4552
240	270	4	2	1	0.1063	0.0786	0.0139	0.3090
300	330	1	1	0	0.0000	.	.	.

The interval printed 120 150 means the interval including 120, and up to but not including 150. The reported survival rate is the survival rate just after the close of the interval.

When you specify more than one number as the argument to interval(), you specify not the widths but the cutoff points themselves.

```
. ltable t died if group==1, interval(120,180,210,240,330)
```

Interval		Beg. Total	Deaths	Lost	Survival	Std. Error	[95% Conf. Int.]	
120	180	19	2	0	0.8947	0.0704	0.6408	0.9726
180	210	17	6	0	0.5789	0.1133	0.3321	0.7626
210	240	11	6	1	0.2481	0.1009	0.0847	0.4552
240	330	4	3	1	0.0354	0.0486	0.0006	0.2245

If any of the underlying failure or censoring times are larger than the last cutoff specified, they are treated as being in the open-ended interval:

```
. ltable t died if group==1, interval(120,180,210,240)
```

Interval		Beg. Total	Deaths	Lost	Survival	Std. Error	[95% Conf. Int.]	
120	180	19	2	0	0.8947	0.0704	0.6408	0.9726
180	210	17	6	0	0.5789	0.1133	0.3321	0.7626
210	240	11	6	1	0.2481	0.1009	0.0847	0.4552
240	.	4	3	1	0.0354	0.0486	0.0006	0.2245

Whether the last interval is treated as open-ended or not makes no difference for survival and failure tables, but it does affect hazard tables. If the interval is open-ended, the hazard is not calculated for it.

◁

▷ Example

The by(*varname*) option specifies that separate tables are to be presented for each value of *varname*. Remember that our rat dataset contains two groups:

(*Continued on next page*)

```
. ltable t died, by(group) interval(30)
                    Beg.                               Std.
      Interval     Total   Deaths   Lost   Survival   Error    [95% Conf. Int.]

group = 1
   120   150        19        1      0     0.9474     0.0512   0.6812    0.9924
   150   180        18        1      0     0.8947     0.0704   0.6408    0.9726
   180   210        17        6      0     0.5789     0.1133   0.3321    0.7626
   210   240        11        6      1     0.2481     0.1009   0.0847    0.4552
   240   270         4        2      1     0.1063     0.0786   0.0139    0.3090
   300   330         1        1      0     0.0000        .        .         .
group = 2
   120   150        21        1      0     0.9524     0.0465   0.7072    0.9932
   150   180        20        2      0     0.8571     0.0764   0.6197    0.9516
   180   210        18        2      1     0.7592     0.0939   0.5146    0.8920
   210   240        15        7      0     0.4049     0.1099   0.1963    0.6053
   240   270         8        2      0     0.3037     0.1031   0.1245    0.5057
   270   300         6        4      0     0.1012     0.0678   0.0172    0.2749
   300   330         2        1      0     0.0506     0.0493   0.0035    0.2073
   330   360         1        0      1     0.0506     0.0493   0.0035    0.2073
```

◁

▷ Example

A failure table is simply a different way of looking at a survival table; failure is $1 -$ survival:

```
. ltable t died if group==1, interval(30) failure
                    Beg.                      Cum.      Std.
      Interval     Total   Deaths   Lost    Failure    Error    [95% Conf. Int.]

   120   150        19        1      0     0.0526     0.0512   0.0076    0.3188
   150   180        18        1      0     0.1053     0.0704   0.0274    0.3592
   180   210        17        6      0     0.4211     0.1133   0.2374    0.6679
   210   240        11        6      1     0.7519     0.1009   0.5448    0.9153
   240   270         4        2      1     0.8937     0.0786   0.6910    0.9861
   300   330         1        1      0     1.0000        .        .         .
```

◁

▷ Example

Selvin (1996, 332) presents follow-up data from Cutler and Ederer (1958) on six cohorts of kidney cancer patients. The goal is to estimate the 5-year survival probability.

Year	Interval	Alive	Deaths	Lost	With-drawn	Year	Interval	Alive	Deaths	Lost	With-drawn
1946	0–1	9	4	1		1948	0–1	21	11	0	
	1–2	4	0	0			1–2	10	1	2	
	2–3	4	0	0			2–3	7	0	0	
	3–4	4	0	0			3–4	7	0	0	7
	4–5	4	0	0		1949	0–1	34	12	0	
	5–6	4	0	0	4		1–2	22	3	3	
1947	0–1	18	7	0			2–3	16	1	0	15
	1–2	11	0	0		1950	0–1	19	5	1	
	2–3	11	1	0			1–2	13	1	1	11
	3–4	10	2	2		1951	0–1	25	8	2	15
	4–5	6	0	0	6						

The following is the Stata dataset corresponding to the table:

```
. use http://www.stata-press.com/data/r8/selvin, clear
. list
```

	year	t	died	pop
1.	1946	.5	1	4
2.	1946	.5	0	1
3.	1946	5.5	0	4
4.	1947	.5	1	7
5.	1947	2.5	1	1

(*output omitted*)

As summary data may often come in the form shown above, it is worth understanding exactly how the data were translated for use with `ltable`. `t` records the time of death or censoring (lost to follow-up or withdrawal). `died` contains 1 if the observation records a death and 0 if it instead records lost or withdrawn patients. `pop` records the number of patients in the category. The first line of the table stated that, in the 1946 cohort, there were 9 patients at the start of the interval 0–1, and during the interval, 4 died, and 1 was lost to follow-up. Thus, we entered in observation 1 that at $t = .5$, 4 patients died and, in observation 2 that at $t = .5$, 1 patient was censored. We ignored the information on the total population because `ltable` will figure that out for itself.

The second line of the table indicated that in the interval 1–2, 4 patients were still alive at the beginning of the interval and, during the interval, 0 died or were lost to follow-up. Since no patients died or were censored, we entered nothing into our data. Similarly, we entered nothing for lines 3, 4, and 5 of the table. The last line for 1946 stated that, in the interval 5–6, 4 patients were alive at the beginning of the interval and that those 4 patients were withdrawn. In observation 3, we entered that there were 4 censorings at $t = 5.5$.

The fact that we chose to record the times of deaths or censoring as midpoints of intervals does not matter; we could just as well have recorded the times as 0.8 and 5.8. By default, `ltable` will form intervals 0–1, 1–2, and so on, and place observations into the intervals to which they belong. We suggest using 0.5 and 5.5 because those numbers correspond to the underlying assumptions made by `ltable` in making its calculations. Using midpoints reminds you of the assumptions.

To obtain the survival rates, we type

```
. ltable t died [freq=pop]
```

Interval		Beg. Total	Deaths	Lost	Survival	Std. Error	[95% Conf. Int.]	
0	1	126	47	19	0.5966	0.0455	0.5017	0.6792
1	2	60	5	17	0.5386	0.0479	0.4405	0.6269
2	3	38	2	15	0.5033	0.0508	0.4002	0.5977
3	4	21	2	9	0.4423	0.0602	0.3225	0.5554
4	5	10	0	6	0.4423	0.0602	0.3225	0.5554
5	6	4	0	4	0.4423	0.0602	0.3225	0.5554

We estimate the 5-year survival rate as .4423 and the 95% confidence interval as .3225 to .5554.

Selvin (1996, 336), in presenting these results, lists the survival in the interval 0–1 as 1, in 1–2 as .597, in 2–3 as .539, and so on. That is, relative to us, he shifted the rates down one row and inserted a 1 in the first row. In his table, the survival rate is the survival rate at the *start* of the interval. In our table, the survival rate is the survival rate at the *end* of the interval (or, equivalently, at the start of the next interval). This is, of course, simply a difference in the way the numbers are presented and not in the numbers themselves.   ◁

▷ Example

The discrete hazard function is the rate of failure—the number of failures occurring within a time interval divided by the width of the interval (assuming there are no censored observations). While the survival and failure tables are meaningful at the "individual" level—with intervals so narrow that each contains only a single failure—that is not true for the discrete hazard. If all intervals contained one death and if all intervals were of equal width, the hazard function would be $1/\Delta t$ and so appear to be a constant!

The empirically determined discrete hazard function can only be revealed by aggregation. Gross and Clark (1975, 37) print data on malignant melanoma at the M. D. Anderson Tumor Clinic between 1944 and 1960. The interval is the time from initial diagnosis:

Interval (years)	Number lost to follow-up	Number with-drawn alive	Number dying
0–1	19	77	312
1–2	3	71	96
2–3	4	58	45
3–4	3	27	29
4–5	5	35	7
5–6	1	36	9
6–7	0	17	3
7–8	2	10	1
8–9	0	8	3
9+	0	0	32

For our statistical purposes, there is no difference between the number lost to follow-up (patients who disappeared) and the number withdrawn alive (patients dropped by the researchers)—both are censored. We have entered the data into Stata; here is a small amount of the data:

```
. use http://www.stata-press.com/data/r8/tumor, clear
. list in 1/6
```

	t	d	pop
1.	.5	1	312
2.	.5	0	19
3.	.5	0	77
4.	1.5	1	96
5.	1.5	0	3
6.	1.5	0	71

We entered each group's time of death or censoring as the midpoint of the intervals and entered the numbers of the table, recording d as 1 for deaths and 0 for censoring. The hazard table is

(*Continued on next page*)

```
. ltable t d [freq=pop], hazard interval(0(1)9)
```

Interval		Beg. Total	Cum. Failure	Std. Error	Hazard	Std. Error	[95% Conf. Int.]	
0	1	913	0.3607	0.0163	0.4401	0.0243	0.3924	0.4877
1	2	505	0.4918	0.0176	0.2286	0.0232	0.1831	0.2740
2	3	335	0.5671	0.0182	0.1599	0.0238	0.1133	0.2064
3	4	228	0.6260	0.0188	0.1461	0.0271	0.0931	0.1991
4	5	169	0.6436	0.0190	0.0481	0.0182	0.0125	0.0837
5	6	122	0.6746	0.0200	0.0909	0.0303	0.0316	0.1502
6	7	76	0.6890	0.0208	0.0455	0.0262	0.0000	0.0969
7	8	56	0.6952	0.0213	0.0202	0.0202	0.0000	0.0598
8	9	43	0.7187	0.0235	0.0800	0.0462	0.0000	0.1705
9	.	32	1.0000	.	.	.	.	.

We specified the `interval()` option as we did and not as `interval(1)` (or omitting the option altogether) to force the last interval to be open-ended. Had we not, and if we had recorded t as 9.5 for observations in that interval (as we did), `ltable` would have calculated a hazard rate for the "interval". In this case, the result of that calculation would have been 2, but no matter what the result, it would have been meaningless since we do not know the width of the interval.

When dealing with the survival or failure function, you are not limited to merely examining a column of numbers. With the `graph` option, you can see the result graphically:

```
. ltable t d [freq=pop], i(0(1)9) graph notab xlab(0(2)10)
```

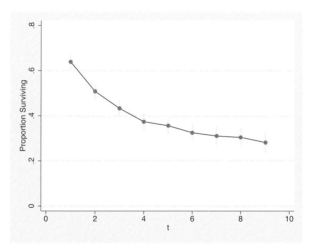

The vertical lines in the graph represent the 95% confidence intervals for the survival function; specifying `noconf` would have suppressed them. Among the options we did specify, although it is not required, `notab` suppressed printing the table, saving us some paper. `xlab()` was passed through to the `graph` command (see [G] *twoway_options*) and was unnecessary, but made the graph look better.

◁

❏ Technical Note

Because a large number of intervals can exist where no failures occur (in which case the hazard estimate is zero), the estimated hazard is best graphically represented using a kernel smooth. Such an estimate is available in `sts graph`; see [ST] **sts graph**.

❏

Methods and Formulas

`ltable` is implemented as an ado-file.

Let τ_i be the individual failure or censoring times. The data are aggregated into intervals given by t_j, $j = 1, \ldots, J$, and $t_{J+1} = \infty$ with each interval containing counts for $t_j \leq \tau < t_{j+1}$. Let d_j and m_j be the number of failures and censored observations during the interval and N_j the number alive at the start of the interval. Define $n_j = N_j - m_j/2$ as the adjusted number at risk at the start of the interval. If the `noadjust` option is specified, $n_j = N_j$.

The product-limit estimate of the survivor function is

$$S_j = \prod_{k=1}^{j} \frac{n_k - d_k}{n_k}$$

(Kalbfleisch and Prentice 2002, 10, 15). Greenwood's formula for the asymptotic standard error of S_j is

$$s_j = S_j \sqrt{\sum_{k=1}^{j} \frac{d_k}{n_k(n_k - d_k)}}$$

(Greenwood 1926; Kalbfleisch and Prentice 2002, 17). s_j is reported as the standard deviation of survival but is not used in generating the confidence intervals since it can produce intervals outside 0 and 1. The "natural" units for the survival function are $\log(-\log S_j)$ and the asymptotic standard error of that quantity is

$$\widehat{s}_j = \sqrt{\frac{\sum d_k / \left\{ n_k(n_k - d_k) \right\}}{\left[\sum \log \left\{ (n_k - d_k)/n_k \right\} \right]^2}}$$

(Kalbfleisch and Prentice 2002, 18). The corresponding confidence intervals are $S_j^{\exp(\pm z_{1-\alpha/2} \widehat{s}_j)}$.

The cumulative failure time is defined as $G_j = 1 - S_j$, and thus the variance is the same as for S_j and the confidence intervals are $1 - S_j^{\exp(\pm z_{1-\alpha/2} \widehat{s}_j)}$.

For purposes of graphing, both S_j and G_j are graphed against t_{j+1}.

Define the within-interval failure rate as $f_j = d_j/n_j$. The maximum likelihood estimate of the (within-interval) hazard is then

(Continued on next page)

$$\lambda_j = \frac{f_j}{(1 - f_j/2)(t_{j+1} - t_j)}$$

The standard error of λ_j is

$$s_{\lambda_j} = \lambda_j \sqrt{\frac{1 - \{(t_{j+1} - t_j)\lambda_j/2\}^2}{d_j}}$$

from which a confidence interval is calculated.

If the noadjust option is specified, the estimate of the hazard is

$$\lambda_j = \frac{f_j}{t_{j+1} - t_j}$$

and its standard error is

$$s_{\lambda_j} = \frac{\lambda_j}{\sqrt{d_j}}$$

The confidence interval is

$$\left[\frac{\lambda_j}{2d_j} \chi^2_{2d_j, \alpha/2}, \ \frac{\lambda_j}{2d_j} \chi^2_{2d_j, 1-\alpha/2} \right]$$

where $\chi^2_{2d_j, q}$ is the qth quantile of the χ^2 distribution with $2d_j$ degrees of freedom (Cox and Oakes 1984, 53–54, 38–40).

For the likelihood-ratio test for homogeneity, let d_g be the total number of deaths in the gth group. Define $T_g = \sum_{i \in g} \tau_i$, where i indexes the individual failure or censoring times. The χ^2 value with $G - 1$ degrees of freedom (where G is the total number of groups) is

$$\chi^2 = 2\left\{ \left(\sum d_g\right) \log\left(\frac{\sum T_g}{\sum d_g}\right) - \sum d_g \log\left(\frac{T_g}{d_g}\right) \right\}$$

(Lawless 1982, 113).

The log-rank test for homogeneity is the test presented by sts test; see [ST] **sts**.

Acknowledgments

ltable is based on the lftbl command by Henry Krakauer and John Stewart (1991). We also thank Michel Henry-Amar, Centre Regional Francois Baclesse, Caen, France for his comments.

References

Armitage, P. and G. Berry. 1994. *Statistical Methods in Medical Research.* 3d ed. Oxford: Blackwell Scientific Publications.

Chiang, C. L. 1984. *The Life Table and Its Applications.* Malabar, FL: Krieger.

Cox, D. R. and D. Oakes. 1984. *Analysis of Survival Data.* London: Chapman and Hall.

Cutler, S. J. and E. Ederer. 1958. Maximum utilization of the life table method in analyzing survival. *Journal of Chronic Diseases* 8: 699–712.

Greenwood, M. 1926. The natural duration of cancer. *Reports on Public Health and Medical Subjects* 33: 1–26. London: His Majesty's Stationery Office.

Gross, A. J. and V. A. Clark. 1975. *Survival Distributions: Reliability Applications in the Biomedical Sciences.* New York: John Wiley & Sons.

Halley, E. 1693. An estimate of the degrees of mortality of mankind, drawn from curious tables of the births and funerals at the city of Breslau, with an attempt to ascertain the price of annuities on lives. *Philosophical Transactions* 17: 596–610. London: The Royal Society.

Kahn, H. A. and C. T. Sempos. 1989. *Statistical Methods in Epidemiology.* New York: Oxford University Press.

Kalbfleisch, J. D. and R. L. Prentice. 2002. *The Statistical Analysis of Failure Time Data.* 2d ed. New York: John Wiley & Sons.

Krakauer, H. and J. Stewart. 1991. ssa1: Actuarial or life-table analysis of time-to-event data. *Stata Technical Bulletin* 1: 23–25. Reprinted in *Stata Technical Bulletin Reprints*, vol. 1, pp. 200–202.

Lawless, J. F. 1982. *Statistical Models and Methods for Lifetime Data.* New York: John Wiley & Sons.

Pagano, M. and K. Gauvreau. 2000. *Principles of Biostatistics.* 2d ed. Pacific Grove, CA: Brooks/Cole.

Pike, M. C. 1966. A method of analysis of a certain class of experiments in carcinogenesis. *Biometrics* 22: 142–161.

Ramalheira, C. 2001. ssa14: Global and multiple causes-of-death life tables from complete or aggregated vital data. *Stata Technical Bulletin* 59: 29–45. Reprinted in *Stata Technical Bulletin Reprints*, vol. 10, pp. 333–355.

Selvin, S. 1996. *Statistical Analysis of Epidemiologic Data.* 2d ed. New York: Oxford University Press.

Also See

Related: [ST] **st**, [ST] **stcox**, [ST] **streg**

Background: [ST] **survival analysis**,
                       Stata Graphics Reference Manual

Title

---
snapspan — Convert snapshot data to time-span data

Syntax

snapspan *idvar* *timevar* *varlist* [, <u>g</u>enerate(*newt0var*) replace]

Description

snapspan converts snapshot data to time-span data. See *Remarks* below for a description of snapshot and time-span data. Time-span data are required for use with survival-analysis commands such as stcox, streg, and stset.

idvar records the subject id; it may be string or numeric.

timevar records the time of the snapshot; it must be numeric and may be recorded on any scale: date, hour, minute, second, etc.

varlist are the "event" variables: the variables that occur at the instant of *timevar*. *varlist* is also to include retrospective variables: variables that are to apply to the time span ending at the time of the current snapshot. The other variables are assumed to be measured at the time of the snapshot and thus apply from the time of the snapshot forward. See *Specifying varlist* below.

Options

<u>g</u>enerate(*newt0var*) adds *newt0var* to the dataset containing the entry time for each converted time-span record.

replace specifies that it is okay to change the data in memory even though the dataset has not been saved on disk in its current form.

Remarks

Snapshot and time-span datasets

snapspan converts a snapshot dataset to a time-span dataset. A snapshot dataset records a subject *id*, a *time*, and then other variables measured at the *time*:

(Continued on next page)

83

Snapshot datasets:

idvar	timevar	x1	x2	...
47	12	5	27	...
47	42	5	18	...
47	55	5	19	...

idvar	datevar	x1	x2	...
122	14jul1998	5	27	...
122	12aug1998	5	18	...
122	08sep1998	5	19	...

idvar	year	x1	x2	...
122	1994	5	27	...
122	1995	5	18	...
122	1997	5	19	...

A time-span dataset records a span of time (*time0*, *time1*]:

```
                                            some variables assumed
                                               to occur at time1
                                                      |
        |<— other variables assumed constant over span —>|
      _____|_____
                                                      |        —> time
        time0                                        time1
```

It is time-span data that are required, for instance, by stset and the st system. The variables assumed to occur at time1 are the failure or event variables. All the other variables are assumed to be constant over the span.

Time-span datasets:

idvar	time0	time1	x1	x2	...	event
47	0	12	5	13	...	0
47	12	42	5	27	...	0
47	42	55	5	18	...	1

idvar	time0	time1	x1	x2	...	event
122	01jan1998	14jul1998	5	13	...	0
122	14jul1998	12aug1998	5	27	...	0
122	12aug1998	08sep1998	5	18	...	1

idvar	time0	time1	x1	x2	...	event
122	1993	1994	5	13	...	0
122	1994	1995	5	27	...	0
122	1995	1997	5	18	...	1

To convert snapshot data to time-span data, you need to distinguish between event and nonevent variables. Event variables happen at an instant.

Say one has a snapshot dataset containing variable e recording an event (e = 1 might record surgery, or death, or becoming unemployed, etc.) and the rest of the variables—call them x1, x2, etc.—recording characteristics (such as sex, birth date, blood pressure, weekly wage, etc.). The same data, in snapshot and time-span form, would be

In snapshot form:						In time-span form:					
id	time	x1	x2	e		id	time0	time	x1	x2	e
1	5	a1	b1	e1		1	.	5	.	.	e1
1	7	a2	b2	e2		1	5	7	a1	b1	e2
1	9	a3	b3	e3		1	7	9	a2	b2	e3
1	11	a4	b4	e4		1	9	11	a3	b3	e4

snapspan converts data from the form on the left to the form on the right:

```
. snapspan id time e
```

The form on the right is suitable for use by stcox and stset and the other survival-analysis commands.

Specifying varlist

The *varlist*—the third variable on—specifies the "event" variables.

In fact, the *varlist* specifies the variables that are to apply to the time span ending at the time of the current snapshot. The other variables are assumed to be measured at the time of the snapshot and thus apply from the time of the snapshot forward.

Thus, *varlist* should include retrospective variables.

For instance, say the snapshot recorded bp, blood pressure; smokes, the answer to the question whether the patient smoked in the last two weeks; and event, a variable recording examination, surgery, etc. Then *varlist* should include smokes and event. The remaining variables, bp and the rest, would be assumed to apply from the time of snapshot forward.

Suppose the snapshot recorded ecs, employment change status (hired, fired, promoted, etc.); wage, the current hourly wage; and ms, current marital status. Then varlist should include esc and ms (assuming snapshot records are not generated for reason of ms change). The remaining variables, wage and the rest, would be assumed to apply from the time of the snapshot forward.

Methods and Formulas

snapspan is implemented as an ado-file.

Also See

Complementary:	[ST] **stset**
Background:	[ST] **st**, [ST] **survival analysis**

Title

st — Survival-time data

Description

The term st refers to survival-time data and the commands—all of which begin with the letters st—for analyzing these data. If you have data on individual subjects with observations recording that this subject came under observation at time t_0, and then, later, at t_1, a failure or censoring was observed, you have what we call survival-time data.

If you have subject-specific data, with observations recording not a span of time, but measurements taken on the subject at that point in time, then you have what we call a snapshot dataset; see [ST] **snapspan**.

If you have data on populations, with observations recording the number of units under test at time t (subjects alive) and the number of subjects that failed or were lost due to censoring, you have what we call count-time data; see [ST] **ct**.

The st commands are

stset	[ST] **stset**	Declare data to be survival-time data
stdes	[ST] **stdes**	Describe survival-time data
stsum	[ST] **stsum**	Summarize survival-time data
stvary	[ST] **stvary**	Report which variables vary over time
stfill	[ST] **stfill**	Fill in by carrying forward values of covariates
stgen	[ST] **stgen**	Generate variables reflecting entire histories
stsplit	[ST] **stsplit**	Split time-span records
stjoin	[ST] **stsplit**	Join time-span records
stbase	[ST] **stbase**	Form baseline dataset
sts	[ST] **sts**	Generate, graph, list, and test the survivor and cumulative hazard functions
stir	[ST] **stir**	Report incidence-rate comparison
stci	[ST] **stci**	Confidence intervals for means and percentiles of survival time
strate	[ST] **strate**	Tabulate failure rate
stptime	[ST] **stptime**	Calculate person-time
stmh	[ST] **strate**	Calculates rate ratios using Mantel–Haenszel method
stmc	[ST] **strate**	Calculates rate ratios using Mantel–Cox method
stcox	[ST] **stcox**	Fit Cox proportional hazards model
stphtest	[ST] **stcox**	Test of Cox proportional hazards assumption
stphplot	[ST] **stphplot**	Graphical assessment of the Cox proportional hazards assumption
stcoxkm	[ST] **stphplot**	Graphical assessment of the Cox proportional hazards assumption
streg	[ST] **streg**	Fit parametric survival models
stcurve	[ST] **streg**	Plot fitted survival functions
sttocc	[ST] **sttocc**	Convert survival-time data to case–control data
sttoct	[ST] **sttoct**	Convert survival-time data to count-time data
cttost	[ST] **cttost**	Convert count-time data to survival-time data
snapspan	[ST] **snapspan**	Convert snapshot data to time-span data
st_*	[ST] **st_is**	Survival analysis subroutines for programmers

The st commands are used for analyzing time-to-absorbing-event (single failure) data and for analyzing time-to-repeated-event (multiple failure) data.

You begin an analysis by `stsetting` your data, which tells Stata the key survival-time variables. This is described in [ST] **stset**. Once you have `stset` your data, you can use the other st commands. If you `save` your data after `stsetting` it, you will not have to re-`stset` it in the future; Stata will remember.

The subsequent st entries are printed in this manual in alphabetical order. You can skip around, but if you want to be an expert on all of Stata's survival analysis capabilities, we suggest the above reading order.

References

Cleves, M. A. 1999. ssa13: Analysis of multiple failure-time data with Stata. *Stata Technical Bulletin* 49: 30–39. Reprinted in *Stata Technical Bulletin Reprints*, vol. 9, pp. 338–349.

Also See

Complementary:	[ST] **stset**
Related:	[ST] **ct**, [ST] **snapspan**
Background:	[ST] **survival analysis**

Title

st_is — Survival analysis subroutines for programmers

Syntax

st_is 2 {full | analysis}

st_show [noshow]

st_ct "[*byvars*]" -> *newtvar newpopvar newfailvar* [*newcensvar* [*newentvar*]]

These commands are for use with survival-time data; see [ST] **st**. You must have stset your data before using these commands.

Description

These commands are provided for programmers—persons wishing to write new st commands.

st_is verifies that the data in memory are survival-time (st) data. If not, it issues the error message "data not st", r(119).

st is currently "release 2", meaning that this is the second design of the system. Programs written for the previous release continue to work. (The previous release of st corresponds to Stata 5.)

Modern programs code st_is 2 full or st_is 2 analysis. The st_is 2 part verifies that the dataset in memory is in release 2 format; if it is in the prior format, it is converted to release 2 format. (Older programs simply code st_is. This verifies that no new features are stset about the data that would cause the old program to break.)

The full and analysis parts indicate whether the dataset may be past, future, or past and future. Code st_is 2 full if the command is suitable for running on the analysis sample and the past and future data (many data-management commands fall into this category). Code st_is 2 analysis if the command is suitable for use only with the analysis sample (most statistical commands fall into this category). See [ST] **stset** for the definitions of past and future.

st_show displays the summary of the survival-time variables or does nothing, depending on what the user specified when stseting the data.

st_ct is a low-level utility that provides risk-group summaries from survival-time data.

Options

noshow requests that st_show display nothing.

Remarks

Remarks are presented under the headings

> *Definitions of characteristics and st variables*
> *Outline of an st command*
> *Using the st_ct utility*
> *Comparison of st_ct with sttoct*
> *Data verification*
> *Data conversion*

Definitions of characteristics and st variables

From a programmer's perspective, st is a set of conventions that specify where certain pieces of information are stored and how to interpret that information together with a few subroutines that make it easier to follow the conventions.

At the lowest level, st is nothing more than a set of Stata characteristics that programmers may access:

char _dta[_dta]	st (marks that the data is st)
char _dta[st_ver]	2 (version number)
char _dta[st_id]	*varname* or nothing; id() variable
char _dta[st_bt0]	*varname* or nothing; t0() variable
char _dta[st_bt]	*varname*; *t* variable from stset t, ...
char _dta[st_bd]	*varname* or nothing; failure() variable
char _dta[st_ev]	list of numbers or nothing; *numlist* from failure(*varname*[==*numlist*])
char _dta[st_enter]	contents of enter() or nothing; *numlist* expanded
char _dta[st_exit]	contents of exit() or nothing; *numlist* expanded
char _dta[st_orig]	contents of origin() or nothing; *numlist* expanded
char _dta[st_bs]	# or 1; scale() value
char _dta[st_o]	_origin or #
char _dta[st_s]	_scale or #
char _dta[st_ifexp]	*exp* or nothing; from stset ... if exp ...
char _dta[st_if]	*exp* or nothing; contents of if()
char _dta[st_ever]	*exp* or nothing; contents of ever()
char _dta[st_never]	*exp* or nothing; contents of never()
char _dta[st_after]	*exp* or nothing; contents of after()
char _dta[st_befor]	*exp* or nothing; contents of before()
char _dta[st_wt]	weight type or nothing; user-specified weight
char _dta[st_wv]	*varname* or nothing; uscr-specified weighting variable
char _dta[st_w]	[*weighttype=weightvar*] or nothing
char _dta[st_show]	noshow or nothing
char _dta[st_t]	_t (for compatibility with release 1)
char _dta[st_t0]	_t0 (for compatibility with release 1)
char _dta[st_d]	_d (for compatibility with release 1)
char _dta[st_n0]	# or nothing; number of st notes
char _dta[st_n1]	text of first note or nothing
char _dta[st_n2]	text of second note or nothing
char _dta[st_set]	text or nothing. If filled in, streset will refuse to execute and present this text as the reason

In addition, all st datasets have the following four variables:

_t0	time of entry (in t units) into risk pool
_t	time of exit (in t units) from risk pool
_d	contains 1 if failure, 0 if censoring
_st	contains 1 if observation is to be used and 0 otherwise

Thus, in a program, you might code

```
display "the failure/censoring base time variable is _t"
display "and its mean in the uncensored subsample is"
summarize _t if _d
```

No matter how simple or complicated the data, these four variables exist and are filled in. For instance, in simple data, _t0 might contain 0 for every observation and _d might always contain 1.

Some st datasets also contain the variables

> _origin   contains evaluated value of origin()
> _scale   contains evaluated value of scale()

The characteristic _dta[st_o] contains the name _origin or it contains a number, often 0. It contains a number when the origin does not vary across observations. _dta[st_s] works the same way with the scale() value. Thus, the origin and scale are, in all cases, '_dta[st_o]' and '_dta[st_s]'. In fact, these characteristics are seldom used because variables _t and _t0 are already adjusted.

Some st datasets have an id() variable that clusters together records on the same subject. The name of the variable varies and the name can be obtained from the characteristic _dta[st_id]. If there is no id() variable, the characteristic contains nothing.

Outline of an st command

If you are writing a new st command, place st_is near the top of your code. This ensures that your command does not execute on inappropriate data. In addition, place st_show following the parsing of your command's syntax; this will display the key st variables. The minimal outline for an st command is

```
program st name
        version 8
        st_is 2 ...
        ... syntax command ...
        ... determined there are no syntax errors ...
        st_show
        ... guts of program ...
end
```

Note that st_is 2 appears even before parsing of input. That is to avoid irritating users when they type a command, get a syntax error, work hard to eliminate the error, and then learn that "data not st".

A fuller outline for an st command, particularly one that performs analysis on the data, is

```
program st name
        version 8
        st_is 2 ...
        syntax ... [, ... noSHow ... ]
        st_show 'show'
        marksample touse
        quietly replace 'touse' = 0 if _st==0
        ... guts of program ...
end
```

Do not forget that all calculations and actions are to be restricted, at the least, to observations for which _st ≠ 0. Observations with _st = 0 are to be ignored.

Using the st_ct utility

st_ct converts the data in memory to observations containing summaries of risk groups. Consider the code

```
st_is 2 analysis
preserve
st_ct "" -> t pop die
```

This would change the data in memory to contain something akin to count-time data. The transformed data would have observations containing

t	time
pop	population at risk at time t
die	number who fail at time t

There would be one record per time t and the data would be sorted by t. The original data are discarded, which is why you should code preserve; see [P] **preserve**.

The above three lines could be used as the basis for calculating the Kaplan–Meier product-limit survivor function estimate; the rest of the code being

```
keep if die
gen double hazard = die/pop
gen double km    = 1-hazard  if _n==1
replace    km    = (1-hazard)*km[_n-1] if _n>1
```

st_ct can be used to obtain risk groups separately for subgroups of the population. The code

```
st_is 2 analysis
preserve
st_ct "race sex" -> t pop die
```

would change the data in memory to contain

race	
sex	
t	time
pop	population at risk at time t
die	number who fail at time t

There will be one observation for each race–sex–t combination and the data will be sorted by race sex t.

With this data, you could calculate the Kaplan–Meier product-limit survivor function estimate for each race–sex group by coding

```
keep if die
gen double hazard = die/pop
by race sex: gen double km    = 1-hazard  if _n==1
by race sex: replace    km    = (1-hazard)*km[_n-1] if _n>1
```

st_ct is a convenient subroutine. The above code fragment works regardless of the complexity of the underlying survival-time data. It does not matter whether there is one record per subject, no censoring, and a single failure per subject, or multiple records per subject, gaps, and recurring failures for the same subject. st_ct forms risk groups that summarize the events recorded by the data.

st_ct can provide the number censored and the number who enter the risk group. The code

```
st_ct "" -> t pop die cens ent
```

creates records containing

t	time
pop	population at risk at time t
die	number who fail at time t
cens	number who are censored at t (after the failures)
ent	number who enter at t (after the censorings)

As before,

```
st_ct "race sex" -> t pop die cens ent
```

would create a similar dataset with records for each race-sex group.

Comparison of st_ct with sttoct

sttoct—see [ST] **sttoct**—is related to st_ct and, in fact, sttoct is implemented in terms of st_ct. The differences between them are

1. sttoct creates ct data, meaning that the dataset is marked as being ct. st_ct merely creates a useful dataset; it does not ctset the data.

2. st_ct creates a total population at risk variable—which is useful in programming—but sttoct creates no such variable.

3. sttoct eliminates thrashings—censorings and re-entries of the same subject as covariates change—if there are no gaps, strata shifting, etc. st_ct does not. Thus, at a particular time, sttoct might show that there are 2 lost to censoring and none entered, whereas st_ct might show 12 censorings and 10 entries. Note that this makes no difference to the calculations of the number at risk and the number who fail, which are the major ingredients in survival calculations.

4. st_ct is faster.

Data verification

As long as you code st_is at the top of your program, you need not verify the consistency of the data. That is, you need not verify that subjects do not fail before they enter, etc.

The dataset is verified when the user stsets it. It is hereby agreed between users and programmers that, if the user makes a substantive change to the data, the user will rerun stset (which can be done by typing stset or streset without arguments) to reverify that all is well.

Data conversion

If you write a program that converts the data from one form of st data to another, or from st data to something else, be sure to issue the appropriate stset command. For instance, a command we have written, stbase, converts the data from st to a simple cross-section in one instance. In our program, we coded stset, clear so that all other st commands would know that this is no longer st data and that making st calculations on it would be inappropriate.

In fact, even if we had forgotten, it would not have mattered. Other st programs would have found many of the key st variables missing and so ended with a "such-and-such not found" error.

Also See

Complementary:	[ST] **stset**, [ST] **sttoct**
Background:	[ST] **st**, [ST] **survival analysis**

Title

stbase — Form baseline dataset

Syntax

> stbase [if *exp*] [in *range*] [, at(*#*) g̲ap(*newvar*) replace n̲opreserve n̲oshow]

stbase is for use with survival-time data; see [ST] **st**. You must stset your data before using this command.

Description

stbase without the at() option converts multiple-record st data to st data with every variable set to its value at baseline, defined as the earliest time at which each subject was observed. stbase without at() does nothing to single-record st data.

stbase, at() converts single- or multiple-record st data to a cross-sectional dataset (not st data) recording the number of failures at the specified time. All variables are given their values at baseline—the earliest time at which each subject was observed. In this form, single-failure data could be analyzed by logistic regression and multiple-failure data by Poisson regression, for instance.

stbase is appropriate for use with single- or multiple-record, single- or multiple-failure, st data.

Options

at(*#*) changes what stbase does. Without the at() option, stbase produces another, related st dataset. With the at() option, stbase produces a related cross-sectional dataset.

gap(*newvar*) is allowed only with at(); it specifies the name of a new variable to be added to the data containing time on gap. If gap() is not specified, the new variable will be named gap or gaptime, depending on which name does not already exist in the data.

replace specifies that it is okay to change the data in memory even though the dataset has not been saved to disk in its current form.

nopreserve is for use by programmers using stbase as a subroutine. It says not to preserve the original dataset so that it can be restored should an error be detected or should the user press *Break*. Programmers would specify this option if, in their program, they had already preserved the original data.

noshow prevents stbase from showing the key st variables. This option is rarely used since most people type stset, show or stset, noshow to reset once and for all whether they want to see these variables mentioned at the top of the output of every st command; see [ST] **stset**.

Remarks

Remarks are presented under the headings

> *stbase without the at() option*
> *stbase with the at() option*
> *Single-failure st data where all subjects enter at 0*
> *Single-failure st data where all subjects enter after time 0*
> *Single-failure st data with gaps and perhaps delayed entry*
> *Multiple-failure st data*

stbase without the at() option

Once you type stbase, you may not streset your data even though the data are st. streset will refuse to run to protect you because the data are changed, and if the original rules were reapplied, they might produce different, incorrect results. The st commands use four key variables:

_t0	the time at which the record came under observation
_t	the time at which the record left observation
_d	1 if the record left under failure, 0 otherwise
_st	whether the observation is to be used (contains 1 or 0)

These variables are adjusted by stbase. The variables _t0 and _t, in particular, were derived from your variables according to options you specified at the time you stset the data, which might include an origin() rule, an entry() rule, and the like. Once intervening observations are eliminated, those rules will not necessarily produce the same results as they did previously.

To illustrate how stbase works, consider multiple-record, time-varying st data, on which you have performed some analysis. You now wish to compare your results with a simpler, non-time-varying analysis. For instance, suppose the variable bp measures blood pressure, and that readings were taken on bp at various times. Perhaps you fitted the model

```
. stcox drug bp sex
```

using these data. You now wish to fit that same model, but this time using the value of bp at baseline. You do this by typing

```
. stbase
. stcox drug bp sex
```

Another way you could perform the analysis is

```
. generate bp0 = bp
. stfill bp0, baseline
. stcox drug bp0 sex
```

See [ST] **stfill**. Which you use makes no difference but, if there were a lot of explanatory variables, stbase would be easier.

stbase changes the data to record the same events, but changes the values of all other variables to their values at the earliest time the subject was observed.

In addition, stbase simplifies the st data where possible. For instance, say one of your subjects has three records in the original data and ends in a failure:

After running stbase, this subject would have a single record in the data:

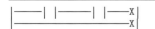

Here are some other examples of how stbase would process records with gaps and multiple failure events:

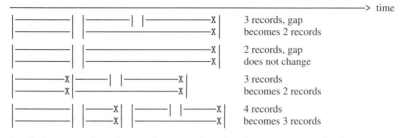

In each of the examples above, the second and subsequent records that stbase creates have the values of the other variables recorded in the first record.

The simplification of the records makes no difference to the st commands; whether a person has

or

makes no difference if all the covariates are the same. The simplification does, however, make it easier for human beings to read the data.

stbase with the at() option

stbase, at() produces a dataset that is not st. The st dataset is converted to a cross-sectional dataset recording the status of each subject at the specified time. Four "new" variables are created:

> the first entry time for the subject,
> the time on gap,
> the time at risk, and
> the number of failures during the time at risk.

The names given to those variables are related to how your data are stset. Pretend that your stset command was

        . stset var1, failure(var2) time0(var3) ...

Then,

the first entry time	will be named	*var3* or time0 or _t0
the time on gap	will be named	gap() or gap or gaptime
the time at risk	will be named	*var1*
the number of (or whether) failures	will be named	*var2* or failure or _d

The ors are mentioned because, for instance, if you did not specify a *var2* variable when you stset your data, stbase, at() looks around for a name.

You need not memorize this; the names are obvious from the output produced by stbase, at().

Let us consider the actions of stbase, at(), with some particular st datasets. Pretend that the command given is

        . stbase, at(5)

thus producing a cross-section at analysis time 5.

Note that the value of time specified with the at() option must correspond to time in the analysis scale, i.e., *t*. See [ST] **stset** for a definition of analysis time.

Single-failure st data where all subjects enter at 0

The result of stbase, at(5) would be one record per subject. Any subject who was censored before time 5 would not appear in the data; the rest would. Those that failed after time 5 will be recorded as having been censored at time 5 (*failvar* = 0); those that failed at time 5 or earlier will have *failvar* = 1.

timevar will contain

	for the failures:	
	time of failure	if failed on or before time 5, or
	5	because the subject has not failed yet
	for the censored:	
	5	because the subject has not failed yet

Among the analyses appropriate to perform with the data would be

1. logistic regression of *failvar* on any of the characteristics; and

2. incidence rate analysis, summing the failures (perhaps within strata) and the time-at-risk *timevar*.

With these data, you could examine 5-year survival probabilities.

Single-failure st data, some subjects enter after time 0

The data produced by stbase, at(5) would be similar to that above except

1. persons who enter on or after time 5 would not be included in the data (because they have not entered yet), and

2. the time-at-risk *timevar* would properly account for when each patient entered.

timevar (the time at risk) will contain

	for the failures:	
	time of failure or less	if failed on or before time 5 (or less because the subject may not have entered at 0); or
	5 or less	because the subject has not failed yet (or less because subject may not have entered at time 0)
	for the censored:	
	5 or less	because the subject has not failed yet (or less because the subject may not have entered at time 0)

Depending on the analysis you are performing, you may have to discard those that enter late. That is easy to do because t0 contains the first time of entry.

Among the things appropriate to do with these data would be

1. logistic regression of *failvar* on any of the characteristics, but only if one restricted the sample to if t0 == 0 because those who entered after time 0 have a lesser risk of failing over the fixed interval.

2. incidence rate analysis, summing the failures (perhaps within stratum) and the time-at-risk *timevar*. In this case, you would have to do nothing differently from what one did in the previous example. The time-at-risk variable already includes the time of entry for each patient.

Single-failure st data with gaps and perhaps delayed entry

These data will be similar to the delayed-entry, no-gap data, but gap will no longer contain 0 in all the observations. It will be 0 for those who have no gap.

If analyzing these data, you could perform

1. logistic regression, but the sample must be restricted to if t0 == 0 & gap == 0; or

2. incidence rate analysis, and nothing would need to be done differently; the time-at-risk *timevar* accounts for late entry and gaps.

Multiple-failure st data

The multiple-failure case parallels the single-failure case except that fail will not solely contain 0 and 1; it will contain 0, 1, 2, ..., depending on the number of failures observed. Regardless of late entry, gaps, etc., the following would be appropriate:

1. Poisson regression of fail, the number of events, but remember to specify exposure(*timevar*); and

2. incidence rate analysis.

Methods and Formulas

stbase is implemented as an ado-file.

Also See

Complementary:	[ST] **stfill**, [ST] **stgen**, [ST] **stset**, [ST] **sttocc**
Background:	[ST] **st**, [ST] **survival analysis**

Title

> **stci** — Confidence intervals for means and percentiles of survival time

Syntax

> stci [if *exp*] [in *range*] [, by(*varlist*) median emean rmean ccorr p(#)
>
> noshow level(#) graph tmax(#) dd(#)]

stci is for use with survival-time data; see [ST] **st** You must stset your data before using this command.
by ... : may be used with stci; see [R] **by**.

Description

stci computes means and percentiles of survival time, standard errors, and confidence intervals. In the case of multiple-event data, survival time is the time until a failure.

stci is appropriate for use with single- or multiple-record, single- or multiple-failure, st data.

Options

by(*varlist*) requests separate summaries for each group along with an overall total. Observations are in the same group if they have equal values of the variables in *varlist*. *varlist* may contain any number of variables, each of which may be string or numeric.

median specifies median survival times. This is the default.

emean and rmean specify mean survival times. If the longest follow-up time is censored, emean (extended mean) computes the mean survival by exponentially extending the survival curve to zero, and rmean (restricted mean) computes the mean survival time restricted to the longest follow-up time. Note that if the longest follow-up time is a failure, the restricted mean survival time and the extended mean survival time are equal.

ccorr specifies that the standard error for the restricted mean survival time be computed using a continuity correction. ccorr is only valid with option rmean.

p(#) specifies the percentile of survival time to be computed. For example, p(25) will compute the 25th percentile of survival times, and p(75) will compute the 75th percentile of survival times. Note that specifying p(50) is the same as specifying the median option.

noshow prevents stci from showing the key st variables. This option is rarely used since most people type stset, show or stset, noshow to reset once and for all whether they want to see these variables mentioned at the top of the output of every st command; see [ST] **stset**.

level(#) specifies the confidence level, in percent, for confidence intervals. The default is level(95) or as set by set level; see [U] **23.6 Specifying the width of confidence intervals**.

graph specifies that the exponentially extended survivor function be plotted. This option is only valid when option emean is also specified, and is not valid in conjunction with the by() option.

tmax(#) is for use with the graph option. It specifies the maximum analysis time to be plotted.

dd(#) specifies the maximum number of decimal digits to be reported for standard errors and confidence intervals. This option affects only how values are reported and not how they are calculated.

Remarks

Single-failure data

Here is an example of stci with single-record survival data:

```
. use http://www.stata-press.com/data/r8/page2
. stset, noshow
. stci
```

	no. of subjects	50%	Std. Err.	[95% Conf. Interval]	
total	40	232	2.562933	213	239

```
. stci, by(group)
```

group	no. of subjects	50%	Std. Err.	[95% Conf. Interval]	
1	19	216	5.171042	190	234
2	21	233	2.179595	232	280
total	40	232	2.562933	213	239

In the above, we obtained the median survival time, by default.

To obtain the 25th or any other percentile of survival time, specify the p(#) option.

```
. stci, p(25)
```

	no. of subjects	25%	Std. Err.	[95% Conf. Interval]	
total	40	198	10.76878	164	220

```
. stci, p(25) by(group)
```

group	no. of subjects	25%	Std. Err.	[95% Conf. Interval]	
1	19	190	8.411659	143	213
2	21	232	14.88531	142	233
total	40	198	10.76878	164	220

Note that the p-percentile of survival time is the analysis time at which $p\%$ of subjects have failed and $1 - p\%$ have not. In the above table, 25% of subjects in group 1 failed by time 190, while 25% of subjects in group 2 failed by time 232, indicating a better survival experience for this group.

We can verify the quantities reported by stci by plotting and examining the Kaplan–Meier survival curves.

. sts graph, by(group)

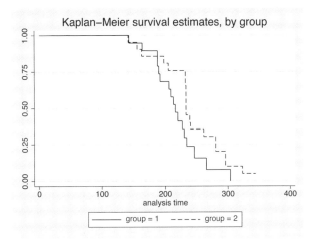

The mean survival time reported by rmean is calculated as the area under the Kaplan–Meier survivor function. If the observation with the largest analysis time is censored, then the survivor function does not go to zero. Consequently, the area under the curve underestimates the mean survival time.

In the above graph, notice that the survival probability for group = 1 goes to 0 at analysis time 344, but the survivor function for group = 2 never goes to 0. For these data, the mean survival time for group = 1 will be properly estimated, but it will be underestimated for group = 2. When we specify the rmean option, Stata will inform us if any of the mean survival times are underestimated.

. stci, rmean by(group)

group	no. of subjects	restricted mean	Std. Err.	[95% Conf. Interval]	
1	19	218.7566	9.122424	200.877	236.636
2	21	241.8571(*)	11.34728	219.617	264.097
total	40	231.3522(*)	7.700819	216.259	246.446

(*) largest observed analysis time is censored, mean is underestimated.

We note that Stata flagged the mean for group = 2 and the overall mean as being underestimated.

If the largest observed analysis time is censored, stci's emean option will extend the survivor function from the last observed time to zero using an exponential function and will compute the area under the entire curve.

. stci, emean

	no. of subjects	extended mean
total	40	234.2557

The resulting area needs to be evaluated with care because it is an ad hoc approximation that can at times be misleading. It is recommended that the extended survivor function be plotted and examined. This is facilitated by the use of stci's graph option.

```
. stci, emean graph
```

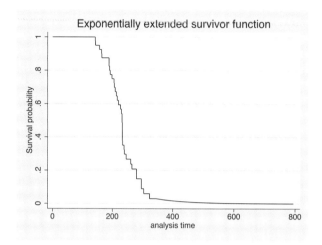

stci also works with multiple-record survival data. Here is a summary of the multiple-record Stanford heart-transplant data introduced in [ST] **stset**:

```
. use http://www.stata-press.com/data/r8/stan3
(Heart transplant data)
. stci
        failure _d:  died
   analysis time _t:  t1
              id:  id
```

	no. of subjects	50%	Std. Err.	[95% Conf. Interval]	
total	103	100	38.64425	69	219

stci with the by() option may produce results with multiple-record data that, at first, you might think are in error:

```
. stci, by(posttran) noshow
```

posttran	no. of subjects	50%	Std. Err.	[95% Conf. Interval]	
0	103	149	22.16591	69	340
1	69	96	38.01968	45	285
total	103	100	38.64425	69	219

Note that, for the number of subjects, $103 + 69 \neq 103$. Variable posttran is not constant for the subjects in this dataset:

```
. stvary posttran, noshow
```

	subjects for whom the variable is				
variable	constant	varying	never missing	always missing	sometimes missing
posttran	34	69	103	0	0

In this dataset, subjects have one or two records. All subjects were eligible for heart transplantation. They have one record if they die or are lost due to censoring before transplantation, and they have two records if the operation was performed. In that case, the first record records their survival up to transplantation, and the second records their subsequent survival. posttran is 0 in the first record and 1 in the second.

Thus, all 103 subjects have records with posttran = 0 and, when stci reported results for this group, it summarized the pretransplantation survival. The median survival time was 149 days.

The posttran = 1 line of stci's output summarizes the post-transplantation survival: 69 patients underwent transplantation and the median survival time was 96 days. For these data, this is not 96 more days, but 96 days in total. That is, the clock was not reset on transplantation. Thus, without attributing cause, we can describe the differences between the groups as an increased hazard of death at early times followed by a decreased hazard later.

Multiple-failure data

If you simply type stci with multiple-failure data, be aware that the reported survival time is the survival time to the first failure under the assumption that the hazard function is not indexed by number of failures.

Here we have some multiple-failure data:

```
. use http://www.stata-press.com/data/r8/mfail2
. st
-> stset t, id(id) failure(d) time0(t0) exit(time .) noshow

               id:  id
    failure event:  d != 0 & d < .
obs. time interval:  (t0, t]
 exit on or before:  time .
. stci
```

	no. of subjects	50%	Std. Err.	[95% Conf. Interval]	
total	926	420	13.42537	394	451

To understand this output, let's also obtain output for each failure separately:

```
. stgen nf = nfailures()
. stci, by(nf)
```

nf	no. of subjects	50%	Std. Err.	[95% Conf. Interval]	
0	926	399	11.50173	381	430
1	529	503	13.68105	425	543
2	221	687	16.83127	549	817
3	58	.	.	.	.
total	926	420	13.42537	394	451

The stgen command added, for each subject, a variable containing the number of previous failures. For a subject, up to and including the first failure, nf is 0. Then nf is 1 up to and including the second failure, and then it is 2, and so on; see [ST] **stgen**.

You should have no difficulty interpreting stci's output. The first line, corresponding to nf = 0, states that among those who had experienced no failures yet, the median time to first failure is 399.

Similarly, the second line, corresponding to nf = 1, is for those who have already experienced one failure. The median time of second failures is 503.

When we simply typed stci, we obtained the same information shown as the total line of the more detailed output. The total survival time distribution is an estimate of the distribution of the time to first failure, under the assumption that the hazard function $h(t)$ is the same across failures—that the second failure is no different from the first failure. This is an odd definition of "same" because the clock t is not reset in $h(t)$ upon failure. What is the hazard of a failure—any failure—at time t? Answer: $h(t)$.

Another definition of same would have it that the hazard of a failure is given by $h(\tau)$, where τ is the time since last failure—that the process resets itself. These definitions are different unless $h()$ is a constant function of t.

Let's examine this multiple-failure data under the process-replication idea. The key variables in this st data are id, t0, t, and d:

```
. st
-> stset t, id(id) failure(d) time0(t0) exit(time .) noshow
                id:  id
     failure event:  d != 0 & d < .
obs. time interval:  (t0, t]
 exit on or before:  time .
```

Our goal is, for each subject, to reset t0 and t to 0 after every failure event. We are going to have to trick Stata, or at least trick stset. stset will not let us set data where the same subject has multiple records summarizing the overlapping periods. The trick is create a new id variable that is different for every id—nf combination (remember, nf is the variable we previously created that records the number of prior failures). Then each of the "new" subjects can have their clock start at time 0:

```
. egen newid = group(id nf)
. sort newid t
. by newid: replace t = t - t0[1]
(808 real changes made)
. by newid: generate newt0 = t0 - t0[1]
. stset t, failure(d) id(newid) time0(newt0)
                id:  newid
     failure event:  d != 0 & d < .
obs. time interval:  (newt0, t]
 exit on or before:  failure

───────────────────────────────────────────────────────────────────
    1734  total obs.
       0  exclusions
───────────────────────────────────────────────────────────────────
    1734  obs. remaining, representing
    1734  subjects
     808  failures in single failure-per-subject data
  435444  total analysis time at risk, at risk from t =           0
                              earliest observed entry t =         0
                                 last observed exit t =         797
```

Note that stset no longer thinks we have multiple-failure data. Whereas, with id, subjects had multiple failures, newid gives a unique identity to each id—nf combination. Each "new" subject has at most one failure.

```
. stci, by(nf)

        failure _d:  d
  analysis time _t:  t
              id:  newid
```

nf	no. of subjects	50%	Std. Err.	[95% Conf. Interval]	
0	926	399	11.22457	381	430
1	529	384	9.16775	359	431
2	221	444	7.406977	325	515
3	58	.	.	.	.
total	1734	404	10.29992	386	430

Compare this table with the one we previously obtained. The number of subjects is the same, but the survival times differ because now we measure the times from one failure to the next, and previously we measured the time from a fixed point. The time between events in these data appears to be independent of event number.

Similarly, we can obtain the mean survival time for these data restricted to the longest follow-up time:

```
. stci, rmean by(nf)

        failure _d:  d
  analysis time _t:  t
              id:  newid
```

nf	no. of subjects	restricted mean	Std. Err.	[95% Conf. Interval]	
0	926	399.1802	8.872794	381.79	416.571
1	529	397.0077(*)	13.36058	370.821	423.194
2	221	397.8051(*)	25.78559	347.266	448.344
3	58	471(*)	0	471	471
total	1734	404.7006	7.021657	390.938	418.463

(*) largest observed analysis time is censored, mean is underestimated.

Saved Results

stci saves in r():

Scalars

r(N_sub)	number of subjects	r(se)	standard error
r(p#)	#th percentile	r(lb)	lower bound of CI
r(rmean)	restricted mean	r(ub)	upper bound of CI
r(emean)	extended mean		

Methods and Formulas

stci is implemented as an ado-file.

The percentiles of survival times are obtained from $S(t)$, the Kaplan–Meier product-limit estimate of the survivor function. The 25th percentile, for instance, is obtained as the maximum value of t such that $S(t) \leq .75$. The restricted mean is obtained as the area under the Kaplan–Meier product-limit survivor curve. The extended mean is obtained by extending the Kaplan–Meier product-limit survivor curve to zero using an exponentially fitted curve and then computing the area under the entire curve. Note that if the longest follow-up time ends in failure, the Kaplan–Meier product-limit survivor curve goes to zero and the restricted mean and extended mean are identical.

The large sample standard error for the pth percentile of the distribution is given by Klein and Moeschberger (1997, 114) as

$$\frac{p\sqrt{\widehat{g}}}{\sqrt{\widehat{S}(t_p)\widehat{f}(t_p)}}$$

where $\widehat{g}$ is the Greenwood pointwise standard error estimate for $\widehat{S}(t_p)$, and $\widehat{f}(t_p)$ is the estimated density function at the pth percentile.

Confidence intervals, however, are not calculated based on this standard error. For a given confidence level, the upper confidence limit for the pth percentile is defined as the first time at which the upper confidence limit for $S(t)$ (based on a $\ln\{-\ln S(t)\}$ transformation) is less than or equal to p, and similarly, the lower confidence limit is defined as the first time at which the lower confidence limit of $S(t)$ is less than or equal to p.

The restricted mean is obtained as the area under the Kaplan–Meier product-limit survivor curve. The extended mean is obtained by extending the Kaplan–Meier product-limit survivor curve to zero by using an exponentially fitted curve and then computing the area under the entire curve. Note that if the longest follow-up time ends in failure, the Kaplan–Meier product-limit survivor curve goes to zero, and the restricted mean and the extended mean are identical.

The standard error for the estimated restricted mean is computed as given by Klein and Moeschberger (1997, 110) and Collett (1994, 295):

$$\widehat{SE} = \sum_{i=1}^{D} \widehat{A}_i \sqrt{\frac{d_i}{R_i(R_i - d_i)}}$$

where the sum is over all distinct failure times, $\widehat{A}_i$ is the estimated area under the curve from time i to the maximum follow-up time, R_i is the number of subjects at risk at time i, and d_i is the number of failures at time i.

The $100(1 - \alpha)\%$ confidence interval for the estimated restricted mean is computed as

$$\widehat{A}_i \pm Z_{1-\alpha/2}\widehat{SE}$$

References

Collett, D. 1994. *Modeling Survival Data in Medical Research*. London: Chapman & Hall.

Klein, J. P. and M. L. Moeschberger. 1997. *Survival Analysis: Techniques for Censored and Truncated data*. New York: Springer.

Also See

Complementary: [ST] **stdes**, [ST] **stgen**, [ST] **stir**, [ST] **sts**, [ST] **stset**, [ST] **stvary**

Background: [ST] **st**, [ST] **survival analysis**

Title

stcox — Fit Cox proportional hazards model

Syntax

stcox [*varlist*] [if *exp*] [in *range*] [, nohr strata(*varnames*) robust

cluster(*varname*) noadjust tvc(*varlist*) texp(*exp*) shared(*varname*)

frailty(gamma) effects(*newvar*) mgale(*newvar*) esr(*newvar(s)*)

schoenfeld(*newvar(s)*) scaledsch(*newvar(s)*) basehc(*newvar*) basechazard(*newvar*)

basesurv(*newvar*) {breslow | efron | exactm | exactp} estimate noshow

offset(*varname*) level(#) *maximize_options*]

stphtest [, km log rank time(*varname*) plot(*varname*) detail *lowess_options*

twoway_options]

stcurve [, cumhaz survival hazard range(# #) at(*varname=#* [*varname=#*...])

[at1(*varname=#* [*varname=#*...]) [at2(*varname=#* [*varname=#*...]) [...]]]

outfile(*filename*, [replace]) alpha1 unconditional kernel(*kernel*) width(#)

line_options twoway_options]

stcox is for use with survival-time data; see [ST] **st**. You must stset your data before using this command.
by ... : may be used with stcox; see [R] **by**.
stphtest may be used after stcox.
stcurve may be used after stcox and streg. See [ST] **streg** for full documentation on stcurve.
stcox shares the features of all estimation commands; see [U] **23 Estimation and post-estimation commands**.
stcox may be used with sw to perform stepwise estimation; see [R] **sw**.

Syntax for predict

predict [*type*] *newvarname* [if *exp*] [in *range*] [, *statistic* nooffset]

where *statistic* is

hr	hazard ratio (relative hazard), predicted $\exp(\mathbf{x}_j\mathbf{b})$ (the default)
xb	linear prediction $\mathbf{x}_j\mathbf{b}$
stdp	standard error of the linear prediction; $\text{SE}(\mathbf{x}_j\mathbf{b})$
* csnell	(partial) Cox–Snell residuals
* deviance	deviance residuals
* ccsnell	cumulative Cox–Snell residuals
* cmgale	cumulative martingale residuals

Unstarred statistics are available both in and out of sample; type predict ... if e(sample) ... if wanted for only the estimation sample. Starred statistics are calculated for only the estimation sample, even when e(sample) is not specified. Note that the (partial) martingale residuals, the efficient score residuals, the Schoenfeld residuals, and the scaled Schoenfeld residuals are available by options specified to the stcox command itself.

Description

stcox fits, via maximum-likelihood, proportional hazards models on st data. stcox is appropriate for use with single- or multiple-record, single- or multiple-failure, st data.

stphtest can be used after stcox to test the proportional hazard assumption based on Schoenfeld residuals. Use of this command requires that you previously specified stcox's schoenfeld() option (if the global test is desired) and/or stcox's scaledsch() option (if the detailed test is desired). For graphical assessment of the proportional hazard assumptions, see [ST] **stphplot**.

Options

Options for stcox

nohr specifies that coefficients rather than exponentiated coefficients are to be displayed or, said differently, coefficients rather than hazard ratios. This option affects only how results are displayed and not how they are estimated. nohr may be specified at estimation time or when redisplaying previously estimated results (which you do by typing stcox without a variable list).

strata(*varnames*) specifies up to 5 strata variables. Observations with equal values of the strata variables are assumed to be in the same stratum. Stratified estimates (equal coefficients across strata but baseline hazard unique to each stratum) are then obtained.

robust specifies that the robust method of calculating the variance–covariance matrix (Lin and Wei 1989) is to be used instead of the conventional inverse-matrix-of-negative-second-derivatives method. If you specify robust, and if you have previously stset an id() variable, the robust calculation will be clustered on the id() variable.

cluster(*varname*) implies robust and specifies a variable on which clustering is to be based. This overrides the default clustering, if any. You seldom need to specify this option, since specifying robust implies cluster() if an id() variable is stset.

noadjust is for use with robust or cluster(). noadjust prevents the estimated variance matrix from being multiplied by $N/(N-1)$ or $g/(g-1)$, where g is the number of clusters. The default adjustment is somewhat arbitrary because it is not always clear how to count observations or clusters. In such cases, however, the adjustment is likely to be biased toward 1, so we would still recommend making it.

tvc(*varlist*) is used to specify those variables that vary continuously with respect to time, i.e., time-varying covariates. This is a convenience option used to speed up calculations and to avoid the need to have to stsplit the data over many failure times.

texp(*exp*) is used in conjunction with tvc(*varlist*) to specify which function of analysis time should be multiplied by the time-varying covariates. For example, specifying texp(ln(_t)) would cause the time-varying covariates to be multiplied by the logarithm of analysis time. If tvc(*varlist*) is used without texp(*exp*), it is understood that you mean texp(_t), and thus the time-varying covariates are multiplied by the analysis time.

Both tvc(*varlist*) and texp(*exp*) are further explained in the section on *Cox regression with continuously time-varying covariates* below.

shared(*varname*) specifies the fitting of a Cox model with shared frailty. The frailties are gamma-distributed latent random effects that enter multiplicatively on the hazard, or equivalently the logarithm of the frailty enters the linear predictor as a random offset. Think of a shared frailty model as a Cox model for panel data. *varname* is a variable in the data which identifies the groups.

Shared frailty models are further discussed in the section *Shared frailty models* below.

frailty(gamma) is for use with shared() and has no effect other than to serve as an analogous syntax for those who fit *parametric* frailty models using streg; see [ST] **streg**. In streg, you may specify either of two frailty distributions (the gamma or the inverse-Gaussian), yet only the gamma is available in stcox.

effects(*newvar*) is for use with shared() and will create *newvar* containing estimates of the log-frailty for each group. The log-frailties are random group-specific offsets to the linear predictor and measure the group effect on the log-relative hazard.

mgale(*newvar*) adds *newvar* containing the partial martingale residuals, which are fully described in *Methods and Formulas* below. If each observation in your data represents a different subject (single-record-per-subject data), then the partial martingale residuals are the martingale residuals.

If you have multiple-records-per-subject data, the value that mgale() stores in each observation is the observation's contribution to the martingale residual, and these partial residuals can be summed within id() to obtain the subject's martingale residual. Say you specify mgale(pmr) and that you have previously stset, id(patid). Then, egen mr = sum(pmr), by(patid) would create the martingale residuals.

esr(*newvar(s)*) adds *newvar(s)* containing the partial efficient score residuals, which are fully described in *Methods and Formulas* below. If each observation in your data represents a different subject (single-record-per-subject data), then the partial efficient score residuals are the efficient score residuals.

If you have multiple-records-per-subject data, then the values esr() stores are each observation's contribution to the score residuals, and these partial residuals can be summed within id() to form the subject's efficient score residuals. This could be accomplished as noted under mgale() above.

One efficient score residual variable is created for each regressor in the model; the first new variable corresponds to the first regressor, the second to the second, and so on.

schoenfeld(*newvar(s)*) adds *newvar(s)* containing the Schoenfeld residuals, which are fully described under *Methods and Formulas*. This option is not available in conjunction with the exactm and exactp options. Schoenfeld residuals are calculated and are reported only at failure times.

One Schoenfeld residual variable is created for each regressor in the model; the first new variable corresponds to the first regressor, the second to the second, and so on.

scaledsch(*newvar(s)*) adds *newvar(s)* containing the scaled Schoenfeld residuals, which are fully described under *Methods and Formulas*. This option is not available if exactm or exactp have been specified. Scaled Schoenfeld residuals are calculated and reported only at failure times.

One scaled Schoenfeld residual variable is created for each regressor in the model; the first new variable corresponds to the first regressor, the second to the second, and so on.

NOTE: The easiest way to specify the preceding three options is, for example, esr(*stub**), where *stub* is a short name of your choosing. Stata then creates variables *stub*1, *stub*2, etc. Alternatively, you may specify each variable explicitly, in which case there must be as many (and no more) variables specified in esr() as regressors in the model.

One caution is necessary for the preceding three options: stcox will drop variables from the model due to collinearity. This is a desirable feature. A side-effect is that the score residual variable, the

Schoenfeld residual variable, and the scaled Schoenfeld residual variable may not align with the regressors in the way you expect. Say you fit a model by typing

```
. stcox x1 x2 x3, esr(r1 r2 r3)
```

Usually, r1 will contain the residual associated with x1, r2 the residual associated with x2, etc.

Now assume that x2 is dropped due to collinearity. In that case, r1 will correspond to x1, r2 to x3, and r3 will contain 0. This happens because, after omitting the collinear variables, there are only two variables in the model: x1 and x3.

basehc(*newvar*) adds *newvar* to the data containing the estimated baseline hazard contributions as described by Kalbfleisch & Prentice (2002, 115). These are used to construct the product-limit type estimator for the baseline survival function generated by basesurv(). If strata() is also specified, baseline estimates for each stratum are provided.

basechazard(*newvar*) adds *newvar* to the data containing the estimated cumulative baseline hazard. If strata() is also specified, cumulative baseline estimates for each stratum are provided.

basesurv(*newvar*) adds *newvar* to the data containing the estimated baseline survival function. Note that, in the null model, this is equivalent to the Kaplan–Meier product-limit estimate. If strata() is also specified, baseline estimates for each stratum are provided.

breslow, efron, exactm, and exactp specify the method for handling tied failures in the calculation of the log partial likelihood (and residuals). breslow is the default. Each method is described in the *Methods and Formulas* section below. Note that efron and the exact methods require substantially more computer time than the default breslow option. exactm and exactp may not be specified with robust or cluster(), or with tvc().

estimate forces the fitting of the null model. All Stata estimation commands redisplay results when the command name is typed without arguments. So does stcox. What if you wish to fit a Cox model on $x_j b$, where $x_j b$ is defined as 0? Logic says you would type stcox. There are no explanatory variables, so there is nothing to type following the command. Unfortunately, stcox looks the same as stcox typed without arguments, which is a request to redisplay results.

To fit the null model, type stcox, estimate.

noshow prevents stcox from showing the key st variables. This option is rarely used since most people type stset, show or stset, noshow to reset once and for all whether they want to see these variables mentioned at the top of the output of every st command; see [ST] **stset**.

offset(*varname*) specifies a variable that is to be entered directly into the linear predictor with coefficient 1.

level(*#*) specifies the confidence level, in percent, for confidence intervals. The default is level(95) or as set by set level; see [U] **23.6 Specifying the width of confidence intervals**.

maximize_options control the maximization process; see [R] **maximize**. You should never have to specify them. Note that if fitting a Cox shared frailty model, *maximize_options* do not apply to the maximization of the Cox partial log-likelihood, but instead to the maximum profile log-likelihood estimation of the frailty parameter.

(*Continued on next page*)

Options for stphtest

km, log, rank, and time() are used to specify the time scaling function.

By default, stphtest performs the tests using the identity function, i.e., analysis time itself.

km specifies that 1 minus the Kaplan–Meier product limit estimate be used.

log specifies that the natural log of analysis time be used.

rank specifies that the rank of analysis time be used.

time(*varname*) specifies a variable containing an arbitrary monotonic transformation of analysis time. It is your responsibility to ensure that *varname* is a monotonic transform.

plot(*varname*) is used to request that a smoothed plot of scaled Schoenfeld residuals versus time be produced for the covariate specified by *varname*. By default, the smoothing is performed using the running mean method implemented in lowess, mean noweight; see [R] **lowess**.

detail specifies that a separate test of the proportional hazard assumption be produced for each covariate in the Cox model. By default, stphtest produces only the global test.

lowess_options are any of the options allowed with the mean and noweight options of the lowess command, except the plot() option; see [R] **lowess**. This particularly includes the width() option, which affects the amount of smoothing.

twoway_options are any of the options documented in [G] ***twoway_options*** excluding by(). These include options for titling the graph (see [G] ***title_options***) and options for saving the graph to disk (see [G] ***saving_option***).

Options for stcurve

See [ST] **streg**.

Options for predict

hr, the default, calculates the relative hazard (hazard ratio); that is, the exponentiated linear prediction, $\exp(\mathbf{x}_j\widehat{\boldsymbol{\beta}})$.

xb calculates the linear prediction from the fitted model. That is, you fit the model by estimating a set of parameters $\beta_1, \beta_2, \ldots, \beta_k$, and the linear prediction is $\widehat{y}_j = \widehat{\beta}_1 x_{1j} + \widehat{\beta}_2 x_{2j} + \cdots + \widehat{\beta}_k x_{kj}$, often written in matrix notation as $\widehat{y}_j = \mathbf{x}_j\widehat{\boldsymbol{\beta}}$.

It is important to understand that the x_{1j}, x_{2j}, ..., x_{kj} used in the calculation are obtained from the data currently in memory and do not have to correspond to the data on the independent variables used in estimating β.

stdp calculates the standard error of the prediction; that is, the standard error of $\widehat{y}_j$.

csnell calculates the (partial) Cox–Snell generalized residuals. If you have a single observation per subject, csnell calculates the usual Cox–Snell residual. Otherwise, csnell calculates the additive contribution of this observation to the subject's overall Cox–Snell residual. In such cases, see option ccsnell below.

In order to specify option csnell, you must have specified 'stcox ..., mgale(*newvar*) ...' when you fitted the model. The name specified for the new variable does not matter, but you must have specified the mgale() option or predict will return an error.

deviance calculates the deviance residual. In the case of multiple-record data, only one value per subject is calculated, and it is placed on the last record for the subject.

In order to specify option deviance, you must have specified 'stcox ..., mgale(*newvar*) ...' when you fitted the model. See the comment for option csnell above.

ccsnell calculates the cumulative Cox–Snell residual in multiple-record data. This is based on calculating the partial Cox–Snell residuals (see option csnell above) and then summing them. Only one value per subject is recorded—the overall sum—and it is placed on the last record for the subject. ccsnell is the same as csnell in single-record survival data.

In order to specify option ccsnell, you must have specified 'stcox ..., mgale(*newvar*) ...' when you fitted the model. See the comment for option csnell above.

cmgale calculates the cumulative martingale residual in multiple-record data. This is based on summing the partial martingale residuals available from stcox. Only one value per subject is recorded—the overall sum—and it is placed on the last record for the subject. cmgale is the same as mgale in single-record survival data.

In order to specify option cmgale, you must have specified 'stcox ..., mgale(*newvar*) ...' when you fitted the model. See the comment for option csnell above.

nooffset is relevant only if you specified offset(*varname*) for stcox. It modifies the calculations made by predict so that they ignore the offset variable; the linear prediction is treated as $\mathbf{x}_j\widehat{\boldsymbol{\beta}}$ rather than $\mathbf{x}_j\widehat{\boldsymbol{\beta}} + \text{offset}_j$.

Remarks

Remarks are presented under the headings

> Cox regression with uncensored data
> Cox regression with censored data
> Treatment of tied failure times
> Cox regression with discrete time-varying covariates
> Cox regression with continuous time-varying covariates
> Robust estimate of variance
> Cox regression with multiple failure data
> Stratified estimation
> Cox regression with shared frailty
> Obtaining baseline function estimates
> Cox regression residuals
> Checking and testing the proportional hazard assumption

What follows is a brief summary of what can be done with stcox and stphtest. For a complete tutorial, see Cleves, Gould, and Gutierrez (2002), who devote three chapters to this topic.

In the Cox proportional hazards model, the hazard is assumed to be

$$h(t) = h_0(t)\exp(\beta_1 x_1 + \cdots + \beta_k x_k)$$

The Cox model provides estimates of $\beta_1, \ldots, \beta_k$ but provides no direct estimate of $h_0(t)$, called the baseline hazard. Formally, the function $h_0(t)$ is not directly estimated, but it is possible to recover an estimate of the cumulative hazard $H_0(t)$ and, from that, an estimate of the baseline survivor function $S_0(t)$.

stcox fits the Cox proportional hazards model, which is to say, it provides estimates of β and its variance–covariance matrix.

stcox, basehc(*newvar*) fits the Cox proportional hazards model and calculates the set of baseline hazard contributions used to estimate the baseline survival function $S_0(t)$.

stcox, basechazard(*newvar*) fits the Cox proportional hazards model and estimates the cumulative baseline hazard function $H_0(t)$.

stcox, basesurv(*newvar*) fits the Cox model and estimates the baseline survival function $S_0(t)$.

The three baseline options may also be used in combination to concurrently produce estimates of their respective functions. In addition, stcox with the strata() option will produce stratified Cox regression estimates. In the stratified estimator, the hazard at time t for a subject in group i is assumed to be

$$h_i(t) = h_{0i}(t) \exp(\beta_1 x_1 + \cdots + \beta_k x_k)$$

That is, the coefficients are assumed to be the same regardless of group, but the baseline hazard is allowed to be group-specific. If you specify the strata() option, the baseline options produce the group-specific estimates of the baseline functions.

Whether or not you specify strata(), stcox can produce either of two variance estimators for β. The default is the conventional, inverse-matrix-of-negative-second-derivatives calculation. The theoretical justification for this estimator is based on likelihood theory.

The robust option instead switches to the robust measure developed by Lin and Wei (1989). This variance estimator is a variant of the estimator discussed in [U] **23.14 Obtaining robust variance estimates**.

Finally, stcox, with the shared() option, will fit a Cox model with shared frailty. A *frailty* is a group-specific latent random effect that multiplies into the hazard function. The distribution of the frailties is gamma with mean one and variance to be estimated by the data. Shared frailty models are used to model within-group correlation—observations within a group are correlated because they share the same frailty.

We give examples below with uncensored, censored, time-varying, and recurring failure data, but it does not matter in terms of what you type. Once you have stset your data, to fit a model you type stcox followed by the names of the explanatory variables. You do this whether your dataset is single- or multiple-record, whether your dataset includes censored observations or delayed entry, and even whether you are analyzing single- or multiple-failure data. In other words, you use stset to describe the properties of the data and then that information is available to stcox—and all the other st commands—so that you do not have to specify it again.

Cox regression with uncensored data

You wish to analyze an experiment testing the ability of emergency generators with a new-style bearing to withstand overloads. For this experiment, the overload protection circuit was disabled and then the generators were run overloaded until they burned up. Here are your data:

(Continued on next page)

```
. use http://www.stata-press.com/data/r8/kva
(Generator experiment)
. list
```

	failtime	load	bearings
1.	100	15	0
2.	140	15	1
3.	97	20	0
4.	122	20	1
5.	84	25	0
6.	100	25	1
7.	54	30	0
8.	52	30	1
9.	40	35	0
10.	55	35	1
11.	22	40	0
12.	30	40	1

Twelve generators, half with the new-style bearings and half with the old, were allocated to this destructive test. The first observation reflects an old-style generator (bearings = 0) under a 15 kVA overload. It stopped functioning after 100 hours. The second generator had new-style bearings (bearings = 1), and under the same overload condition lasted 140 hours. Paired experiments were also performed under 20, 25, 30, 35, and 40 kVA overloads.

You wish to fit a Cox proportional hazards model in which the failure rate depends on the amount of overload and the style of the bearings. That is, you assume that bearings and load do not affect the shape of the overall hazard function, but they do affect the relative risk of failure. To fit this model, you type

```
. stset failtime
(output omitted)
. stcox load bearings
           failure _d:  1 (meaning all fail)
     analysis time _t:  failtime

Iteration 0:   log likelihood = -20.274897
Iteration 1:   log likelihood = -10.515114
Iteration 2:   log likelihood = -8.8700259
Iteration 3:   log likelihood = -8.5915211
Iteration 4:   log likelihood = -8.5778991
Iteration 5:   log likelihood =  -8.577853
Refining estimates:
Iteration 0:   log likelihood =  -8.577853

Cox regression -- Breslow method for ties

No. of subjects =          12                  Number of obs   =        12
No. of failures =          12
Time at risk    =         896
                                               LR chi2(2)      =     23.39
Log likelihood  =    -8.577853                 Prob > chi2     =    0.0000
```

_t	Haz. Ratio	Std. Err.	z	P>|z|	[95% Conf. Interval]	
load	1.52647	.2188172	2.95	0.003	1.152576	2.021653
bearings	.0636433	.0746609	-2.35	0.019	.0063855	.6343223

You find that after controlling for overload, the new-style bearings result in a lower hazard and therefore a longer survival time.

Once an stcox model has been fitted, typing stcox without arguments redisplays the previous results. Options that affect the display, such as nohr—which requests that coefficients rather than hazard ratios be displayed—can be specified upon estimation or when results are redisplayed:

```
. stcox, nohr

Cox regression -- Breslow method for ties

No. of subjects =           12              Number of obs   =          12
No. of failures =           12
Time at risk    =          896
                                            LR chi2(2)      =       23.39
Log likelihood  =    -8.577853             Prob > chi2     =      0.0000
```

_t	Coef.	Std. Err.	z	P>|z|	[95% Conf. Interval]
load	.4229578	.1433485	2.95	0.003	.1419999	.7039157
bearings	-2.754461	1.173115	-2.35	0.019	-5.053723	-.4551981

❏ Technical Note

stcox's iteration log looks like a standard Stata iteration log right up to the point where it says "Refining estimates". The Cox proportional hazards likelihood function is indeed a difficult function, both conceptually and numerically. Up until Stata says "Refining estimates", it maximizes the Cox likelihood in the standard way using double-precision arithmetic. Then, just to be sure that the answers are accurate, Stata switches to quad-precision routines (which is to say, double double precision) and completes the maximization procedure from its current location on the likelihood.

❏

Cox regression with censored data

You have data on 48 participants in a cancer drug trial. Of these 48, 28 received treatment (drug = 1) and 20 receive a placebo (drug = 0). The participants range in age from 47 to 67 years. You wish to analyze time until death, measured in months. Your data include one observation for each patient. The variable studytime records either the month of their death or the last month that they were known to be alive. Some of the patients still live, so together with studytime is died, indicating their health status. Persons known to have died—"noncensored" in the jargon—have died = 1, whereas the patients who are still alive—"right-censored" in the jargon—have died = 0.

Here is an overview of your data:

```
. use http://www.stata-press.com/data/r8/drugtr, clear
(Patient Survival in Drug Trial)
. st
-> stset studytime, failure(died)

     failure event:  died != 0 & died < .
obs. time interval:  (0, studytime]
 exit on or before:  failure
```

```
. summarize
    Variable |       Obs        Mean    Std. Dev.       Min        Max
-------------+--------------------------------------------------------
   studytime |        48        15.5    10.25629          1         39
        died |        48    .6458333    .4833211          0          1
        drug |        48    .5833333    .4982238          0          1
         age |        48      55.875    5.659205         47         67
         _st |        48           1           0          1          1
-------------+--------------------------------------------------------
          _d |        48    .6458333    .4833211          0          1
          _t |        48        15.5    10.25629          1         39
         _t0 |        48           0           0          0          0
```

You typed `stset studytime, failure(died)` previously; that is how `st` knew about this dataset. To fit the Cox model, you type

```
. stcox drug age

         failure _d:  died
   analysis time _t:  studytime

Iteration 0:   log likelihood = -99.911448
Iteration 1:   log likelihood = -83.551879
Iteration 2:   log likelihood = -83.324009
Iteration 3:   log likelihood = -83.323546
Refining estimates:
Iteration 0:   log likelihood = -83.323546

Cox regression -- Breslow method for ties

No. of subjects =           48                  Number of obs    =         48
No. of failures =           31
Time at risk    =          744
                                                LR chi2(2)       =      33.18
Log likelihood  =    -83.323546                 Prob > chi2      =     0.0000

------------------------------------------------------------------------------
          _t | Haz. Ratio   Std. Err.      z    P>|z|     [95% Conf. Interval]
-------------+----------------------------------------------------------------
        drug |   .1048772   .0477017    -4.96   0.000     .0430057    .2557622
         age |   1.120325   .0417711     3.05   0.002     1.041375     1.20526
------------------------------------------------------------------------------
```

You find that the drug results in a lower hazard and therefore a longer survival time controlling for age. Older patients are more likely to die. The model as a whole is statistically significant.

The hazard ratios reported correspond to a one-unit change in the corresponding variable. It is more typical to report relative risk for 5-year changes in age. To obtain such a hazard ratio, you create a new age variable such that a one-unit change means a 5-year change:

(Continued on next page)

```
. replace age = age/5
age was int now float
(48 real changes made)

. stcox drug age, nolog

         failure _d:  died
   analysis time _t:  studytime

Cox regression -- Breslow method for ties

No. of subjects =          48                    Number of obs    =          48
No. of failures =          31
Time at risk    =         744
                                                 LR chi2(2)       =       33.18
Log likelihood  =   -83.323544                   Prob > chi2      =      0.0000
```

_t	Haz. Ratio	Std. Err.	z	P>|z|	[95% Conf. Interval]	
drug	.1048772	.0477017	-4.96	0.000	.0430057	.2557622
age	1.764898	.3290196	3.05	0.002	1.224715	2.543338

Treatment of tied failure times

The proportional hazards model assumes that the hazard function is continuous, and thus that there are no tied survival times. Due to the way that time is recorded, however, tied events do occur in survival data. In such cases, the partial likelihood needs to be modified. See the *Methods and Formulas* section for further details on the methods described below.

Stata provides four methods for handling tied failures in the calculation of the Cox partial likelihood through the options breslow, efron, exactm, and exactp. If there are no ties in the data, the results are identical regardless of the method selected.

When there are tied failure times, we must decide how to handle the calculation of the risk pools for these tied observations. Assume that there are two observations that fail in succession. In the calculation involving the second observation, the first observation is not in the risk pool since failure has already occurred. If the two observations have the same failure time, then how to calculate the risk pool for the second observation and in which order to calculate the two observations are at issue.

There are two views of time. In the first, time is continuous, so ties should not occur. If they have occurred, the likelihood reflects the marginal probability that the tied failure events occurred before the nonfailure events in the risk pool (the order that they occurred is not important). This is called the exact marginal likelihood (option exactm).

In the second view, time is discrete, so ties are expected. The likelihood is changed to reflect this discreteness and calculates the conditional probability that the observed failures are the ones that fail in the risk pool given the observed number of failures. This is called the exact partial likelihood (option exactp).

Let's assume that there are 5 subjects, e_1, e_2, e_3, e_4, e_5, in the risk pool and that subjects e_1 and e_2 fail. It could be that had we been able to observe the events at a better resolution, e_1 failed from risk pool $e_1 + e_2 + e_3 + e_4 + e_5$ and then e_2 failed from risk pool $e_2 + e_3 + e_4 + e_5$. Alternatively, it could be that e_2 failed first from risk pool $e_1 + e_2 + e_3 + e_4 + e_5$ and then e_1 failed from risk pool $e_1 + e_3 + e_4 + e_5$.

The Breslow method (option breslow) for handling tied values simply says that since we do not know the order, we will use the largest risk pool for each of the tied failure events. In other words, it assumes that e_1 failed and e_2 failed, both from risk pool $e_1 + e_2 + e_3 + e_4 + e_5$. This approximation is very fast and is the default method for handling ties. If there are many ties in the dataset, this approximation will not be very good as the risk pools include too many observations. The Breslow method is an approximation of the exact marginal likelihood.

The Efron method (option `efron`) for handling tied values says that the first risk pool must be $e_1 + e_2 + e_3 + e_4 + e_5$ and then that the second risk pool is either $e_2 + e_3 + e_4 + e_5$ or $e_1 + e_3 + e_4 + e_5$. From this, Efron noted that the e_1 and e_2 terms were in the second risk pool with probability 1/2 and so used for the second risk pool $.5(e_1 + e_2) + e_3 + e_4 + e_5$. Efron's approximation is a more accurate approximation of the exact marginal likelihood than Breslow's but takes longer to calculate.

The exact marginal method (option `exactm`) is a misnomer in the sense that the calculation performed is also an *approximation* of the exact marginal likelihood. It is an approximation because the evaluation of the likelihood (and derivatives) is accomplished using 15-point Gauss–Laguerre quadrature. For small to moderate samples, this is slower than the Efron approximation, but the difference in execution time diminishes when samples become larger. You may want to consider the quadrature when deciding to use this method. If the number of tied deaths is large (on average), the quadrature approximation of the function is not well-behaved. A small amount of empirical checking suggests that if the number of tied deaths is larger (on average) than 30, the quadrature does not approximate the function well.

Viewing time as discrete, the exact partial method (option `exactp`) is the final method available. This approach is equivalent to computing conditional logistic regression where the groups are defined by the risk sets and the outcome is given by the death variable. This is the slowest method to use and can take a significant amount of time if the risk sets and the number of tied failures are large.

Cox regression with discrete time-varying covariates

In [ST] **stset** we introduce the Stanford heart transplant data—data for which there are one or two records per patient depending on whether they received a new heart.

This dataset (Crowley and Hu 1977) consists of 103 patients admitted to the Stanford Heart Transplantation Program. Patients were admitted into the program after review by a committee and then waited for an available donor heart. While waiting, some died or were transferred out of the program, but 67% received a transplant. The dataset includes the year the patient was accepted into the program along with the patient's age, whether the patient had other heart surgery previously, and whether the patient received a transplant.

In the data, `posttran` becomes 1 when a patient receives a new heart, and so is a time-varying covariate. That does not matter in terms of what we type to fit the model:

```
. use http://www.stata-press.com/data/r8/stan3, clear
(Heart transplant data)

. stset t1, failure(died) id(id)
  (output omitted)

. stcox age posttran surg year

         failure _d:  died
   analysis time _t:  t1
                 id:  id
Iteration 0:   log likelihood = -298.31514
Iteration 1:   log likelihood =  -289.7344
Iteration 2:   log likelihood = -289.53498
Iteration 3:   log likelihood = -289.53378
Iteration 4:   log likelihood = -289.53378
Refining estimates:
Iteration 0:   log likelihood = -289.53378
```

```
Cox regression -- Breslow method for ties

No. of subjects =          103                   Number of obs   =        172
No. of failures =           75
Time at risk    =      31938.1
                                                 LR chi2(4)      =      17.56
Log likelihood  =   -289.53378                   Prob > chi2     =     0.0015
```

_t	Haz. Ratio	Std. Err.	z	P>|z|	[95% Conf. Interval]	
age	1.030224	.0143201	2.14	0.032	1.002536	1.058677
posttran	.9787243	.3032597	-0.07	0.945	.5332291	1.796416
surgery	.3738278	.163204	-2.25	0.024	.1588759	.8796
year	.8873107	.059808	-1.77	0.076	.7775022	1.012628

We find that older patients have higher hazards, that patients tend to do better over time, and that patients with prior surgery do better. Whether a patient ultimately receives a transplant does not seem to make much difference.

Cox regression with continuous time-varying covariates

The basic proportional hazards regression assumes the relationship

$$h(t) = h_0(t) \exp(\beta_1 x_1 + \cdots + \beta_k x_k)$$

where $h_0(t)$ is the baseline hazard function. For most purposes this model is sufficient, but sometimes we may wish to introduce variables of the form $z_i(t) = z_i g(t)$, which vary continuously with time so that

$$h(t) = h_0(t) \exp \left\{ \beta_1 x_1 + \cdots + \beta_k x_k + g(t)(\gamma_1 z_1 + \cdots + \gamma_m z_m) \right\}$$

where $(z_1, \ldots, z_m)$ are the time-varying covariates and where estimation has the net effect of estimating, say, a regression coefficient γ_i for a covariate $g(t)z_i$, which is a function of the current time.

The time-varying covariates $(z_1, \ldots, z_m)$ are specified using the tvc(*varlist*) option, and $g(t)$ is specified using the texp(*exp*) option, where t in $g(t)$ is analysis time. For example, if we want $g(t) = \ln(t)$, we would use texp(ln(_t)) since _t stores the analysis time once the data are stset.

Since the calculations in Cox regression only concern themselves with the times at which failures occur, the above could also be achieved by stsplitting the data at the observed failure times and manually generating the time-varying covariates. In cases where this is feasible, the above merely represents a more convenient way to accomplish this. However, for large datasets with many distinct failure times, using stsplit may produce datasets that are too large to fit in memory, and even if this were not so, the estimation would take far longer to complete. It is for these reasons that the options described above were introduced.

▷ Example

Consider a dataset consisting of 45 observations on recovery time from walking pneumonia. Recovery time (in days) is recorded in the variable time, and there exist measurements on the covariates age, drug1, and drug2, where drug1 and drug2 interact a choice of treatment with initial dosage level. The study was terminated after 30 days, and thus those who had not recovered by that time were censored (cured = 0).

```
. use http://www.stata-press.com/data/r8/drugtr2, clear

. list in 1/12, separator(0)
```

	age	drug1	drug2	time	cured
1.	36	0	50	20.6	1
2.	14	0	50	6.8	1
3.	43	0	125	8.6	1
4.	25	100	0	10	1
5.	50	100	0	30	0
6.	26	0	100	13.6	1
7.	21	150	0	5.4	1
8.	25	0	100	15.4	1
9.	32	125	0	8.6	1
10.	28	150	0	8.5	1
11.	34	0	100	30	0
12.	40	0	50	30	0

Patient 1 took 50 mg of drug number 2 and was cured after 20.6 days, while Patient 5 took 100 mg of drug number 1 and had yet to recover when the study ended, and thus was censored at 30 days.

We run a standard Cox regression after stsetting the data.

```
. stset time, failure(cured)

     failure event:  cured != 0 & cured < .
obs. time interval:  (0, time]
 exit on or before:  failure

        45  total obs.
         0  exclusions

        45  obs. remaining, representing
        36  failures in single record/single failure data
     677.9  total analysis time at risk, at risk from t =         0
                              earliest observed entry t =          0
                                  last observed exit t =          30

. stcox age drug1 drug2

          failure _d:  cured
    analysis time _t:  time

Iteration 0:   log likelihood = -116.54385
Iteration 1:   log likelihood = -102.77311
Iteration 2:   log likelihood = -101.92794
Iteration 3:   log likelihood = -101.92504
Iteration 4:   log likelihood = -101.92504
Refining estimates:
Iteration 0:   log likelihood = -101.92504

Cox regression -- Breslow method for ties

No. of subjects =          45                    Number of obs   =          45
No. of failures =          36
Time at risk    =  677.9000034
                                                 LR chi2(3)      =       29.24
Log likelihood  =   -101.92504                   Prob > chi2     =      0.0000
```

_t	Haz. Ratio	Std. Err.	z	P>|z|	[95% Conf. Interval]	
age	.8759449	.0253259	-4.58	0.000	.8276873	.9270162
drug1	1.008482	.0043249	1.97	0.049	1.000041	1.016994
drug2	1.00189	.0047971	0.39	0.693	.9925323	1.011337

The output includes p-values for the tests of the null hypotheses that each regression coefficient is zero or, equivalently, that each hazard ratio is one. That all hazard ratios are apparently close to one is a matter of scale; however, we can see that drug number 1 significantly increases the risk of being cured, and so is an effective drug, while drug number 2 is ineffective (given the presence of age and drug number 1 in the model).

Suppose now that we wish to fit a model in which we account for the effect that as time goes by, the actual level of the drug remaining in the body diminishes, say, at an exponential rate. If it is known that the half-life of both drugs is close to 2 days, then we can say that the actual concentration level of the drug in the patient's blood is proportional to the initial dosage times $\exp(-0.35t)$, where t is analysis time. We now fit a model that reflects this change.

```
. stcox age, tvc(drug1 drug2) texp(exp(-0.35*_t)) nolog
        failure _d:  cured
   analysis time _t:  time

Cox regression -- Breslow method for ties

No. of subjects =           45                  Number of obs   =         45
No. of failures =           36
Time at risk    =  677.9000034
                                                LR chi2(3)      =      36.98
Log likelihood  =   -98.052763                  Prob > chi2     =     0.0000

------------------------------------------------------------------------------
         _t | Haz. Ratio   Std. Err.      z    P>|z|     [95% Conf. Interval]
-----------+------------------------------------------------------------------
rh         |
       age |   .8614636    .028558    -4.50   0.000     .8072706    .9192948
-----------+------------------------------------------------------------------
t          |
     drug1 |   1.304744    .1135967    3.06   0.002     1.100059    1.547514
     drug2 |   1.200613    .1113218    1.97   0.049     1.001103    1.439882
------------------------------------------------------------------------------
```

Note: Second equation contains variables that continuously vary with respect to time; variables are interacted with current values of exp(-0.35*_t).

The first equation, rh, reports the results (hazard ratios) for the covariates that do not vary over time; the second equation, t, reports the results for the time-varying covariates.

As the level of drug in the blood system decreases, the drug's effectiveness will diminish. Accounting for this serves to unmask the effects of both drugs in that we now see increased effects on both. In fact, the effect on recovery time of drug number 2 now becomes significant.

❏ Technical Note

The interpretation of hazard ratios requires careful consideration here. For the first model, the hazard ratio for, say, drug1 is interpreted as the proportional change in hazard when the dosage level of drug1 is increased by one unit. For the second model, the hazard ratio for drug1 is the proportional change in hazard when the blood concentration level, i.e., drug1*$\exp(-0.35t)$, increases by one.

❏

Since the number of observations in our data is relatively small, for illustrative purposes we can stsplit the data at each recovery time, manually generate the blood concentration levels, and re-fit the second model.

```
. generate id=_n
. streset, id(id)
(output omitted)
```

```
. stsplit, at(failures)
(31 failure times)
(812 observations (episodes) created)

. generate drug1emt = drug1*exp(-0.35*_t)

. generate drug2emt = drug2*exp(-0.35*_t)

. stcox age drug1emt drug2emt

        failure _d:  cured
   analysis time _t:  time
              id:  id
Iteration 0:   log likelihood = -116.54385
Iteration 1:   log likelihood = -99.321912
Iteration 2:   log likelihood =  -98.07369
Iteration 3:   log likelihood =  -98.05277
Iteration 4:   log likelihood = -98.052763
Refining estimates:
Iteration 0:   log likelihood = -98.052763

Cox regression -- Breslow method for ties

No. of subjects =          45               Number of obs   =         857
No. of failures =          36
Time at risk    =  677.9000034
                                            LR chi2(3)      =       36.98
Log likelihood  =    -98.052763            Prob > chi2     =      0.0000
```

_t	Haz. Ratio	Std. Err.	z	P>|z|	[95% Conf. Interval]	
age	.8614636	.028558	-4.50	0.000	.8072706	.9192948
drug1emt	1.304744	.1135967	3.06	0.002	1.100059	1.547514
drug2emt	1.200613	.1113218	1.97	0.049	1.001103	1.439882

Note that we get the same answer. However, this required more work for both Stata and for the user.

◁

The full functionality of stcox is available with time-varying covariates, including the generation of residuals and baseline functions. The only exception to this rule is when the exactm or exactp options are specified for handling ties, in which case the tvc(*varlist*) option is currently not supported. In those cases, you must use the stsplit approach outlined above.

❏ Technical Note

Finally, it should be noted that the specification of $g(t)$ via the texp(*exp*) option is intended for functions of analysis time, _t only, with the default being texp(_t) if left unspecified. However, specifying any other valid Stata expression will not produce a syntax error, yet in most cases will not yield the anticipated output. For example, specifying texp(*varname*) will not generate interaction terms. This mainly has to do with how the calculations are carried out—by careful summations over risk pools at each failure time.

❏

Robust estimate of variance

By default, stcox produces the conventional estimate for the variance–covariance matrix of the coefficients (and hence, the reported standard errors). If, however, you specify the robust option, stcox switches to the robust variance estimator (Lin and Wei 1989).

The key to the robust calculation is using the efficient score residuals for each of the subjects in the data for the variance calculation. Even in simple single-record, single-failure survival data, the same subjects appear repeatedly in the risk pools, and the robust calculation tries to account for that.

▷ Example

Refitting the Stanford heart transplant data model with robust standard errors, we obtain

```
. use http://www.stata-press.com/data/r8/stan3, clear
(Heart transplant data)
. stset t1, failure(died) id(id)
 (output omitted)
. stcox age posttran surg year, robust

        failure _d:  died
   analysis time _t:  t1
              id:  id
Iteration 0:   log pseudo-likelihood = -298.31514
Iteration 1:   log pseudo-likelihood =  -289.7344
Iteration 2:   log pseudo-likelihood = -289.53498
Iteration 3:   log pseudo-likelihood = -289.53378
Iteration 4:   log pseudo-likelihood = -289.53378
Refining estimates:
Iteration 0:   log pseudo-likelihood = -289.53378

Cox regression -- Breslow method for ties

No. of subjects      =          103          Number of obs   =        172
No. of failures      =           75
Time at risk         =      31938.1
                                             Wald chi2(4)    =      19.68
Log pseudo-likelihood =   -289.53378         Prob > chi2     =     0.0006

                    (standard errors adjusted for clustering on id)
```

_t	Haz. Ratio	Robust Std. Err.	z	P>|z|	[95% Conf. Interval]	
age	1.030224	.0148771	2.06	0.039	1.001474	1.059799
posttran	.9787243	.2961736	-0.07	0.943	.5408498	1.771104
surgery	.3738278	.1304912	-2.82	0.005	.1886013	.7409665
year	.8873107	.0613176	-1.73	0.084	.7749139	1.01601

Note the word Robust above Std. Err. in the table and the phrase "standard errors adjusted for clustering on id" above the table.

The hazard ratio estimates are the same as before, but the standard errors are slightly different.

◁

❑ Technical Note

In the previous example, stcox knew to specify cluster(id) for you when you specified robust.

To see the importance of cluster(id), consider simple single-record, single-failure survival data, a piece of which is

```
      t0         t        died         x
       0         5          1          1
       0         9          0          1
       0         8          0          0
```

and then consider the absolutely equivalent multiple-record survival data:

id	t0	t	died	x
1	0	3	0	1
1	3	5	1	1
2	0	6	0	1
2	6	9	0	1
3	0	3	0	0
3	3	8	0	0

Both of these datasets record the same underlying data, so both should produce the same numerical results. This should be true whether robust is specified.

In the second dataset, were one to ignore id, it would appear that there are six observations on six subjects. The key ingredients in the robust calculation are the efficient score residuals, and viewing the data as six observations on six subjects produces different score residuals. Let us call the six score residuals s_1, s_2, ..., s_6 and the three score residuals that would be generated by the first dataset S_1, S_2, and S_3. It turns out that $S_1 = s_1 + s_2$, $S_2 = s_3 + s_4$, and $S_3 = s_5 + s_6$.

That residuals sum is the key to understanding the cluster() option. When you specify cluster(id), Stata makes the robust calculation based not on the overly detailed s_1, s_2, ..., s_6, but on $s_1 + s_2$, $s_3 + s_4$, and $s_5 + s_6$. That is, Stata sums residuals within clusters before entering them into subsequent calculations (where they are squared), and that is why results estimated from the second dataset are equal to those estimated from the first. In more complicated datasets with time-varying regressors, delayed entry, and gaps, it is this action of summing within cluster that, in effect, treats the cluster (which is typically a subject) as a unified whole.

Because we had stset an id() variable, stcox knew to specify cluster(id) for us when we specified robust. You may, however, override the default clustering by specifying cluster() with a different variable than what you used in stset, id(). This is useful in analyzing multiple-failure data, where you need to stset a pseudo-id establishing the time from last failure as the onset of risk.

❏

Cox regression with multiple failure data

In [ST] **stsum**, we introduce a multiple-failure dataset:

```
. use http://www.stata-press.com/data/r8/mfail, clear

. stdes
```

| Category | total | | per subject | | |
		mean	min	median	max
no. of subjects	926				
no. of records	1734	1.87257	1	2	4
(first) entry time		0	0	0	0
(final) exit time		470.6857	1	477	960
subjects with gap	0				
time on gap if gap	0	.	.	.	.
time at risk	435855	470.6857	1	477	960
failures	808	.8725702	0	1	3

This dataset contains two variables—x1 and x2—which we believe affect the hazard of failure.

If our interest is simply in analyzing these multiple-failure data as if the baseline hazard remains unchanged as events occur (that is, the hazard may change with time, but time is measured from 0 and is independent of when the last failure occurred), we can type

```
. stcox x1 x2, robust
Iteration 0:   log pseudo-likelihood = -5034.9569
 (output omitted )
Iteration 3:   log pseudo-likelihood = -4978.1914
Refining estimates:
Iteration 0:   log pseudo-likelihood = -4978.1914

Cox regression -- Breslow method for ties

No. of subjects      =         926          Number of obs   =      1734
No. of failures      =         808
Time at risk         =      435855
                                             Wald chi2(2)    =    152.13
Log pseudo-likelihood =    -4978.1914        Prob > chi2     =    0.0000

                     (standard errors adjusted for clustering on id)
```

	Haz. Ratio	Robust Std. Err.	z	P>|z|	[95% Conf. Interval]	
x1	2.273456	.1868211	9.99	0.000	1.935259	2.670755
x2	.329011	.0523425	-6.99	0.000	.2408754	.4493951

We chose to fit this model with robust standard errors—we specified robust—but you could have estimated conventional standard errors if you wished.

In [ST] **stsum**, we discuss analyzing this dataset as time since last failure. We wished to assume that the hazard function remained unchanged with failure except that one restarted the same hazard function. To that end, we made the following changes to our data:

```
. stgen nf = nfailures()
. egen newid = group(id nf)
. sort newid t
. by newid: replace t = t - t0[1]
(808 real changes made)
. by newid: gen newt0 = t0 - t0[1]
. stset t, id(newid) failure(d) time0(newt0) noshow
            id:  newid
   failure event:  d != 0 & d < .
obs. time interval:  (newt0, t]
 exit on or before:  failure

     1734  total obs.
        0  exclusions

     1734  obs. remaining, representing
     1734  subjects
      808  failures in single failure-per-subject data
   435444  total analysis time at risk, at risk from t =          0
                         earliest observed entry t =          0
                          last observed exit t =        797
```

That is, we took each subject and made numerous newid subjects out of each, with each subject entering at time 0 (now meaning the time of the last failure). id still identifies real subject, but Stata thinks the identifier variable is newid because we stset, id(newid). If we were to fit a model using robust, we would get

```
. stcox x1 x2, robust nolog
```

Cox regression -- Breslow method for ties

No. of subjects	=	1734	Number of obs	=	1734
No. of failures	=	808			
Time at risk	=	435444			
			Wald chi2(2)	=	88.51
Log pseudo-likelihood =		-5062.5815	Prob > chi2	=	0.0000

(standard errors adjusted for clustering on newid)

_t	Haz. Ratio	Robust Std. Err.	z	P>|z|	[95% Conf. Interval]	
x1	2.002547	.1936906	7.18	0.000	1.656733	2.420542
x2	.2946263	.0569167	-6.33	0.000	.2017595	.4302382

Note carefully the message concerning the clustering: standard errors have been adjusted for clustering on `newid`. We, however, want the standard errors adjusted for clustering on `id`, so we must specify the `cluster()` option:

```
. stcox x1 x2, robust cluster(id) nolog
```

Cox regression -- Breslow method for ties

No. of subjects	=	1734	Number of obs	=	1734
No. of failures	=	808			
Time at risk	=	435444			
			Wald chi2(2)	=	93.66
Log pseudo-likelihood =		-5062.5815	Prob > chi2	=	0.0000

(standard errors adjusted for clustering on id)

_t	Haz. Ratio	Robust Std. Err.	z	P>|z|	[95% Conf. Interval]	
x1	2.002547	.1920151	7.24	0.000	1.659452	2.416576
x2	.2946263	.0544625	-6.61	0.000	.2050806	.4232709

That is, if you are using `robust`, you must remember to specify `cluster()` for yourself when

1. you are analyzing multiple-failure data, and

2. you have played a trick on Stata to reset time to time-since-last-failure, so what Stata considers the subjects are really subsubjects.

Stratified estimation

When you type

```
. stcox xvars, strata(svars)
```

you are allowing the baseline hazard functions to differ for the groups identified by *svars*. Said differently, this is equivalent to fitting separate Cox proportional hazards models under the constraint that the coefficients, but not the baseline hazard functions, are equal.

▷ Example

   Pretend that in the Stanford heart experiment data there was a change in treatment for all patients,
pre- and post-transplant, in 1970 and then again in 1973. Further assume that the proportional hazards
assumption is not reasonable for these changes in treatment—perhaps the changes result in short-run
but little expected long-run benefit. Your interest in the data is not in the effect of these treatment
changes but in the effect of transplantation, for which you still find the proportional hazards assumption
reasonable. One way you might fit your model to account for these fictional changes is

```
. use http://www.stata-press.com/data/r8/stan3, clear
(Heart transplant data)

. generate pgroup = year

. recode pgroup min/69=1 70/72=2 73/max=3
(pgroup: 172 changes made)

. stcox age posttran surg year, strata(pgroup) nolog

         failure _d:  died
   analysis time _t:  t1
                id:   id

Stratified Cox regr. -- Breslow method for ties

No. of subjects =          103                    Number of obs   =        172
No. of failures =           75
Time at risk    =      31938.1
                                                   LR chi2(4)      =      20.67
Log likelihood  =    -213.35033                    Prob > chi2     =     0.0004
```

_t	Haz. Ratio	Std. Err.	z	P>|z|	[95% Conf. Interval]	
age	1.027406	.0150188	1.85	0.064	.9983874	1.057268
posttran	1.075476	.3354669	0.23	0.816	.583567	1.982034
surgery	.2222415	.1218386	-2.74	0.006	.0758882	.6508429
year	.5523966	.1132688	-2.89	0.004	.3695832	.825638

```
                                                         Stratified by pgroup
```

Of course, you could obtain the robust estimate of variance by also including the robust option.

◁

Cox regression with shared frailty

   A shared frailty model is the survival-data analog to regression models with random effects. A
frailty is a latent random effect that enters multiplicatively on the hazard function. In the context of a
Cox model, the data are organized as $i = 1, \ldots, n$ groups with $j = 1, \ldots, n_i$ observations in group
i. For the jth observation in the ith group, the hazard is

$$h_{ij}(t) = h_0(t)\alpha_i \exp(\mathbf{x}_{ij}\boldsymbol{\beta})$$

where α_i is the group-level frailty. The frailties are unobservable positive quantities and are assumed
to have mean one and variance θ, to be estimated from the data. One fits a Cox shared frailty model
by specifying shared(*varname*), where *varname* defines the groups over which frailties are shared.
stcox, shared() treats the frailties as being gamma-distributed, but this is mainly an issue of
computational convenience; see *Methods and Formulas*. Theoretically, any distribution with positive
support, mean one, and finite variance, may be used to model frailty.

Shared frailty models are used to model within-group correlation; observations within a group are correlated because they share the same frailty. The estimate of θ is used to measure the degree of within-group correlation, and the shared frailty model reduces to standard Cox when $\theta = 0$.

For $\nu_i = \log \alpha_i$, the hazard can also be expressed as

$$h_{ij}(t) = h_0(t) \exp(\mathbf{x}_{ij}\boldsymbol{\beta} + \nu_i)$$

and thus the log-frailties, ν_i, are analogous to random effects in standard linear models. In fact, one can retrieve estimates of the ν_i by specifying effects(*newvar*).

▷ Example

Consider the data from a study of 38 kidney dialysis patients, as described in McGilchrist and Aisbett (1991). The study is concerned with the prevalence of infection at the catheter insertion point. Two recurrence times (in days) are measured for each patient, and each recorded time is the time from initial insertion (onset of risk) to infection or censoring.

```
. use http://www.stata-press.com/data/r8/catheter, clear
(Kidney data, McGilchrist and Aisbett, Biometrics, 1991)
. list in 1/10
```

	patient	time	infect	age	female
1.	1	16	1	28	0
2.	1	8	1	28	0
3.	2	13	0	48	1
4.	2	23	1	48	1
5.	3	22	1	32	0
6.	3	28	1	32	0
7.	4	318	1	31.5	1
8.	4	447	1	31.5	1
9.	5	30	1	10	0
10.	5	12	1	10	0

Each patient (patient) has two recurrence times (time) recorded, with each catheter insertion resulting in either infection (infect==1) or right-censoring (infect==0). Among the covariates measured are age and sex (female==1 if female, female==0 if male).

One subtlety to note concerns the use of the generic term, "subjects". In this example, the subjects are taken to be the individual catheter insertions, and not the patients themselves. This is a function of how the data were recorded—the onset of risk occurs at catheter insertion (of which there are two for each patient), and not (say) at the time of admission of the patient into the study. Thus, we have two subjects (insertions) within each group (patient).

It is reasonable to assume independence of patients but unreasonable to assume that recurrence times within each patient are independent. One solution would be to fit a standard Cox model, adjusting the standard errors of the estimated hazard ratios to account for the possible correlation by specifying cluster(patient).

Alternatively, one can model the correlation by assuming that the correlation is the result of a latent patient-level effect, or frailty. That is, rather than fitting a standard model and specifying cluster(patient), fit a frailty model by specifying shared(patient).

```
. stset time, fail(infect)
(output omitted)
```

```
. stcox age female, shared(patient)

        failure _d:  infect
   analysis time _t:  time

Fitting comparison Cox model:

Estimating frailty variance:

Iteration 0:   log profile likelihood = -182.06713
Iteration 1:   log profile likelihood =  -181.9791
Iteration 2:   log profile likelihood = -181.97453
Iteration 3:   log profile likelihood = -181.97453

Fitting final Cox model:

Iteration 0:   log likelihood = -199.05599
Iteration 1:   log likelihood = -183.72296
Iteration 2:   log likelihood = -181.99509
Iteration 3:   log likelihood = -181.97455
Iteration 4:   log likelihood = -181.97453
Refining estimates:
Iteration 0:   log likelihood = -181.97453

Cox regression --
        Breslow method for ties          Number of obs    =       76
        Gamma shared frailty             Number of groups =       38
Group variable: patient

No. of subjects =            76          Obs per group: min =        2
No. of failures =            58                         avg =        2
Time at risk    =          7424                         max =        2

                                         Wald chi2(2)     =    11.66
Log likelihood  =   -181.97453           Prob > chi2      =   0.0029
```

_t	Haz. Ratio	Std. Err.	z	P>|z|	[95% Conf. Interval]	
age	1.006202	.0120965	0.51	0.607	.9827701	1.030192
female	.2068678	.095708	-3.41	0.001	.0835376	.5122756
theta	.4754497	.2673107				

```
Likelihood-ratio test of theta=0: chibar2(01) =    6.27 Prob>=chibar2 = 0.006
```
Note: Standard errors of hazard ratios are conditional on theta.

From the output, we obtain $\widehat{\theta} = 0.475$, and given the standard error of $\widehat{\theta}$ and likelihood-ratio test of $H_o : \theta = 0$, we find a significant frailty effect, meaning that the correlation within patient cannot be ignored. Contrast this with the analysis of the same data in [ST] **streg**. In that analysis, both Weibull and lognormal shared frailty models were considered. In the case of Weibull, there was significant frailty; in the case of lognormal there wasn't.

The estimated ν_i are not displayed in the coefficient table, but may be retrieved at estimation by specifying effects(*newvar*).

```
. qui stcox age female, shared(patient) effects(nu)

. sort nu

. list patient nu in 1/2
```

	patient	nu
1.	21	-2.4487068
2.	21	-2.4487068

```
. list patient nu in 75/l
```

	patient	nu
75.	7	.51871588
76.	7	.51871588

From the above we estimate that the least frail patient is Patient 21, with $\widehat{\nu}_{21} = -2.45$, and that the most frail patient is Patient 7, with $\widehat{\nu}_7 = 0.52$.

◁

In shared frailty Cox models, the estimation consists of two layers. In the outer layer, the optimization is in terms of θ only. For fixed θ, the inner layer consists of fitting a standard Cox model via penalized log likelihood, with the ν_i introduced as estimable coefficients of dummy variables identifying the groups. The penalty term in the penalized log likelihood is a function of θ; see *Methods and Formulas*. The final estimate of θ is taken to be the one that maximizes the penalized log likelihood. Once the optimal θ is obtained, it is held fixed and a final penalized Cox model is fit. As a result, the standard errors of the main regression parameters (or hazard ratios, if displayed as such) are treated as conditional on θ fixed at its optimal value.

❏ Technical Note

With gamma-distributed frailty, hazard ratios decay over time in favor of the *frailty effect*, and thus the displayed "Haz. Ratio" in the above output is actually the hazard ratio only for $t = 0$. The degree of decay depends on θ. Should the estimated θ be close to zero, the hazard ratios do regain their usual interpretation; see Gutierrez (2002) for details.

❏

❏ Technical Note

The likelihood-ratio test of $\theta = 0$ is a boundary test, and thus requires careful consideration concerning the calculation of its p-value. In particular, the null distribution of the likelihood-ratio test statistic is not the usual χ_1^2, but rather is a 50:50 mixture of a χ_0^2 (point mass at zero) and a χ_1^2, denoted as $\bar{\chi}_{01}^2$. See Gutierrez et al. (2001) for more details.

❏

❏ Technical Note

In [ST] **streg**, shared frailty models are compared and contrasted with *unshared* frailty models. Unshared frailty models are used to model heterogeneity, and the frailties are integrated out of the conditional survival function to produce an unconditional survival function, which serves as a basis for all likelihood calculations.

Given the nature of Cox regression (the baseline hazard remains unspecified), there exists no Cox regression analog to the unshared parametric frailty model as fitted using **streg**. That is not to say that one cannot fit a shared frailty model with one observation per group (which is possible as long as you do not fit a null model). Note, however, that there are subtle differences in singleton-group data between shared and unshared frailty models; see Gutierrez (2002).

❏

Obtaining baseline function estimates

When you specify options basechazard(*newvar*) and basesurv(*newvar*)—which you may do together or separately—you obtain estimates of the baseline cumulative hazard and survival functions. When you specify the option basehc(*newvar*), you obtain estimates of the baseline hazard contribution at each failure time, which are factors used to develop both the product-limit estimator for the survival function generated by basesurv(*newvar*) and the estimator of the cumulative hazard function generated by basechazard(*newvar*).

Although in theory $S_0(t) = \exp\{-H_0(t)\}$, where $S_0(t)$ is the baseline survival function and $H_0(t)$ is the baseline cumulative hazard, the estimates produced by basechazard() and basesurv() do not exactly correspond in this manner, although they closely do. The reason is that stcox uses different estimation schemes for each; the exact formulas are given in the *Methods and Formulas* section.

When the model is fitted with the strata() option, you obtain estimates of the baseline functions for each stratum.

Let us first understand how stcox stores the results.

Mathematically, the baseline hazard contribution $h_i = (1 - \alpha_i)$ (see Kalbfleisch and Prentice 2002, 115) is defined at every analytic time t_i at which a failure occurs and is undefined (or, if you prefer, 0) at other times. Stata stores h_i in observations where a failure occurred and missing values in the other observations. For instance, here are some data on which we have fitted a Cox model and specified the option basehc(h):

```
. use http://www.stata-press.com/data/r8/stan3, clear
(Heart transplant data)

. generate age40 = age-40

. generate year70 = year-70

. stcox age40 posttran surg year70, basehc(h)
(output omitted)

. list id _t0 _t _d h in 1/10
```

	id	_t0	_t	_d	h
1.	1	0	50	1	.01503465
2.	2	0	6	1	.02035303
3.	3	0	1	0	.
4.	3	1	16	1	.03339642
5.	4	0	36	0	.
6.	4	36	39	1	.01365406
7.	5	0	18	1	.01167142
8.	6	0	3	1	.02875689
9.	7	0	51	0	.
10.	7	51	675	1	.06215003

Here is the interpretation: At time _t = 50, the hazard contribution h_1 is .0150. At time _t = 6, the hazard contribution h_2 is .0204.

In observation 3, no hazard contribution is stored. Observation 3 contains a missing because observation 3 did not fail at time 1.

All of which is to say that values of the hazard contributions are stored only in observations that are marked as failing.

The baseline survivor function $S_0(t)$ is defined at all values of t: its estimate changes its value when failures occur and, at times when no failures occur, the estimated $S_0(t)$ is equal to its value at the time of the last failure.

Here are some data in which we specified both basehc(h) and basesurv(s):

```
. list id _t0 _t _d h s in 1/10
```

	id	_t0	_t	_d	h	s
1.	1	0	50	1	.01503465	.68100303
2.	2	0	6	1	.02035303	.89846438
3.	3	0	1	0	.	.99089681
4.	3	1	16	1	.03339642	.84087361
5.	4	0	36	0	.	.7527663
6.	4	36	39	1	.01365406	.73259264
7.	5	0	18	1	.01167142	.82144038
8.	6	0	3	1	.02875689	.93568733
9.	7	0	51	0	.	.6705895
10.	7	51	675	1	.06215003	.26115633

At time $\_t = 50$, the baseline survivor function is .6810 or, more precisely, $S_0(50 + 0) = .6810$. What we mean by $S_0(t)$ is $S(t + 0)$, the probability of surviving just beyond time t. This is done to clarify that the probability does not include failure at precisely time t.

Understanding what is stored is easier if we sort by _t:

```
. generate notd = -_d

. sort _t notd

. drop notd

. sort _t

. list id _t0 _t _d h s in 1/18
```

	id	_t0	_t	_d	h	s
1.	15	0	1	1	.00910319	.99089681
2.	45	0	1	0	.	.99089681
3.	3	0	1	0	.	.99089681
4.	20	0	1	0	.	.99089681
5.	43	0	2	1	.02775802	.96339147
6.	61	0	2	1	.02775802	.96339147
7.	75	0	2	1	.02775802	.96339147
8.	39	0	2	0	.	.96339147
9.	95	0	2	0	.	.96339147
10.	46	0	2	0	.	.96339147
11.	54	0	3	1	.02875689	.93568733
12.	42	0	3	1	.02875689	.93568733
13.	6	0	3	1	.02875689	.93568733
14.	60	0	3	0	.	.93568733
15.	23	0	3	0	.	.93568733
16.	68	0	3	0	.	.93568733
17.	94	0	4	0	.	.93568733
18.	72	0	4	0	.	.93568733

Note that the baseline hazard contribution is stored on every failure record—and if multiple failures occur at a time, the value of the hazard contribution is repeated—and the baseline survival is stored on every record. (More correctly, baseline values are stored on records that meet the criterion and which were used in estimation. If some observations are explicitly or implicitly excluded from the estimation, their baseline values will be set to missing no matter what.)

With this listing, we get a better indication as to how the hazard contributions are used to calculate the survival function. Since the patient with $id = 15$ died at time $t_1 = 1$, his hazard contribution is $h_{15} = .00910319$. Since that was the only death at $t_1 = 1$, the estimated survival function at this time is $S_0(1) = 1 - h_{15} = 1 - .00910319 = .99089681$. The next death occurs at time $t_1 = 2$, and the hazard contribution at this time for patient 61 is $h_{61} = .02775802$. Multiplying the previous survival function value by $1 - h_{61}$ gives the new survival function at $t_1 = 2$ as $S_0(2) = .96339147$. The other survival function values are then calculated in succession, using this method at each failure time. At times when no failures occur, the survival function remains unchanged.

If we had fitted a stratified model—if we had specified the `strata()` option—the recorded baseline hazard contribution and survival on each record would be for the stratum of the record.

❏ Technical Note

If you want the baseline hazard contribution stored on every record for which it is defined and not just for the failure records, you would do the following:

```
. sort _t _d
. by _t: replace h = h[_N]
```

The above assumes that you specified `basehc(h)` when fitting the Cox model. If you also specified the `strata()` option, say `strata(group)`, the instructions would be

```
. sort group _t _d
. by group _t: replace h = h[_N]
```

In both of these examples, all we did was place the data in time order: we put the failures at the end of each time group, and then copied the last value of h within each time to all the observations for that time.

It is a useful test of your understanding to consider obtaining the estimate of $S_0(t)$ from the h_i's for yourself. One way you could do that is

```
. sort _t _d
. by _t: keep if _d & _n==_N
. generate s = 1-h
. replace s = s[_n-1]*s if _n>1
```

If you had obtained stratified estimates, the equivalent code would be

```
. sort group _t _d
. by group _t: keep if _d & _n==_N
. generate s = 1-h
. by group: replace s = s[_n-1]*s if _n>1
```

❏

▷ Example

One thing to do with the baseline functions is to graph them. Remember, baseline functions refer to the values of the functions when all covariates are set to 0. Let's graph the survival curve for the heart transplant model we have been fitting and, to make the baseline curve reasonable, let us do that at `age = 40` and `year = 70`. (We did previously without explanation, but the following technical note provides important information on why baseline values should be chosen to be reasonable, which is to say, in the range of the data.)

Thus, we will begin by creating variables that, when 0, correspond to the baseline values we desire, and then refit our model, specifying the basesurv() option:

```
. use http://www.stata-press.com/data/r8/stan3, clear
(Heart transplant data)
. generate age40 = age-40
. generate year70 = year-70
. stcox age40 posttran surg year70, bases(s) nolog
        failure _d:  died
   analysis time _t:  t1
              id:  id
Cox regression -- Breslow method for ties
No. of subjects =          103                  Number of obs   =        172
No. of failures =           75
Time at risk    =      31938.1
                                                LR chi2(4)      =      17.56
Log likelihood  =    -289.53378                 Prob > chi2     =     0.0015
```

_t	Haz. Ratio	Std. Err.	z	P>|z|	[95% Conf. Interval]	
age40	1.030224	.0143201	2.14	0.032	1.002536	1.058677
posttran	.9787243	.3032597	-0.07	0.945	.5332291	1.796416
surgery	.3738278	.163204	-2.25	0.024	.1588759	.8796
year70	.8873107	.059808	-1.77	0.076	.7775022	1.012628

```
. summarize s
```

Variable	Obs	Mean	Std. Dev.	Min	Max
s	172	.629187	.2530009	.130666	.9908968

Note first that our re-centering of age and year did not affect the estimation. Also, for your information, the s variable we have just generated is the s variable we summarized above.

Here is a graph of the baseline survival curve:

```
. line s _t, sort c(J)
```

There is an easier way to graph estimated survival, cumulative hazard, and hazard functions, both at baseline and at specified covariate values. Use stcurve, as we demonstrate later.

◁

❏ Technical Note

If you specify the basechazard() or basesurv() options, for numerical accuracy reasons it is important that the baseline functions correspond to something reasonable in your data. Remember, the baseline functions correspond to all covariates equal to 0 in your Cox model.

Consider, for instance, a Cox model that included the variable calendar year among the covariates. Say year varied between 1980 and 1996. Then the baseline functions would correspond to year 0, almost 2,000 years in the past. Say the estimated coefficient on year was $-.2$ (meaning the hazard ratio for one year to the next is a reasonable 0.82).

Think carefully about the contribution to the predicted log cumulative hazard: it would be approximately $-.2 \times 2,000 = -400$. Now $e^{-400} \approx 10^{-173}$ which, on a digital computer, is 1. There is simply no hope that $H_0(t)e^{-400}$ will produce an accurate estimate of $H(t)$.

Even with less extreme numbers, problems arise even in the calculation of the baseline survivor function. Baseline hazard contributions very near 1 produce baseline survivor functions with the steps differing by tens of orders of magnitudes because the calculation of the survivor function is cumulative. Producing a meaningful graph of such a survivor function is hopeless, and adjusting the survivor function (as opposed to the hazard contribution) to other values of the covariates is too much work.

For these reasons, it is important that covariate values of 0 be meaningful if you are going to specify the basechazard() or basesurv() options. As the baseline values move to absurdity, the first problem you will encounter is a baseline survivor function that is too hard to interpret, even though the baseline hazard contributions are estimated accurately. Further out, the procedure Stata uses to estimate the baseline hazard contributions will break down—it will produce results that are exactly 1.

This, in fact, occurs with the Stanford heart transplant data:

```
. use http://www.stata-press.com/data/r8/stan3, clear
(Heart transplant data)
. stcox age posttran surg year, basec(ch) bases(s)
 (output omitted)
. summarize ch s
```

Variable	Obs	Mean	Std. Dev.	Min	Max
ch	172	745.1134	682.8671	11.88239	2573.637
s	172	1.45e-07	9.43e-07	0	6.24e-06

The hint that there are problems is that the values of ch are huge and the values of s are close to zero. In this dataset, it is age (which ranges from 8 to 64 with a mean value of 40) and year (which ranges from 67 to 74) that are the problems. The baseline functions correspond to a newborn at the turn of the century on the waiting list for a heart transplant!

To obtain accurate estimates of the baseline functions, you should type

```
. drop ch s
. generate age40 = age-40
. generate year70 = year-70
. stcox age40 posttran surg year70, basec(ch) bases(s)
 (output omitted)
. summarize ch s
```

Variable	Obs	Mean	Std. Dev.	Min	Max
ch	172	.5685743	.521076	.0090671	1.963868
s	172	.629187	.2530009	.130666	.9908968

Adjusting the variables makes no difference in terms of the coefficient (and hence hazard ratio) estimates, but it changes the values at which the baseline functions are estimated to be within the range of the data.

❏

❏ Technical Note

When used with shared frailty models, the options `basehc()`, `basesurv()`, and `basechazard()` will produce estimates of baseline quantities that are based on the last-step penalized Cox model fit. Thus, the term *baseline* means that not only are the covariates are set to zero, but ν_i as well, and thus *baseline* corresponds to $\alpha_l = 1$ (and zero covariates).

❏

Cox regression residuals

Stata can calculate score residuals, efficient score residuals (esr), martingale residuals, Schoenfeld residuals, and scaled Schoenfeld residuals with options specified to the `stcox` command. Cox–Snell and deviance residuals can be calculated with options to `predict` and, in the case of time-varying covariates and multiple observations per subject, cumulative martingale and cumulative Cox–Snell residuals can also be calculated using `predict`.

Although the uses of residuals vary and depend on the data and user preferences, traditional and suggested uses are the following: Cox–Snell residuals are useful in assessing overall model fit; martingale residuals are useful in determining the functional form of covariates to be included in the model and are occasionally useful in assessing lack of fit; Schoenfeld and score residuals are useful for checking and testing the proportional hazard assumption, examining leverage points, and identifying outliers; and deviance residuals are useful in examining model accuracy and identifying outliers.

▷ Example

Let us first examine the use of Cox–Snell residuals. Using the cancer data, we first perform a Cox regression requesting that martingale residuals be calculated. Martingale residuals must be requested because these are used by `predict` to calculate the Cox–Snell residuals. If we forget, `predict` will issue an error message reminding us of this.

```
. use http://www.stata-press.com/data/r8/drugtr, clear
(Patient Survival in Drug Trial)
. stset studytime, failure(died)
 (output omitted )
. stcox age drug, mgale(mg)
 (output omitted )
. predict double cs, csnell
```

The first command performs the Cox regression, calculates the martingale residuals, and saves them in the variable `mg`. In the second command, the `csnell` option tells `predict` to output to a new variable, `cs`, the Cox–Snell residuals. If the Cox regression model fits the data, then these residuals should have a standard censored exponential distribution with hazard ratio 1. We can verify the model's fit by calculating, based, for example, on the Kaplan–Meier estimated survival function or the Aalen–Nelson estimator, an empirical estimate of the cumulative hazard function, using the Cox–Snell residuals as the time variable and the data's original censoring variable. If the model fits the data, then the plot of the cumulative hazard versus `cs` should be a straight line with slope 1.

To do this, we first `stset` the data, specifying cs as our new failure time variable and `died` as the failure/censoring indicator. We then use the `sts generate` command to generate the variable km containing the Kaplan–Meier survival estimates. Lastly, we generate the cumulative hazard H using the relationship $H = -\ln(\text{km})$ and plot it against cs.

```
. stset cs, failure(died)
(output omitted)
. sts generate km=s

. generate double H=-ln(km)
(1 missing value generated)

. line H cs cs, sort ytitle("") clstyle(. refline) legend(nodraw)
```

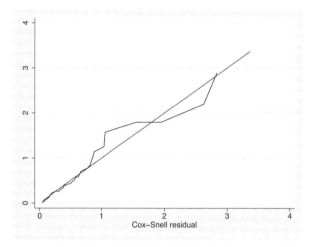

We specified cs twice in the `graph` command so that a reference 45° line is plotted. Comparing the jagged line with the reference line, we observe that the Cox model does not fit these data too badly.

◁

❑ Technical Note

The statement "if the Cox regression model fits the data, then the Cox–Snell residuals have a standard censored exponential distribution with hazard ratio 1" only holds if the true parameters β and the true cumulative baseline hazard function, $H_0(t)$, are used in the calculation of the residuals. Since we use estimates $\widehat{\beta}$ and $\widehat{H}_0(t)$, deviations from the 45° line in the above plots could be partially due to uncertainty about these estimates. This is particularly important for small sample sizes and in the right-hand tail of the distribution, where the baseline hazard is more variable due to the reduced effective sample caused by prior failures and censoring.

❑

▷ Example

Let us now examine the martingale residuals. Martingale residuals are useful in assessing the functional form of a covariate to be entered into a Cox model. Sometimes the covariate may need transforming so that the transformed variable will satisfy the assumptions of the proportional hazards model. The procedure for finding the appropriate functional form of a variable is to fit a Cox model excluding the variable, and then to plot a `lowess` smooth of the martingale residuals against some transformation of the variable in question. If the transformation is appropriate, then the smooth should be approximately linear.

We apply this procedure to our Cancer data in order to find an appropriate transformation of `age` (or to verify that `age` need not be transformed).

```
. use http://www.stata-press.com/data/r8/drugtr, clear
(Patient Survival in Drug Trial)
. stset studytime, failure(died)
(output omitted)
. stcox drug, mgale(mg)
(output omitted)
. lowess mg age, mean noweight title("") note("") m(o)
```

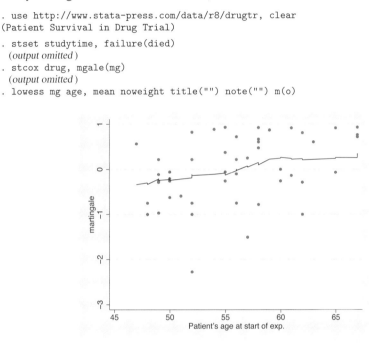

We used the `lowess` command with the options `mean` and `noweight` to obtain a plot of the running mean smoother to ease interpretation. Alternatively, a lowess smoother or other smoother could be used; see [R] **lowess**. The smooth appears nearly linear, supporting the inclusion of the untransformed version of `age` in our Cox model. Had the smooth not been linear, then we would have tried smoothing the martingale residuals against various transformations of `age` until we found one that produced a near-linear smooth.

◁

Martingale residuals can also be interpreted as the difference over time of the observed number of failures minus that predicted by the model. Thus, a plot of the martingale residuals versus the linear predictor may be used in the detection of outliers.

Plots of martingale residuals are sometimes difficult to interpret, however, because these residuals are skewed, taking values in $(-\infty, 1)$. For this reason, deviance residuals are preferred for examining model accuracy and outlier identification.

▷ Example

Using our cancer data, we can calculate the deviance residuals using `predict`, but we first need to refit the Cox model with `age` (untransformed) included and regenerate the martingale residuals for this model. It is necessary that martingale residuals be calculated and saved when fitting the Cox model since `predict, deviance` requires that. Deviance residuals are a rescaling of the martingale residuals so that they are symmetric about zero, and thus are more like residuals obtained from linear regression. Plots of these residuals against the linear predictor, survival time, rank order of survival, or observation number can be useful in identifying aberrant observations and assessing model fit.

```
. drop mg
. stcox drug age, mgale(mg)
  (output omitted)
. predict double xb, xb
. scatter mg xb
```

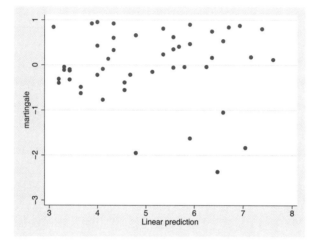

```
. predict double dev, deviance
. scatter dev xb
```

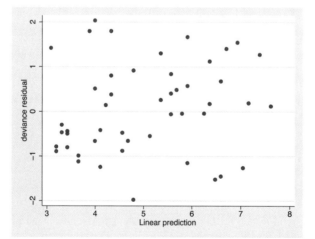

We first plotted the martingale residuals versus the linear predictor, and then the deviance residuals versus the linear predictor. Given their symmetry about zero, deviance residuals are easier to interpret, although both graphs yield the same information. With uncensored data, deviance residuals should resemble white noise if the fit is adequate. Censored observations would be represented as clumps of deviance residuals near zero (Klein and Moeschberger 1997, 356). Given what we see above, there do not appear to be any outliers.   ◁

In evaluating the adequacy of the fitted model, it is important to determine if any one or any group of observations has a disproportionate influence on the estimated parameters. This is known as influence or leverage analysis. The preferred method of performing influence or leverage analysis is to compare the estimated parameter, $\widehat{\beta}$, obtained from the full data, with estimated parameters $\widehat{\beta}_{(i)}$, obtained by fitting the model to the $n - 1$ observations remaining after the ith observation is removed. If $\widehat{\beta} - \widehat{\beta}_{(i)}$ is close to zero, then the ith observation has little influence on the estimate. The process is repeated for all observations included in the original model. To compute these differences for a dataset with n observations, we would have to execute `stcox` n additional times, which could be impractical for large datasets. In such cases, an approximation to $\widehat{\beta} - \widehat{\beta}_{(i)}$ based on the efficient score residuals can be calculated as

$$\Delta' \mathbf{V}(\widehat{\beta})$$

where $\mathbf{V}(\widehat{\beta})$ is the variance–covariance matrix and Δ' is the matrix of efficient score residuals. The difference $\widehat{\beta} - \widehat{\beta}_{(i)}$ is commonly referred to as `dfbeta` in the literature; see [R] **regression diagnostics**.

▷ Example

We now perform an influence analysis using the cancer data. To do this, we first obtain the efficient score residuals by typing `stcox age drug, esr(esr*)`. This command generates two new variables, `esr1` and `esr2`, corresponding to the two covariates in the model. The first variable, `esr1`, contains the efficient score residuals corresponding to the first covariate listed in the model (`age`), and the second variable, `esr2`, contains the efficient score residuals corresponding to the second covariate listed in the model (`drug`). We then use the `mkmat` command to create a matrix of the efficient score residuals, we obtain the variance–covariance matrix, and then we multiply the two matrices.

```
. use http://www.stata-press.com/data/r8/drugtr, clear
(Patient Survival in Drug Trial)
. stset studytime, fail(died)
 (output omitted )
. stcox age drug, esr(esr*)
 (output omitted )
. set matsize 100
. mkmat esr1 esr2, matrix(esr)
. mat V = e(V)
. mat Inf = esr*V
. svmat Inf, names(s)
```

The last command saves the estimates of `dfbeta` $= \widehat{\beta} - \widehat{\beta}_{(i)}$ in the variables `s1` and `s2`. We can now label these new variables and plot them versus time or observation number to identify observations with disproportionate influence.

```
. label var s1 "dfbeta - Age"
. label var s2 "dfbeta - Drug"
```

(Continued on next page)

. scatter s1 studytime, ylab(-.1(.1).2) yline(0)

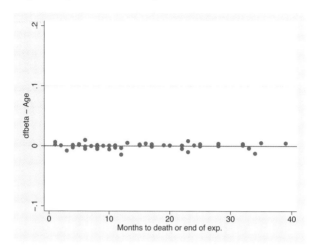

. scatter s2 studytime, ylab(-.1(.1).2) yline(0)

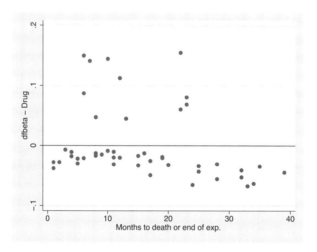

◁

Checking and testing the proportional hazard assumption

The most important assumption of the Cox proportional hazards model is that the hazard ratio is proportional over time. Using an example from Garrett (1997), suppose a group of cancer patients on an experimental treatment are followed for 10 years. If the hazard of dying for the nontreated group is twice the rate as that for the treated group (HR = 2.0), the proportional hazards assumption implies that this ratio is the same at one year, two years, or at any point on the time scale. Because the Cox model is constrained to follow this assumption, it is important to evaluate its validity. If the assumption fails, alternative modeling choices would be more appropriate (e.g., a stratified Cox model).

We provide three programs, `stphtest`, `stphplot` and `stcoxkm`, which can be used to evaluate the proportional hazards assumption. Both `stphplot` and `stcoxkm` are graphical methods documented separately. See [ST] **stphplot** for complete description of these tests.

The test of proportional hazards implemented in `stphtest` is based on the generalization by Grambsch and Therneau (1994). They showed that many of the popular tests for proportional hazards are in fact tests of nonzero slope in a generalized linear regression of the scaled Schoenfeld residuals on functions of time. The `stphtest` command tests, for individual covariates and globally, the null hypothesis of zero slope in the appropriate regression. The test of zero slope is equivalent to testing that the log hazard ratio function is constant over time; thus, rejection of the null hypothesis of a zero slope indicates deviation from the proportional hazards assumption. The choice of time function depends on the problem at hand. The `stphtest` command allows three common choices as options, `log`, `rank` and `km`, and allows the use of any user-defined function of time through the `time()` option. Additionally, when no option is specified the tests are performed using analysis time without further transformation.

▷ Example

This example uses data from a leukemia remission study (Garrett 1997). The data consist of 42 patients who are followed over time to examine how long (`weeks`) before they go out of remission (`relapse`: 1 = yes, 0 = no). Half the patients received a new experimental drug, and the other half received a standard drug (`treatment1`: 1 = drug A, 0 = standard). White blood cell count, a strong indicator of the presence of leukemia, is divided into three categories (`wbc3cat`: 1 = normal, 2 = moderate, 3 = high).

```
. use http://www.stata-press.com/data/r8/leukemia, clear
(Leukemia Remission Study)

. describe
Contains data from http://www.stata-press.com/data/r8/leukemia.dta
  obs:            42                          Leukemia Remission Study
  vars:            8                          23 Sep 2002 15:25
  size:          504 (99.9% of memory free)
```

variable name	storage type	display format	value label	variable label
weeks	byte	%8.0g		Weeks in Remission
relapse	byte	%8.0g	yesno	Relapse
treatment1	byte	%8.0g	trt1lbl	Treatment I
treatment2	byte	%8.0g	trt2lbl	Treatment II
wbc3cat	byte	%9.0g	wbclbl	White Blood Cell Count
wbc1	byte	%8.0g		wbc3cat==Normal
wbc2	byte	%8.0g		wbc3cat==Moderate
wbc3	byte	%8.0g		wbc3cat==High

```
Sorted by:  weeks
```

In this example, we examine whether the proportional hazards assumption holds for a model with covariates `wbc2`, `wbc1`, and `treatment1`. After `stsetting` the data, we first run `stcox`, saving both the scaled and the nonscaled Schoenfeld residuals, and then use `stphtest`. `stphtest` requires that we request scaled Schoenfeld residuals if we want a separate test for each covariate and that we request the unscaled residuals if we want the global test. If we forget at estimation time, `stphtest` will remind us.

```
. stset weeks, failure(relapse)

     failure event:  relapse != 0 & relapse < .
obs. time interval:  (0, weeks]
 exit on or before:  failure

─────────────────────────────────────────────────────────────
     42  total obs.
      0  exclusions
─────────────────────────────────────────────────────────────
     42  obs. remaining, representing
     30  failures in single record/single failure data
    541  total analysis time at risk, at risk from t =          0
                          earliest observed entry t =          0
                           last observed exit t =             35
. stcox treatment1 wbc2 wbc3, scaledsch(sca*) schoenfeld(sch*) nolog

         failure _d:  relapse
   analysis time _t:  weeks

Cox regression -- Breslow method for ties

No. of subjects =          42                   Number of obs    =          42
No. of failures =          30
Time at risk    =         541
                                                LR chi2(3)       =       33.02
Log likelihood  =   -77.476905                  Prob > chi2      =      0.0000
```

_t	Haz. Ratio	Std. Err.	z	P>|z|	[95% Conf. Interval]	
treatment1	.2834551	.1229874	-2.91	0.004	.1211042	.6634517
wbc2	3.637825	2.201306	2.13	0.033	1.111134	11.91015
wbc3	10.92214	7.088783	3.68	0.000	3.06093	38.97284

```
. stphtest, rank detail

     Test of proportional hazards assumption

     Time:  Rank(t)
```

	rho	chi2	df	Prob>chi2
treatment1	-0.02802	0.02	1	0.8755
wbc2	-0.10665	0.32	1	0.5735
wbc3	-0.02238	0.02	1	0.8987
global test		0.51	3	0.9159

Because we saved both the Schoenfeld residuals and the scaled Schoenfeld residuals by specifying the detail option on the stphtest command, both covariate-specific and global tests were produced. We can see that there is no evidence that the proportional hazards assumption has been violated. When we saved the residuals using stcox, it did not matter what we named them, it just mattered that we did save them.

Another variable on this dataset measures a different drug (treatment2: 1 = drug B, 0 = standard). We now wish to examine the proportional hazards assumption for the previous model by substituting treatment2 for treatment1.

After dropping the previous Schoenfeld and scaled Schoenfeld residuals, we fit a new Cox model and perform the test for proportional hazards.

```
. drop sca* sch*

. stcox treatment2 wbc2 wbc3, scaledsch(sca*) schoenfeld(sch*) nolog

        failure _d:  relapse
   analysis time _t:  weeks

Cox regression -- Breslow method for ties

No. of subjects =          42              Number of obs    =          42
No. of failures =          30
Time at risk    =         541
                                           LR chi2(3)       =       23.93
Log likelihood  =   -82.019053            Prob > chi2      =      0.0000
```

_t	Haz. Ratio	Std. Err.	z	P>|z|	[95% Conf. Interval]	
treatment2	.8483777	.3469054	-0.40	0.688	.3806529	1.890816
wbc2	3.409628	2.050784	2.04	0.041	1.048905	11.08353
wbc3	14.0562	8.873693	4.19	0.000	4.078529	48.44314

```
. stphtest, rank detail
        Test of proportional hazards assumption

     Time:  Rank(t)
```

	rho	chi2	df	Prob>chi2
treatment2	-0.63673	15.47	1	0.0001
wbc2	-0.17380	0.90	1	0.3426
wbc3	-0.08294	0.21	1	0.6481
global test		15.76	3	0.0013

treatment2 clearly violates the proportional hazards assumption. A single hazard ratio describing the effect of this drug is inappropriate.

◁

❑ Technical Note

The test of the proportional hazards assumption is based on the principle that the assumption restricts $\beta_j(t) = \beta$ for all t, which implies that a plot of $\beta_j(t)$ versus time will have a slope of zero. Grambsch and Therneau (1994) showed that $E(s_j^*) + \widehat{\beta} \approx \beta(t_j)$, where s_j^* is the scaled Schoenfeld residual and $\widehat{\beta}$ is the estimated coefficient from the Cox model. Thus, a plot of $s_j^* + \beta$ versus some function of time provides a graphical assessment of the assumption.

❑

❑ Technical Note

The tests of the proportional hazards assumption assume homogeneity of variance across risk sets. This allows the use of the estimated overall (pooled) variance–covariance matrix in the equations. Although these tests have been shown by Grambsch and Therneau (1994) to be fairly robust to departures from this assumption, care must be exercised where this assumption may not hold, particularly when performing a stratified Cox analysis. In such cases, we recommend that the proportional hazards assumption be checked separately for each stratum.

❑

stcurve

After fitting a Cox model, stcurve may be used to plot the estimated hazard, cumulative hazard, and survival functions.

```
. use http://www.stata-press.com/data/r8/drugtr, clear
(Patient Survival in Drug Trial)
. stcox age drug, basesurv(s) basehc(h)
(output omitted)
. stcurve, survival
```

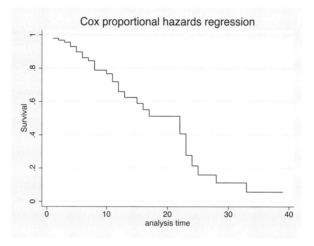

By default, the curve is evaluated at the mean values of all the predictors, but you are free to specify other values if you wish.

```
. stcurve, survival at1(drug=0) at2(drug=1)
```

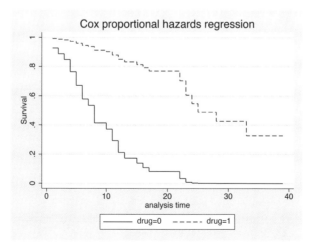

In this example, we asked for two plots, one for the placebo group and one for the treatment group. For both groups, the value of age was held at its mean value for the overall estimation sample.

stcurve can also be used to plot estimated hazard functions. The hazard function is estimated by a kernel smooth of the estimated hazard contributions, available to us after specifying basehc(h) to stcox; see [ST] **sts graph** for details. As such, we can customize the smooth as we would any other; see [R] **kdensity** for details.

. stcurve, hazard at1(drug=0) at2(drug=1) kernel(gauss) yscale(log)

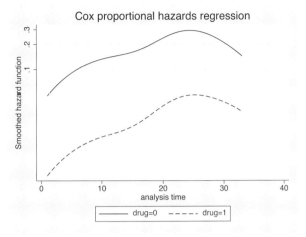

In the case of the hazard plot, we plotted on a log scale to demonstrate the proportionality of hazards under this model; see the technical note below on smoothed hazards.

When we ran stcox above, we specified the options basesurv(s) and basehc(h). If plotting the hazard, you must specify option basehc() to stcox—that is how stcurve retrieves the necessary information in order to estimate the hazard. If plotting the cumulative hazard, you must specify basechazard() to stcox. If plotting the survival function, you must specify basesurv(). If you forget, stcurve will complain.

❑ Technical Note

In the case of survival or cumulative hazard estimation, stcurve works by first estimating the baseline function and then modifying it to adhere to the specified (or by default, mean) covariate patterns. As such, as mentioned previously it is important that *baseline* (when all covariates are equal to zero) correspond to something meaningful, and preferably in the range of your data. Otherwise, stcurve could encounter numerical difficulties. We ignored our own advise above and left age unchanged. Had we encountered numerical problems, or funny looking graphs, we would have known to try shifting age so that age==0 was in the range of our data.

In the case of hazard estimation, stcurve works by first transforming the estimated hazard contributions (as produced by basehc()) to adhere to the necessary covariate pattern, then applying the smooth. If plotting multiple curves, each is smoothed independently, although the same bandwidth is used for each.

The smoothing takes place in the hazard scale and not in the log-hazard scale. As a result, note that the resulting curves will look nearly, but not exactly, parallel when plotted on a log scale. This inexactitude is an issue of the smoothing, and should not be interpreted as a deviation from the proportional hazards assumption; stcurve (after stcox) assumes proportionality of hazards and will reflect this in the produced plots. If smoothing were a perfect science, the curves would be exactly parallel when plotted on a log scale. If you should encounter estimated hazards exhibiting severe disproportionality, this may signal a numerical problem as described above. Try re-centering your covariates so that baseline is more reasonable.

❑

Acknowledgments

We thank Peter Sasieni of Cancer Research UK for his statistical advice and guidance in implementing the robust variance estimator for this command. We would also like to thank Joanne M. Garrett of the University of North Carolina at Chapel Hill for her contributions to the stphtest command.

Saved Results

stcox saves in e():

Scalars

e(N)	number of observations	e(N_g)	number of groups
e(N_sub)	number of subjects	e(g_max)	largest group size
e(N_fail)	number of failures	e(g_min)	smallest group size
e(risk)	total time at risk	e(g_avg)	average group size
e(N_clust)	number of clusters	e(ll)	log likelihood
e(df_m)	model degrees of freedom	e(ll_0)	log likelihood, constant-only model
e(r2_p)	pseudo R-squared	e(chi2)	χ^2
e(theta)	frailty parameter	e(se_theta)	standard error of θ
e(chi2_c)	χ^2, comparison model	e(ll_c)	log likelihood, comparison model
e(p_c)	significance, comparison model		

Macros

e(cmd)	cox or stcox_fr	e(chi2type)	Wald or LR; type of model χ^2 test
e(cmd2)	stcox	e(offset)	offset
e(depvar)	_t	e(mgale)	variable containing partial martingale residuals
e(t0)	_t0		
e(clustvar)	name of cluster variable	e(vl_esr)	variables containing partial efficient score residuals
e(shared)	frailty grouping variable		
e(re_var)	variable containing estimated random effects	e(vl_sch)	variables containing Schoenfeld residuals
e(method)	requested estimation method	e(vl_ssc)	variables containing scaled Schoenfeld residuals
e(ties)	method used for handling ties		
e(texp)	function used for time-varying covariates	e(bases)	variable containing baseline survival function
e(vcetype)	covariance estimation method	e(basec)	variable containing baseline cumulative haz. function
e(predict)	program used to implement predict		
e(crittype)	optimization criterion	e(basehc)	variable containing baseline hazard contributions

Matrices

e(b)	coefficient vector	e(V)	variance–covariance matrix of the estimators

Functions

e(sample)	marks estimation sample

stphtest saves in r():

Scalars

r(df)	global test degrees of freedom	r(chi2)	global test χ^2

Methods and Formulas

`stcox` and `stphtest` are implemented as ado-files.

The proportional hazards model with time-dependent explanatory variables was first suggested by Cox (1972). For an introductory explanation, see, for example, Hosmer and Lemeshow (1999, Chapters 3, 4, and 7), Kahn and Sempos (1989, 193–198), and Selvin (1996, 391–422). For a comprehensive review of the methods in this entry, see Klein and Moeschberger (1997). For a detailed development, see Kalbfleisch and Prentice (2002). For more Stata-specific insight, see Cleves, Gould, & Gutierrez (2002).

Let $\mathbf{x}_i$ be the row vector of covariates for the time interval $(t_{0i}, t_i]$ for the ith observation in the dataset ($i = 1, \ldots, N$). `stcox` obtains parameter estimates $\widehat{\boldsymbol{\beta}}$ by maximizing the partial log-likelihood function

$$L = \sum_{j=1}^{D} \left[\sum_{k \in D_j} \mathbf{x}_k \boldsymbol{\beta} - d_j \ln \left\{ \sum_{i \in R_j} \exp(\mathbf{x}_i \boldsymbol{\beta}) \right\} \right]$$

where j indexes the ordered failure times $t_{(j)}$ ($j = 1, \ldots, D$), D_j is the set of d_j observations that fail at $t_{(j)}$, d_j is the number of failures at $t_{(j)}$, and R_j is the set of observations k that are at risk at time $t_{(j)}$ (i.e., all k such that $t_{0k} < t_{(j)} \leq t_k$). This formula for L is for unweighted data and handles ties using the Peto–Breslow approximation (Peto 1972; Breslow 1974), which is the default method of handling ties in `stcox`.

The variance of $\widehat{\boldsymbol{\beta}}$ is estimated by the conventional inverse matrix of (negative) second derivatives of L unless `robust` is specified, in which case the method of Lin and Wei (1989) is used. If `cluster()` is specified, the efficient score residuals are summed within cluster before application of the sandwich (robust) estimator.

See [R] **maximize** for a description of the maximization algorithm and for the translation of estimated results to hazard ratios when `hr` is specified.

`mgale()` stores in each observation the observation's contribution to the martingale residual, which we call the "partial" martingale residual—partial is our terminology. The derivative of L can be written as

$$\frac{\partial L}{\partial \boldsymbol{\beta}} = \sum_{j=1}^{D} \sum_{i=1}^{N} \mathbf{x}_i \, dM_i(t_{(j)})$$

where

$$dM_i(t_{(j)}) = \delta_{ij} - I(t_{0i} < t_{(j)} \leq t_i) \frac{d_j \exp(\mathbf{x}_i \boldsymbol{\beta})}{\sum_{\ell \in R_j} \exp(\mathbf{x}_\ell \boldsymbol{\beta})}$$

with $\delta_{ij} = 1$ if observation i fails at $t_{(j)}$ and 0 otherwise. $I(\cdot)$ is the indicator function. $dM_i(t_{(j)})$ is the increment of the martingale residual for the ith observation due to the failures at time $t_{(j)}$. The `mgale()` option saves the partial martingale residuals ΔM_i, which are the sum of the $dM_i(t_{(j)})$ over all failure times $t_{(j)}$ such that $t_{0i} < t_{(j)} \leq t_i$. For single-record data, the partial martingale residual is the martingale residual. For multiple-record data, the martingale residual for an individual can be obtained by summing the partial martingale residuals over all observations belonging to the individual.

For a discussion of martingale residuals, see, for instance, Fleming and Harrington (1991, 163–197).

The increments of the efficient score residuals are

$$d\mathbf{F}_{ij} = \mathbf{x}_{ij}^* \, dM_i(t_{(j)})$$

where

$$\mathbf{x}_{ij}^* = \mathbf{x}_i - \frac{\sum_{\ell \in R_j} \mathbf{x}_\ell \exp(\mathbf{x}_\ell \boldsymbol{\beta})}{\sum_{\ell \in R_j} \exp(\mathbf{x}_\ell \boldsymbol{\beta})}$$

When the esr() option is specified, stcox creates $p = \dim(\mathbf{x})$ new variables containing $\Delta \mathbf{F}_i$, which are the sum of $d\mathbf{F}_{ij}$ over all failure times $t_{(j)}$ such that $t_{0i} < t_{(j)} \leq t_i$. The efficient score residuals for an individual can be obtained by summing $\Delta \mathbf{F}_i$ over all observations i belonging to the individual.

The estimated baseline hazard contribution, if requested, is obtained as $h_j = 1 - \widehat{\alpha}_j$, where $\widehat{\alpha}_j$ is the solution of

$$\sum_{k \in D_j} \frac{\exp(\mathbf{x}_k \boldsymbol{\beta})}{1 - \widehat{\alpha}_j^{\exp(\mathbf{x}_k \boldsymbol{\beta})}} = \sum_{l \in R_j} \exp(\mathbf{x}_l \boldsymbol{\beta})$$

(Kalbfleisch and Prentice 2002, equation (4.34), 115).

The estimated baseline survivor function, if requested, is obtained as

$$\widehat{S}_0(t_{(j)}) = \prod_{h=0}^{j-1} \widehat{\alpha}_h$$

where $\widehat{\alpha}_0 = 1$.

The estimated baseline cumulative hazard function, if requested, is related to the baseline survivor function calculation, yet the values of $\widehat{\alpha}_j$ are set at their starting values and are not iterated. Equivalently,

$$\widehat{H}_0(t_{(j)}) = \sum_{h=0}^{j-1} \frac{d_j}{\sum_{l \in R_j} \exp(\mathbf{x}_l \boldsymbol{\beta})}$$

Tied values are handled using one of four approaches. The log likelihoods corresponding to the four approaches are given with weights (exactp does not allow weights) and offsets by

(Continued on next page)

$$L_{\text{breslow}} = \sum_{j=1}^{D} \sum_{i \in D_j} \left[w_i(\mathbf{x}_i\boldsymbol{\beta} + \text{offset}_i) - w_i \ln \left\{ \sum_{\ell \in R_j} w_\ell \exp(\mathbf{x}_\ell\boldsymbol{\beta} + \text{offset}_\ell) \right\} \right]$$

$$L_{\text{efron}} = \sum_{j=1}^{D} \sum_{k=1}^{d_j} \left\{ \frac{1}{d_j} \sum_{i \in D_j} w_i(\mathbf{x}_i\boldsymbol{\beta} + \text{offset}_i) - \right.$$

$$\left. \left(\frac{1}{d_j} \sum_{i \in D_j} w_i \right) \ln \sum_{\ell \in R_j} w_\ell f_{kj\ell} \exp(\mathbf{x}_\ell\boldsymbol{\beta} + \text{offset}_\ell) \right\}$$

$$f_{kj\ell} = \begin{cases} \dfrac{d_j - k + 1}{d_j} & \text{if } \delta_{\ell j} = 1 \\ 1 & \text{otherwise} \end{cases}$$

$$L_{\text{exactm}} = \sum_{j=1}^{D} \ln \int_0^\infty \prod_{\ell \in D_j} \left\{ 1 - \exp\left(-\frac{e_\ell}{s}t\right) \right\}^{w_\ell} \exp(-t)dt$$

$$e_\ell = \exp(\mathbf{x}_\ell\boldsymbol{\beta} + \text{offset}_\ell)$$

$$s = \sum_{\substack{k \in R_j \\ k \notin D_j}} w_k \exp(\mathbf{x}_k\boldsymbol{\beta} + \text{offset}_k) = \text{sum of weighted nondeath risk scores}$$

$$L_{\text{exactp}} = \sum_{j=1}^{D} \left\{ \sum_{i \in R_j} \delta_{ij}(\mathbf{x}_i\boldsymbol{\beta} + \text{offset}_i) - \ln f(r_j, d_j) \right\}$$

$$f(r, d) = f(r - 1, d) + f(r - 1, d - 1) \exp(\mathbf{x}_k\boldsymbol{\beta} + \text{offset}_k)$$

$$k = r^{\text{th}} \text{ observation in the set } R_j$$

$$r_j = \text{cardinality of the set } R_j$$

$$f(r, d) = \begin{cases} 0 & \text{if } r < d \\ 1 & \text{if } d = 0 \end{cases}$$

Calculations for the exact marginal log-likelihood (and associated derivatives) are obtained with 15-point Gauss–Laguerre quadrature. The `breslow` and `efron` options both provide approximations of the exact marginal log-likelihood. The `efron` approximation is a better (closer) approximation, but the `breslow` is a much faster approximation. The choice of which approximation to use in a given situation should generally be driven by the proportion of ties in the data.

Note that weights are not allowed with the `exactp` method.

Cox–Snell residuals are calculated using

$$r_{C_i} = \delta_{i\cdot} - r_{M_i}$$

where r_{M_i} are the martingale residuals, and $\delta_{i\cdot}$ is a censoring indicator equal to 1 if the ith observation is a failure and 0 otherwise. The martingale residuals are nonnegative values following an exponential distribution with mean 1. Modified Cox–Snell residuals are defined similarly, except that the residual is augmented by some positive constant if the observation is censored. The most common modifications are to add either 1 or the Crowley–Hu adjustment of $\ln 2 \approx .693$ (which is the median of an exponential distribution with parameter 1).

Martingale residuals are calculated using

$$r_{M_i} = \delta_{i\cdot} - \sum_{j=1}^{D} I(t_{0i} < t_{(j)} \le t_i) \frac{d_j \exp(\mathbf{x}_i \boldsymbol{\beta} + \text{offset}_i)}{\displaystyle\sum_{\ell \in R_j} \exp(\mathbf{x}_\ell \boldsymbol{\beta} + \text{offset}_\ell)}$$

These residuals are in $(-\infty, 1)$.

Using the Breslow–Peto approach, we have

$$r_{M_i} = 1 - \exp(\mathbf{x}_i \boldsymbol{\beta} + \text{offset}_i) \sum_{k:t_k \le t_i} \frac{w_k \delta_{k\cdot}}{\sum_{\ell \in R_i} w_\ell \exp(\mathbf{x}_\ell \boldsymbol{\beta} + \text{offset}_\ell)}$$

Using the Efron approach, we have

$$r_{M_i} = \left\{ w_i - \frac{1}{d_i} \left(\sum_{j=1}^{d_i} w_i \right) \frac{\exp(\mathbf{x}_i \boldsymbol{\beta} + \text{offset}_i)}{\sum_{\ell \in R_i} w_\ell f_{ij\ell} \exp(\mathbf{x}_\ell \boldsymbol{\beta} + \text{offset}_\ell)} \right\} \bigg/ w_i$$

Deviance residuals are calculated using

$$r_{D_i} = \text{sign}(r_{M_i}) \left[-2 \left\{ r_{M_i} + \delta_{i\cdot} \log(\delta_{i\cdot} - r_{M_i}) \right\} \right]^{1/2}$$

These are expected to be symmetric about zero but do not necessarily sum to zero.

Schoenfeld residuals are calculated using the Breslow–Peto approach as

$$r_{uS_i} = \sum_{j=1}^{D} \delta_{ij} \left\{ x_{ui} - \frac{\displaystyle\sum_{\ell \in R_j} x_{u\ell} \exp(\mathbf{x}_\ell \boldsymbol{\beta} + \text{offset}_\ell)}{\displaystyle\sum_{\ell \in R_j} \exp(\mathbf{x}_\ell \boldsymbol{\beta} + \text{offset}_\ell)} \right\}$$

Using the Efron approach, we have

$$r_{uS_i} = \sum_{j=1}^{D} \delta_{ij} \left\{ x_{ui} - \frac{\displaystyle\sum_{\ell \in R_j} x_{u\ell} f_{ij\ell} \exp(\mathbf{x}_\ell \boldsymbol{\beta} + \text{offset}_\ell)}{\displaystyle\sum_{\ell \in R_j} f_{ij\ell} \exp(\mathbf{x}_\ell \boldsymbol{\beta} + \text{offset}_\ell)} \right\}$$

Note that these are for each of the covariates, $u = 1, \dots, p$.

Grambsch and Therneau (1994) presented a scaled adjustment for the Schoenfeld residuals that permits the interpretation of the smoothed residuals as a nonparametric estimate of the log hazard ratio function. These are defined at each death event as

$$r_{uS_i^*} = \widehat{\beta}_u + \Delta(\mathbf{S}\widehat{\mathbf{V}}^{-1})_{ui}$$

where $\Delta = \sum_{i=1}^{N} \delta_{i.}$ is the total number of deaths, $\mathbf{S}$ is the matrix of Schoenfeld residuals, and $\widehat{\mathbf{V}}$ is the variance matrix estimate. These residuals are centered at $\widehat{\beta}_u$ for each of the covariates and should have slope zero when plotted against functions of time. The stphtest command uses these residuals, tests the null hypothesis that the slope is equal to zero for each covariate in the model, and performs the global test proposed by Grambsch and Therneau (1994). The test of zero slope is equivalent to testing that the log hazard ratio function is constant over time.

For a specified function of time, $g(t)$, the statistic for testing individual covariates is (for $\overline{g}(t) = \Delta^{-1} \sum_{i=1}^{N} \delta_{i.} g(t_i)$)

$$\frac{\left[\sum_{i=1}^{N} \left\{\delta_{i.} g(t_i) - \overline{g}(t)\right\} r_{uS_i^*}\right]^2}{\Delta \widehat{\mathbf{V}}_{uu} \sum_{i=1}^{N} \left\{\delta_{i.} g(t_i) - \overline{g}(t)\right\}^2}$$

which is asymptotically distributed as a χ^2 random variate with 1 degree of freedom.

The statistic for the global tests is calculated as

$$\left[\sum_{i=1}^{N} \left\{\delta_{i.} g(t_i) - \overline{g}(t)\right\} \mathbf{r}_i\right]' \left[\frac{\Delta \widehat{\mathbf{V}}}{\sum_{i=1}^{N} \left\{\delta_{i.} g(t_i) - \overline{g}(t)\right\}^2}\right] \left[\sum_{i=1}^{N} \left\{\delta_{i.} g(t_i) - \overline{g}(t)\right\} \mathbf{r}_i\right]$$

for $\mathbf{r}_i$ a vector of the p (unscaled) Schoenfeld residuals for the ith observation. The global test statistic is asymptotically distributed as a χ^2 random variate with p degrees of freedom.

The equations for the scaled Schoenfeld residuals and the two test statistics just described assume homogeneity of variance across risk sets. Although these tests are fairly robust to deviations from this assumption, care must be exercised, particularly when dealing with a stratified Cox model.

For shared frailty models, the data are organized into G groups with the ith group consisting of n_i observations, $i = 1, \ldots, G$. Following Therneau and Grambsch (2002, 253–255), estimation of θ takes place via maximum profile log-likelihood. For fixed θ, estimates of $\boldsymbol{\beta}$ and $\nu_1, \ldots, \nu_G$ are obtained by maximizing

$$L(\theta) = L_{\text{Cox}}(\boldsymbol{\beta}, \nu_1, \ldots, \nu_G) + \sum_{i=1}^{G} \left[\frac{1}{\theta}\left\{\nu_i - \exp(\nu_i)\right\} + \right.$$
$$\left. \left(\frac{1}{\theta} + D_i\right)\left\{1 - \ln\left(\frac{1}{\theta} + D_i\right)\right\} - \frac{\ln \theta}{\theta} + \ln \Gamma\left(\frac{1}{\theta} + D_i\right) - \ln \Gamma\left(\frac{1}{\theta}\right)\right]$$

where D_i is the number of death events in group i, and $L_{\text{Cox}}(\boldsymbol{\beta}, \nu_1, \ldots, \nu_G)$ is the standard Cox partial log-likelihood, with the ν_i treated as the coefficients of indicator variables identifying the groups. That is, the jth observation in the ith group has log-relative hazard $\mathbf{x}_{ij}\boldsymbol{\beta} + \nu_i$. The estimate of the frailty parameter, $\widehat{\theta}$, is chosen as that which maximizes $L(\theta)$. The final estimates of $\boldsymbol{\beta}$ are obtained by maximizing $L(\widehat{\theta})$ in $\boldsymbol{\beta}$ and the ν_i; the ν_i are not reported in the coefficient table but are available optionally. The estimated variance–covariance matrix of $\widehat{\beta}$ is obtained as the appropriate submatrix of the variance matrix of $(\widehat{\beta}, \widehat{\nu}_1, \ldots, \widehat{\nu}_G)$, and that matrix is obtained as the inverse of the negative hessian of $L(\widehat{\theta})$. As such, standard errors and inference based on $\widehat{\beta}$ should be treated as conditional on $\theta = \widehat{\theta}$.

The likelihood-ratio test statistic for testing $H_o : \theta = 0$ is calculated as minus twice the difference between the log-likelihood for a Cox model without shared frailty and $L(\widehat{\theta})$ evaluated at the final $(\widehat{\beta}, \widehat{\nu}_1, \ldots, \widehat{\nu}_G)$. All residuals and predictions are calculated treating the estimated ν_i as fixed offsets to the linear predictor. Estimated baseline functions are interpreted as that for which $\mathbf{x}$ *and* ν equal zero.

References

Breslow, N. E. 1974. Covariance analysis of censored survival data. *Biometrics* 30: 89–99.

Cleves, M. A. 1999. ssa13: Analysis of multiple failure-time data with Stata. *Stata Technical Bulletin* 49: 30–39. Reprinted in *Stata Technical Bulletin Reprints*, vol. 9, pp. 338–349.

Cleves, M. A., W. W. Gould, and R. G. Gutierrez. 2002. *An Introduction to Survival Analysis Using Stata*. College Station, TX: Stata Press.

Cox, D. R. 1972. Regression models and life-tables (with discussion). *Journal of the Royal Statistical Society*, Series B 34: 187–220.

——. 1975. Partial likelihood. *Biometrika* 62: 269–276.

Cox, D. R. and D. Oakes. 1984. *Analysis of Survival Data*. London: Chapman & Hall.

Cox, D. R. and E. J. Snell. 1968. A general definition of residuals (with discussion). *Journal of the Royal Statistical Society B* 30: 248–275.

Crowley, J. and M. Hu. 1977. Covariance analysis of heart transplant survival data. *Journal of the American Statistical Association* 72: 27–36.

Dupont, W. D. 2003. *Statistical Modeling for Biomedical Researchers*. Cambridge: Cambridge University Press. (*Forthcoming in first quarter 2003.*)

Fleming, T. R. and D. P. Harrington. 1991. *Counting Processes and Survival Analysis*. New York: John Wiley & Sons.

Garrett, J. M. 1997. gr23: Graphical assessment of the Cox model proportional hazards assumption. *Stata Technical Bulletin* 35: 9–14. Reprinted in *Stata Technical Bulletin Reprints*, vol. 6, pp. 38–44.

Grambsch, P. M. and T. M. Therneau. 1994. Proportional hazards tests and diagnostics based on weighted residuals. *Biometrika* 81: 515–526.

Gutierrez, R. G. 2002. Parametric frailty and shared frailty survival models. *The Stata Journal* 2: 22–44.

Gutierrez, R. G., S. L. Carter, and D. M. Drukker. 2001. On boundary-value likelihood ratio tests. *Stata Technical Bulletin* 60: 15–18. Reprinted in *Stata Technical Bulletin Reprints*, vol. 10, pp. 269–273.

Hills, M. and B. L. De Stavola. 2002. *A Short Introduction to Stata for Biostatistics*. London: Timberlake Consultants Press.

Hosmer, D. W., Jr., and S. Lemeshow. 1999. *Applied Survival Analysis*. New York: John Wiley & Sons.

Jenkins, S. P. 1997. sbe17: Discrete time proportional hazards regression. *Stata Technical Bulletin* 39: 22–32. Reprinted in *Stata Technical Bulletin Reprints*, vol. 7, pp. 109–121.

Kahn, H. A. and C. T. Sempos. 1989. *Statistical Methods in Epidemiology*. New York: Oxford University Press.

Kalbfleisch, J. D. and R. L. Prentice. 2002. *The Statistical Analysis of Failure Time Data*. 2d ed. New York: John Wiley & Sons.

Klein, J. P. and M. L. Moeschberger. 1997. *Survival Analysis: Techniques for Censored and Truncated data*. New York: Springer.

Lin, D. Y. and L. J. Wei. 1989. The robust inference for the Cox proportional hazards model. *Journal of the American Statistical Association* 84: 1074–1078.

McGilchrist, C. A. and C. W. Aisbett. 1991. Regression with frailty in survival analysis. *Biometrics* 47: 461–466.

Newman, S. C. 2001. *Biostatistical Methods in Epidemiology.* New York: John Wiley & Sons.

Peto, R. 1972. Contribution to the discussion of paper by D. R. Cox. *Journal of the Royal Statistical Society*, Series B 34: 205–207.

Rogers, W. H. 1994. ssa4: *Ex post* tests and diagnostics for a proportional hazards model. *Stata Technical Bulletin* 19: 23–27. Reprinted in *Stata Technical Bulletin Reprints*, vol. 4, pp. 186–191.

Royston, J. P. 2001. *Flexible alternatives to the Cox model. The Stata Journal* 1: 1–28.

Schoenfeld, D. 1982. Partial residuals for the proportional hazards regression model. *Biometrika* 69: 239–241.

Selvin, S. 1996. *Statistical Analysis of Epidemiologic Data.* 2d ed. New York: Oxford University Press.

Sterne, J. A. C. and K. Tilling. 2002. G-estimation of causal effects, allowing for time-varying confounding. *The Stata Journal* 2: 164–182.

Therneau, T. M. and P. M. Grambsch. 2000. *Modeling Survival Data: Extending the Cox Model.* New York: Springer.

Also See

Complementary:	[ST] **stphplot**, [ST] **sts**, [ST] **stset**,
	[R] **adjust**, [R] **lincom**, [R] **linktest**, [R] **lrtest**, [R] **mfx**, [R] **nlcom**,
	[R] **predict**, [R] **predictnl**, [R] **sw**, [R] **test**, [R] **testnl**, [R] **vce**
Related:	[ST] **streg**
Background:	[U] **16.5 Accessing coefficients and standard errors**,
	[U] **23 Estimation and post-estimation commands**,
	[U] **23.14 Obtaining robust variance estimates**,
	[ST] **st**, [ST] **survival analysis**,
	[R] **maximize**

Title

> **stdes** — Describe survival-time data

Syntax

stdes [if *exp*] [in *range*] [, weight no<u>sh</u>ow]

stdes is for use with survival-time data; see [ST] **st**. You must stset your data before using this command. by ... : may be used with stdes; see [R] **by**.

Description

stdes presents a brief description of the st data in a computer or data-based sense rather than in an analytical or statistical sense.

stdes is appropriate for use with single- or multiple-record, single- or multiple-failure, st data.

Options

weight specifies that you wish the description to use weighted rather than unweighted statistics. weight does nothing unless you specified a weight when you stset the data. The weight option, and ignoring the weights, is unique to stdes. The purpose of stdes is to describe the data in a computer sense—the number of records, etc.—and for that purpose, the weights are best ignored.

noshow prevents stdes from showing the key st variables. This option is rarely used since most people type stset, show or stset, noshow to reset once and for all whether they want to see these variables mentioned at the top of the output of every st command; see [ST] **stset**.

Remarks

Here is an example of stdes with single-record survival data:

```
. use http://www.stata-press.com/data/r8/page2
. stdes
        failure _d:  dead
   analysis time _t:  time
```

Category	total	mean	min	median	max
			per subject		
no. of subjects	40				
no. of records	40	1	1	1	1
(first) entry time		0	0	0	0
(final) exit time		227.95	142	231	344
subjects with gap	0				
time on gap if gap	0				
time at risk	9118	227.95	142	231	344
failures	36	.9	0	1	1

In this dataset, there is one record per subject. The purpose of this summary is not analysis—it is to describe how the data are arranged. Looking at this, one can quickly see that there is one record per subject (the number of subjects equals the number of records but, if there is any doubt, we can also see that the minimum and maximum number of records per subject is 1), that all the subjects entered at time 0, that the subjects exited between times 142 and 344 (median 231), that there are no gaps (as there could not be if there is only one record per subject), that the total time at risk is 9,118 (distributed reasonably evenly across the subjects), and that the total number of failures is 36 (with a maximum of 1 failure per subject).

Here is a description of the multiple-record Stanford heart transplant data that we introduced in [ST] **stset**:

```
. use http://www.stata-press.com/data/r8/stan3, clear
(Heart transplant data)
. stdes
        failure _d:  died
  analysis time _t:  t1
               id:  id
```

| Category | total | | per subject | | |
		mean	min	median	max
no. of subjects	103				
no. of records	172	1.669903	1	2	2
(first) entry time		0	0	0	0
(final) exit time		310.0786	1	90	1799
subjects with gap	0				
time on gap if gap	0	.	.	.	.
time at risk	31938.1	310.0786	1	90	1799
failures	75	.7281553	0	1	1

In this dataset, patients have one or two records. Although not revealed by the output, a patient has one record if the patient never received a heart transplant and two if the patient did receive a transplant; the first reflecting the patient's survival up to the time of transplantation and the second their subsequent survival:

```
. stset, noshow          /* to not show the st marker variables */
. stdes if !transplant
```

| Category | total | | per subject | | |
		mean	min	median	max
no. of subjects	34				
no. of records	34	1	1	1	1
(first) entry time		0	0	0	0
(final) exit time		96.61765	1	21	1400
subjects with gap	0				
time on gap if gap	0	.	.	.	.
time at risk	3285	96.61765	1	21	1400
failures	30	.8823529	0	1	1

(Continued on next page)

```
. stdes if transplant
```

| | | per subject | | | |
Category	total	mean	min	median	max
no. of subjects	69				
no. of records	138	2	2	2	2
(first) entry time		0	0	0	0
(final) exit time		415.2623	5.1	207	1799
subjects with gap	0				
time on gap if gap	0	.	.	.	.
time at risk	28653.1	415.2623	5.1	207	1799
failures	45	.6521739	0	1	1

Finally, here is the stdes from multiple-failure data:

```
. use http://www.stata-press.com/data/r8/mfail2
. stdes
```

| | | per subject | | | |
Category	total	mean	min	median	max
no. of subjects	926				
no. of records	1734	1.87257	1	2	4
(first) entry time		0	0	0	0
(final) exit time		470.6857	1	477	960
subjects with gap	6				
time on gap if gap	411	68.5	16	57.5	133
time at risk	435444	470.2419	1	477	960
failures	808	.8725702	0	1	3

Note that the maximum number of failures per subject observed is 3, although 50% had just one failure, and that 6 of the subjects have a gap in their histories.

Saved Results

stdes saves in r():

Scalars

r(N_sub)	number of subjects	r(gap)	total gap, if gap
r(N_total)	number of records	r(gap_min)	minimum gap, if gap
r(N_min)	minimum number of records	r(gap_mean)	mean gap, if gap
r(N_mean)	mean number of records	r(gap_med)	median gap, if gap
r(N_med)	median number of records	r(gap_max)	maximum gap, if gap
r(N_max)	maximum number of records	r(tr)	total time at risk
r(t0_min)	minimum first entry time	r(tr_min)	minimum time at risk
r(t0_mean)	mean first entry time	r(tr_mean)	mean time at risk
r(t0_med)	median first entry time	r(tr_med)	median time at risk
r(t0_max)	maximum first entry time	r(tr_max)	maximum time at risk
r(t1_min)	minimum final exit time	r(N_fail)	number of failures
r(t1_mean)	mean final exit time	r(f_min)	minimum number of failures
r(t1_med)	median final exit time	r(f_mean)	mean number of failures
r(t1_max)	maximum final exit time	r(f_med)	median number of failures
r(N_gap)	number of subjects with gap	r(f_max)	maximum number of failures

Methods and Formulas

stdes is implemented as an ado-file.

References

Cleves, M. A., W. W. Gould, and R. G. Gutierrez. 2002. *An Introduction to Survival Analysis Using Stata.* College Station, TX: Stata Press.

Also See

Complementary:	[ST] **stset**, [ST] **stsum**, [ST] **stvary**
Background:	[ST] **st**, [ST] **survival analysis**

Title

stfill — Fill in by carrying forward values of covariates

Syntax

stfill *varlist* [if *exp*] [in *range*] , { <u>b</u>aseline | <u>f</u>orward } [<u>nosh</u>ow]

stfill is for use with survival-time data; see [ST] **st**. You must stset your data before using this command.

Description

stfill is intended for use with multiple-record st data; data for which id() has been stset. stfill may be used with single-record data, but it literally does nothing. That is, stfill is appropriate for use with multiple-record, single- or multiple-failure, st data.

stfill, baseline changes variables to contain the value at the earliest time each subject was observed, making the variable constant over time. stfill, baseline changes all subsequent values of the specified variables to equal the first value whether they originally contained missing or not.

stfill, forward fills in missing values of each variable with that of the most recent time at which the variable was last observed. stfill, forward changes only missing values.

You must specify either the baseline or the forward option.

if *exp* and in *range* operate slightly differently from their usual definitions to work as you would expect. if and in restrict where changes can be made to the data but, no matter what, all stset observations are used to provide the values to be carried forward.

Options

baseline specifies that values are to be replaced with the values at baseline, the earliest time at which the subject was observed. All values of the specified variables are replaced, missing and nonmissing.

forward specifies that values are to be carried forward, and that previously observed, nonmissing values are to be used to fill in later values that are missing in the specified variables.

noshow prevents stsum from showing the key st variables. This option is rarely used since most people type stset, show or stset, noshow to reset once and for all whether they want to see these variables mentioned at the top of the output of every st command; see [ST] **stset**.

Remarks

stfill serves two purposes:

1. to assist in fixing data errors, and

2. to make baseline analyses easier.

Let us begin with repairing broken data.

You have a multiple-record st dataset which, due to how it was constructed, has a problem with the gender variable:

```
. use http://www.stata-press.com/data/r8/mrecord
. stvary sex
        failure _d:  myopic
  analysis time _t:  t
               id:  id

              subjects for whom the variable is
                                              never    always   sometimes
   variable |  constant    varying          missing   missing    missing

        sex |      131          1               22         0        110
```

Note that for 110 subjects, sex is sometimes missing and, for one more subject, the value of sex changes over time! The sex change is an error, but the missing values occurred because sometimes the subject's sex was not filled in on the revisit forms. We will assume that you have checked the changing-sex subject and determined that the baseline record is correct in that case, too.

```
. stfill sex, baseline
        failure _d:  myopic
  analysis time _t:  t
               id:  id

replace all values with value at earliest observed:
             sex:  221 real changes made
. stvary sex
        failure _d:  myopic
  analysis time _t:  t
               id:  id

              subjects for whom the variable is
                                              never    always   sometimes
   variable |  constant    varying          missing   missing    missing

        sex |      132          0              132         0          0
```

The sex variable is now completely filled in.

In this same dataset, there is another variable—bp, blood pressure—that is not always filled in because readings were not always taken.

```
. stvary bp
        failure _d:  myopic
  analysis time _t:  t
               id:  id

              subjects for whom the variable is
                                              never    always   sometimes
   variable |  constant    varying          missing   missing    missing

         bp |       18        114                9         0        123
```

(bp is constant for 18 patients because it was only taken once—at baseline.) Anyway, you decide it will be good enough, when bp is missing, to use the previous value of bp:

```
. stfill bp, forward noshow
replace missing values with previously observed values:
          bp:   263 real changes made
. stvary bp, noshow
             subjects for whom the variable is
                                       never    always   sometimes
    variable │  constant    varying    missing  missing  missing
   ──────────┼─────────────────────────────────────────────────────
          bp │      18        114        132        0        0
```

So much for data repair and fabrication.

Much later, deep in analysis, you are concerned about the bp variable and decide to compare results with a model that simply includes blood pressure at baseline. You are undecided on the issue and want to have both variables in your data:

```
. generate bp0 = bp
. stfill bp0, baseline
replace all values with value at earliest observed:
          bp0:   406 real changes made
. stvary bp bp0
             subjects for whom the variable is
                                       never    always   sometimes
    variable │  constant    varying    missing  missing  missing
   ──────────┼─────────────────────────────────────────────────────
          bp │      18        114        132        0        0
         bp0 │     132          0        132        0        0
```

Methods and Formulas

stfill is implemented as an ado-file.

Also See

Complementary:	[ST] **stbase**, [ST] **stgen**, [ST] **stset**, [ST] **stvary**
Background:	[ST] **st**, [ST] **survival analysis**

Title

> **stgen** — Generate variables reflecting entire histories

Syntax

> stgen [*type*] *newvar* = *function*

where *function* is

> ever(*exp*)
> never(*exp*)
> always(*exp*)
> min(*exp*)
> max(*exp*)
> when(*exp*)
> when0(*exp*)
> count(*exp*)
> count0(*exp*)
> minage(*exp*)
> maxage(*exp*)
> avgage(*exp*)
> nfailures()
> ngaps()
> gaplen()
> hasgap()

stgen is for use with survival-time data; see [ST] **st**. You must stset your data before using this command.

Description

stgen provides a convenient way to generate new variables reflecting entire histories—variables you could create for yourself using generate (and especially, generate with the by *varlist*: prefix), but that would require too much thought and there would be too much chance of making a mistake.

These functions are intended for use with multiple-record survival data, but may be used with single-record data. With single-record data, each function reduces to a single generate, and generate would be a more natural way to approach the problem.

stgen is appropriate for use with multiple-record, single- or multiple-failure, st data.

If your interest is in generating calculated values such as the survivor function, etc., see [ST] **sts**.

Functions

In the description of the functions below, note that when we say time units, we mean the same units as *timevar* from stset *timevar*, ... For instance, if *timevar* is the number of days since 01 January 1960 (a Stata date), then time units are days. If *timevar* is in years—years since 1960 or years since diagnosis or whatever—then time units are years.

When we say variable X records a "time" we mean a variable recording when something occurred recorded in the same units and with the same base as *timevar*. If *timevar* is a Stata date, then "time" is correspondingly a Stata date.

163

When we say t units, or analysis time units, we mean a variable in the units *timevar*/scale() from stset *timevar*, scale(...) .... If you did not specify a scale(), then t units are the same as time units. Alternatively, say *timevar* is recorded as a Stata date and you specified scale(365.25). Then t units are years. Caution: if you specified a nonconstant scale—scale(myvar), where myvar varies from subject to subject—then t units are different for every subject.

When we say "an analysis time", we mean when something occurred recorded in the units (*timevar*-origin())/scale(). We only speak about analysis time in terms of the beginning and end of each time-span record.

Although in the *Description* we said stgen creates variables reflecting entire histories, it is important to understand that stgen restricts itself to the stset observations, and entire history thus means the entire history as it is currently stset. If you really want to use entire histories as recorded in the data, type streset, past or streset, past future before using stgen and then type streset to reset to the original analysis sample.

The functions are

ever(*exp*) creates *newvar* containing 1 (true) if the expression is ever true (nonzero) and 0 otherwise. For instance,

> . stgen everlow = ever(bp<100)

would create everlow containing, for each subject, uniformly 1 or 0. Every record for a subject would contain everlow = 1 if, on any stset record for the subject, bp < 100, and otherwise everlow would be 0.

never(*exp*) is the reverse of ever(); it creates *newvar* containing 1 (true) if the expression is always false (0) and 0 otherwise. For instance,

> . stgen neverlow = never(bp<100)

would create neverlow containing, for each subject, uniformly 1 or 0. Every record for a subject would contain neverlow = 1 if, on every stset record for the subject, bp < 100 is false.

always(*exp*) creates *newvar* containing 1 (true) if the expression is always true (nonzero) and 0 otherwise. For instance,

> . stgen lowlow = always(bp<100)

would create lowlow containing, for each subject, uniformly 1 or 0. Every record for a subject would contain lowlow = 1 if, on every stset record for a subject, bp < 100.

min(*exp*) and max(*exp*) create newvar containing the minimum or maximum nonmissing value of *exp* within id(). min() and max() are often used with variables recording "a time" (see definition above), such as min(visitdat).

when(*exp*) and when0(*exp*) create newvar containing the "time" when *exp* first became true within the previously stset id(). Note that the result is in time, not t units; see definition above.

when() and when0() differ about when the *exp* became true. Records record time spans (*time0*, *time1*]. when() assumes the expression became true at the end of the time span, *time1*. when0() assumes the expression became true at the beginning of the time span, *time0*.

For example, assume you previously 'stset myt, failure(*eventvar*=...) ...'. when() would be appropriate for use with *eventvar*, and presumably when0() would be appropriate for use with the remaining variables.

count(*exp*) and count0(*exp*) create *newvar* containing the number of occurrences when *exp* is true within id().

count() and count0() differ in when they assume *exp* occurs. count() assumes *exp* corresponds to the end of the time-span record. Thus, even if *exp* is true in this record, the count would remain unchanged until the next record.

count0() assumes *exp* corresponds to the beginning of the time-span record. Thus, if *exp* is true in this record, the count is immediately updated.

For example, assume you previously 'stset myt, failure(*eventvar*=...) ...'. count() would be appropriate for use with *eventvar*, and presumably count0() would be appropriate for use with the remaining variables.

minage(*exp*), maxage(*exp*), and avgage(*exp*) return the elapsed time, in time units, since *exp* at the beginning, end, or middle of the record, respectively. *exp* is expected to evaluate to a time in time units. minage(), maxage(), and avgage() would be appropriate for use with the result of when(), when0(), min(), and max(), for instance.

Also see [ST] **stsplit**; stsplit will divide the time-span records into new time-span records that record specified intervals of ages.

nfailures() creates *newvar* containing the cumulative number of failures for each subject as of the entry time for the observation. nfailures() is intended for use with multiple-failure data; with single-failure data, nfailures() is always 0. In multiple-failure data,

    . stgen nfail = nfailures()

might create, for a particular subject, the following:

id	time0	time1	fail	x	nfail
93	0	20	0	1	0
93	20	30	1	1	0
93	30	40	1	2	1
93	40	60	0	1	2
93	60	70	0	2	2
93	70	80	1	1	2

Note that the total number of failures for this subject is 3, and yet the maximum of the new variable nfail is 2. At time 70, the beginning of the last record, there had been 2 failures previously, and there were 2 failures up to but not including time 80.

ngaps() creates *newvar* containing the cumulative number of gaps for each subject as of the entry time for the record. Delayed entry (an opening gap) is not considered a gap. For example,

    . stgen ngap = ngaps()

might create, for a particular subject, the following:

id	time0	time1	fail	x	ngap
94	10	30	0	1	0
94	30	40	0	2	0
94	50	60	0	1	1
94	60	70	0	2	1
94	82	90	1	1	2

gaplen() creates *newvar* containing the time on gap, measured in analysis time units, for each subject as of the entry time for the observation. Delayed entry (an opening gap) is not considered a gap. Continuing with the previous example,

```
. stgen gl = gaplen()
```

would produce

id	time0	time1	fail	x	ngap	gl
94	10	30	0	1	0	0
94	30	40	0	2	0	0
94	50	60	0	1	1	10
94	60	70	0	2	1	0
94	82	90	1	1	2	12

hasgap() creates *newvar* containing uniformly 1 if the subject ever has a gap, and 0 otherwise. Delayed entry (an opening gap) is not considered a gap.

Remarks

stgen does nothing you cannot do in other ways, but it is convenient.

It is worth considering how you would obtain results like those created by stgen should you need something that stgen will not create for you. Pretend you have an st dataset for which you have previously

```
. stset t, failure(d) id(id)
```

Assume these are some of the data:

id	t	d	bp
27	30	0	90
27	50	0	110
27	60	1	85
28	11	0	120
28	40	1	130

If we were to type

```
. stgen everlow = ever(bp<100)
```

then the new variable everlow would contain for these two subjects

id	t	d	bp	everlow
27	30	0	90	1
27	50	0	110	1
27	60	1	85	1
28	11	0	120	0
28	40	1	130	0

Variable everlow is 1 for subject 27 because in two of the three observations, $bp < 100$, and everlow is 0 for subject 28 because everlow is never less than 100 in either of the two observations.

Here is one way we could have created everlow for ourselves:

```
. generate islow = bp<100
. sort id
. by id: generate sumislow = sum(islow)
. by id: generate everlow = sumislow[_N]>0
. drop islow sumislow
```

The generic term for code like the above is explicit subscripting, and if you want to know more, see [U] **16.7 Explicit subscripting**.

Anyway, that is what stgen did for us although, internally, stgen used denser code that was equivalent to

```
. by id, sort: generate everlow=sum(bp<100)
. by id: replace everlow = everlow[_N]>0
```

Obtaining things like the time on gap is no more difficult. When we `stset` the data, `stset` created variable `_t0` to record the entry time. Then `stgen`'s `gaplen` function is equivalent to

```
. sort id _t
. by id: generate gaplen = _t0-_t[_n-1]
```

Seeing this, you should realize that if all you wanted was the cumulative length of the gap before the current record, you could type

```
. sort id _t
. by id: generate curgap = sum(_t0-_t[_n-1])
```

and if, instead, you wanted a variable that was 1 if there were a gap just before this record and 0 otherwise, you could type

```
. sort id _t
. by id: generate iscurgap = (_t0-_t[_n-1])>0
```

All of this is to convince you that, despite the usefulness of `stgen`, understanding [U] **16.7 Explicit subscripting** really is worth your while.

▷ Example

Let us use the `stgen` commands to real effect. We have a multiple-record, multiple-failure dataset.

```
. use http://www.stata-press.com/data/r8/mrmf
. st
-> stset t, id(id) failure(d) time0(t0) exit(time .) noshow
                  id:  id
        failure event:  d != 0 & d < .
   obs. time interval:  (t0, t]
    exit on or before:  time .
. stdes
```

| | | ┌─────────── per subject ───────────┐ | | | |
Category	total	mean	min	median	max
no. of subjects	926				
no. of records	1734	1.87257	1	2	4
(first) entry time		0	0	0	0
(final) exit time		470.6857	1	477	960
subjects with gap	6				
time on gap if gap	411	68.5	16	57.5	133
time at risk	435444	470.2419	1	477	960
failures	808	.8725702	0	1	3

Also in this dataset are two covariates, x1 and x2. We wish to fit a Cox model on these data but wish to assume that the baseline hazard for first failures is different from that for second and subsequent failures.

Note that our data contain 6 subjects with gaps. Since it is possible that failures might have occurred during the gap, we begin by dropping those 6 subjects:

```
. stgen hg = hasgap()
. drop if hg
(14 observations deleted)
```

The 6 subjects had 14 records among them. That done, we can create variable nf containing the number of failures and, from that, variable group which will be 0 when subjects have experienced no previous failures and 1 thereafter:

```
. stgen nf = nfailures()
. generate byte group = nf>0
```

We can now fit our stratified model:

```
. stcox x1 x2, strata(group) robust
Iteration 0:   log pseudo-likelihood = -4499.9966
Iteration 1:   log pseudo-likelihood = -4444.7797
Iteration 2:   log pseudo-likelihood = -4444.4596
Iteration 3:   log pseudo-likelihood = -4444.4596
Refining estimates:
Iteration 0:   log pseudo-likelihood = -4444.4596

Stratified Cox regr. -- Breslow method for ties

No. of subjects    =          920          Number of obs   =       1720
No. of failures    =          800
Time at risk       =       432153
                                            Wald chi2(2)    =     102.78
Log pseudo-likelihood =   -4444.4596        Prob > chi2     =     0.0000
                          (standard errors adjusted for clustering on id)
```

		Robust				
_t	Haz. Ratio	Std. Err.	z	P>|z|	[95% Conf.	Interval]
x1	2.087903	.1961725	7.84	0.000	1.736738	2.510074
x2	.2765613	.052277	-6.80	0.000	.1909383	.4005806

```
                                               Stratified by group
```
◁

Methods and Formulas

stgen is implemented as an ado-file.

Also See

Complementary: [ST] **stci**, [ST] **sts**, [ST] **stset**, [ST] **stvary**

Background: [U] **16.7 Explicit subscripting**,
                    [ST] **st**, [ST] **survival analysis**

Title

stir — Report incidence-rate comparison

Syntax

stir *exposedvar* [if *exp*] [in *range*] [, <u>strata</u>(*varname*) <u>nosh</u>ow *ir_options*]

stir is for use with survival-time data; see [ST] **st**. You must stset your data before using this command.

by ... : may be used with stir; see [R] **by**.

Description

stir reports point estimates and confidence intervals for the incidence rate ratio and difference. stir is an interface to the ir command; see [ST] **epitab**.

By the logic of ir, *exposedvar* should be a 0/1 variable, 0 meaning unexposed and 1 meaning exposed. stir, however, will allow any two-valued coding and even allow *exposedvar* to be a string variable.

stir may not be used with pweighted data.

stir is appropriate for use with single- or multiple-record, single- or multiple-failure, st data.

Options

<u>strata</u>(*varname*) specifies that the calculation is to be stratified on *varname*, which may be a numeric or a string variable. Within-stratum statistics are shown and then combined with Mantel–Haenszel weights.

<u>nosh</u>ow prevents stir from showing the key st variables. This option is rarely used since most people type stset, show or stset, noshow to reset once and for all whether they want to see these variables mentioned at the top of the output of every st command; see [ST] **stset**.

ir_options refers to the options of ir including, most importantly, level(#), estandard, istandard, and standard(*varname*) — see [ST] **epitab** — but you can specify any of the options ir allows except by() (which is stir's strata() option).

(*Continued on next page*)

Remarks

stir examines the incidence rate and time at risk.

```
. use http://www.stata-press.com/data/r8/page2
. stset, noshow
. stir group
note:  Exposed <-> group==2 and Unexposed <-> group==1
```

	group Exposed	Unexposed	Total
Failure	19	17	36
Time	5023	4095	9118

| Incidence Rate | .0037826 | .0041514 | .0039482 | |

	Point estimate	[95% Conf. Interval]	
Inc. rate diff.	−.0003688	−.002974	.0022364	
Inc. rate ratio	.9111616	.4484366	1.866047	(exact)
Prev. frac. ex.	.0888384	−.8660469	.5515634	(exact)
Prev. frac. pop	.04894			

(midp) Pr(k<=19) =		0.3900	(exact)
(midp) 2*Pr(k<=19) =		0.7799	(exact)

Saved Results

stir saves in r():

Scalars

r(p)	one-sided p-value	r(ub_irr)	upper bound of CI for irr
r(ird)	incidence rate difference	r(afe)	attributable (prev.) fraction among exposed
r(lb_ird)	lower bound of CI for ird	r(lb_afe)	lower bound of CI for afe
r(ub_ird)	upper bound of CI for ird	r(ub_afe)	upper bound of CI for afe
r(irr)	incidence rate ratio	r(afp)	attributable fraction for the population
r(lb_irr)	lower bound of CI for irr		

Methods and Formulas

stir is implemented as an ado-file.

stir simply accumulates numbers of failures and time at risk by exposed and unexposed (by strata, if necessary) and passes the calculation to ir; see [ST] **epitab**.

References

Dupont, W. D. 2003. *Statistical Modeling for Biomedical Researchers.* Cambridge: Cambridge University Press. (*Forthcoming in first quarter 2003.*)

Also See

Complementary:	[ST] **stset**, [ST] **stsum**
Background:	[ST] **st**, [ST] **survival analysis**; [ST] **epitab**

Title

> **stphplot** — Graphical assessment of the Cox proportional hazards assumption

Syntax

stphplot $\left[$ if $exp\right]$, $\left\{$ by(*varname*) | strata(*varname*)$\right\}$

$\left[$ adjust(*varlist*) nolntime nonegative zero noshow plot(*plot*) *connected_options* *twoway_options* $\right]$

stcoxkm $\left[$ if $exp\right]$, by(*varname*) $\left[$ separate

ties(breslow | efron | exactp | exactm)

noshow plot(*plot*) *connected_options* *twoway_options* $\right]$

stphplot and stcoxkm are for use with survival-time data; see [ST] **st**. You must stset your data before using these commands.

Description

stphplot plots $-\ln\left(-\ln(\text{survival})\right)$ curves for each category of a nominal or ordinal covariate versus ln(analysis time). These are often referred to as "log-log" plots. Optionally, these estimates can be adjusted for covariates. The proportional hazards assumption is not violated when the curves are parallel.

stcoxkm plots Kaplan–Meier observed survival curves and compares them with the Cox predicted curves for the same variable. The closer the observed values are to the predicted, the less likely it is that the proportional hazards assumption has been violated. Do not run stcox before this command; stcoxkm will execute stcox itself to estimate the model and obtain predicted values.

Options

by(*varname*) specifies the nominal or ordinal covariate. You must specify by() with stcoxkm and you must specify either by() or strata() with stphplot.

strata(*varname*) is an alternative to by() for stphplot. Rather than separate Cox models being estimated for each value of *varname*, a single, stratified Cox model is estimated. You must also specify adjust(*varlist*) with the strata(*varname*) option; see [ST] **sts graph**.

adjust(*varlist*) adjusts the estimates to that for *average* values of the *varlist* specified. Alternatively the estimates can be adjusted to *zero* values of *varlist* by specifying the zero option. adjust(*varlist*) can be specified with by(); it is required with strata(*varname*).

nolntime specifies that curves be plotted against analysis time instead of against ln(analysis time).

nonegative specifies that $\ln\left\{-\ln(\text{survival})\right\}$ curves be plotted instead of $-\ln\left\{-\ln(\text{survival})\right\}$.

zero is used with adjust() to specify that the estimates be adjusted to the 0 values of the *varlist* rather than average values.

noshow prevents stphplot and stcoxkm from showing the key st variables. This option is rarely used since most people type stset, show or stset, noshow to reset once and for all whether they want to see these variables mentioned at the top of the output of every st command; see [ST] **stset**.

separate is used with stcoxkm to produce separate plots of predicted and observed for each value of the variable specified with by().

ties(breslow | efron | exactm | exactp) is used to specify one of the methods available to stcox for handling tied failures. If not specified, ties(breslow) is assumed; see [ST] **stcox**.

plot(*plot*) provides a way to add other plots to the generated graph; see [G] *plot_option*.

connected_options affect the rendition of the plotted points connected by lines; see
[G] **graph twoway connected**.

twoway_options are any of the options documented in [G] *twoway_options* excluding by(). These include options for titling the graph (see [G] *title_options*) and options for saving the graph to disk (see [G] *saving_option*).

Remarks

What follows is paraphrased from Garrett (1997). Any errors are ours.

The most important assumption of the Cox proportional hazards model is that the hazard ratio is proportional over time. For example, suppose a group of cancer patients on an experimental treatment are followed for 10 years. If the hazard of dying for the nontreated group is twice the rate as that of the treated group (HR = 2.0), the proportional hazards assumption implies that this ratio is the same at one year, at two years, or at any point on the time scale. Because the Cox model, by definition, is constrained to follow this assumption, it is important to evaluate its validity before modeling. If the assumption fails, alternative modeling choices would be more appropriate (e.g., a stratified Cox model).

We provide three programs, stphplot, stcoxkm, and stphtest, which can be used to evaluate the proportional hazards assumption. Both stphplot and stcoxkm are graphical methods, whereas stphtest performs statistical tests based on the distribution of Schoenfeld residuals and is described in [ST] **stcox**.

The two commands stphplot and stcoxkm provide graphical methods for assessing violations of the proportional hazards assumption. Although using graphs to assess the validity of the assumption is subjective, it can be a helpful tool.

stphplot plots $-\ln\left(-\ln(\text{survival})\right)$ curves for each category of a nominal or ordinal covariate versus ln(analysis time). These are often referred to as "log–log" plots. Optionally, these estimates can be adjusted for covariates. If the plotted lines are reasonably parallel, the proportional hazards assumption has not been violated, and it would be appropriate to base the estimate for that variable on a single baseline survivor function.

Another graphical method of evaluating the proportional hazards assumption, though less common, is to plot the Kaplan–Meier observed survival curves and compare them with the Cox predicted curves for the same variable. This plot is produced with stcoxkm. When the predicted and observed curves are close together, the proportional hazards assumption has not been violated.

▷ Example

These examples use data from a leukemia remission study (Garrett 1997). The data consist of 42 patients who are followed over time to see how long (weeks) it takes them to go out of remission (relapse: 1 = yes, 0 = no). Half the patients receive a new experimental drug and the other half receive a standard drug (treatment1: 1 = drug A, 0 = standard). White blood cell count, a strong indicator of the presence of leukemia, is divided into three categories (wbc3cat: 1 = normal, 2 = moderate, 3 = high).

```
. use http://www.stata-press.com/data/r8/leukemia
(Leukemia Remission Study)
. describe
Contains data from http://www.stata-press.com/data/r8/leukemia.dta
  obs:            42                          Leukemia Remission Study
  vars:            8                          23 Sep 2002 02:25
  size:           504 (99.9% of memory free)
```

variable name	storage type	display format	value label	variable label
weeks	byte	%8.0g		Weeks in Remission
relapse	byte	%8.0g	yesno	Relapse
treatment1	byte	%8.0g	trt1lbl	Treatment I
treatment2	byte	%8.0g	trt2lbl	Treatment II
wbc3cat	byte	%9.0g	wbclbl	White Blood Cell Count
wbc1	byte	%8.0g		wbc3cat==Normal
wbc2	byte	%8.0g		wbc3cat==Moderate
wbc3	byte	%8.0g		wbc3cat==High

```
Sorted by:  weeks
. stset weeks, failure(relapse) noshow
     failure event:  relapse != 0 & relapse < .
obs. time interval:  (0, weeks]
 exit on or before:  failure
```

```
     42  total obs.
      0  exclusions
```

```
     42  obs. remaining, representing
     30  failures in single record/single failure data
    541  total analysis time at risk, at risk from t =          0
                          earliest observed entry t =          0
                            last observed exit t =            35
```

In this example, we examine whether the proportional hazards assumption holds for drug A versus the standard drug (treatment1). First, we will use stphplot, followed by stcoxkm.

(Continued on next page)

```
. stphplot, by(treatment1)
```

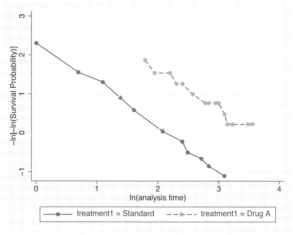

Figure 1.

```
. stcoxkm, by(treatment1) c(l l l l) m(O O S T)
```

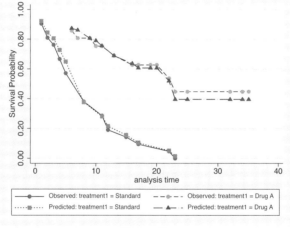

Figure 2.

Figure 1 (stphplot) displays lines that are parallel, implying that the proportional hazards assumption for treatment1 has not been violated. This is confirmed in Figure 2 (stcoxkm) where the observed values and predicted values are close together.

The graph in Figure 3 is the same as Figure 1, adjusted for white blood cell count (using two dummy variables). The adjustment variables were centered temporarily by stphplot before the adjustment was made.

(Continued on next page)

```
. stphplot, strata(treatment1) c(l l) adj(wbc2 wbc3)
```

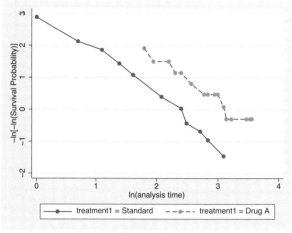

Figure 3.

Note that the lines in Figure 3 are still parallel, although somewhat closer together. Examining the proportional hazards assumption on a variable without adjustment for covariates is usually adequate as a diagnostic tool before using the Cox model. However, if it is known that adjustment for covariates in a final model is necessary, one may wish to re-examine whether the proportional hazards assumption still holds.

There is another variable in this dataset which measures a different drug (treatment2: 1 = drug B, 0 = standard). We wish to examine the proportional hazards assumption for this variable.

```
. stphplot, by(treatment2) c(l l)
```

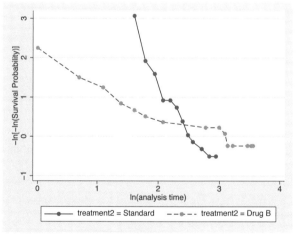

Figure 4.

```
. stcoxkm, by(treatment2) c(l l l l) m(i i i i) separate
```

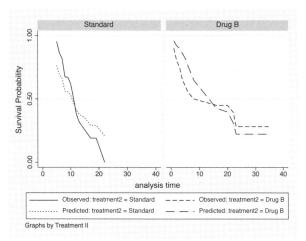

Figure 5.

Clearly this variable violates the proportion hazards assumption. In figure 4, we see that not only are the lines nonparallel, they cross in the data region. And in figure 5, we see that there are considerable differences between the observed and predicted values. We have overestimated the positive effect of drug B for the first half of the study and have underestimated it in the later weeks. A single hazard ratio describing the effect of this drug would be inappropriate. We definitely would want to stratify this variable in our Cox model.

◁

Methods and Formulas

stphplot and stcoxkm are implemented as ado-files.

For a single covariate, x, the Cox proportional hazard model reduces to

$$h(t; x) = h_0(t) \exp(\mathbf{x}\boldsymbol{\beta})$$

where $h_0(t)$ is the baseline hazard function from the Cox model. Let $S_0(t)$ and $H_0(t)$ be the corresponding Cox baseline survivor and baseline cumulative hazard functions respectively.

The proportional hazards assumption implies that

$$H(t) = H_0(t) \exp(\mathbf{x}\boldsymbol{\beta})$$

or

$$\ln H(t) = \ln H_0(t) + \mathbf{x}\boldsymbol{\beta}$$

where $H(t)$ is the cumulative hazard function. Thus, under the proportional hazards assumption, the logs of the cumulative hazard functions at each level of the covariate have equal slope. This is the basis for the method implemented in stphplot.

The proportional hazards assumption also implies that

$$S(t) = S_0(t)^{\exp(\mathbf{x}\boldsymbol{\beta})}$$

Let $\widehat{S}(t)$ be the estimated survivor function based on the Cox model. This function is a step function like the Kaplan–Meier estimate and, in fact, reduces to the Kaplan–Meier estimate when $\mathbf{x} = \mathbf{0}$. Thus, for each level of the covariate of interest, we can assess violations of the proportional hazards assumption by comparing these survival estimates with estimates calculated independently of the model. See, for example, Kalbfleisch and Prentice (2002) or Hess (1995).

stcoxkm plots Kaplan–Meier estimated curves for each level of the covariate together with the Cox model predicted baseline survival curve. The closer the observed values are to the predicted, the less likely it is that the proportional hazards assumption has been violated.

Acknowledgment

The original versions of stphplot and stcoxkm were written by Joanne M. Garrett, University of North Carolina at Chapel Hill.

References

Garrett, J. M. 1997. gr23: Graphical assessment of the Cox model proportional hazards assumption. *Stata Technical Bulletin* 35: 9–14. Reprinted in *Stata Technical Bulletin Reprints*, vol. 6, pp. 38–44.

——. 1998. ssa12: Predicted survival curves for the Cox proportional hazards model. *Stata Technical Bulletin* 44: 37–41. Reprinted in *Stata Technical Bulletin Reprints*, vol. 8, pp. 285–290.

Hess, K. R. 1995. Graphical methods for assessing violations of the proportional hazards assumption in Cox regression. *Statistics in Medicine* 14: 1707–1723.

Kalbfleisch, J. D. and R. L. Prentice. 2002. *The Statistical Analysis of Failure Time Data.* 2d ed. New York: John Wiley & Sons.

Also See

Related: [ST] **stcox**, [ST] **stset**

Background: [ST] **st**, [ST] **survival analysis**

Title

> **stptime** — Calculate person-time, incidence rates, and SMR

Syntax

$$\texttt{stptime} \; \left[\texttt{if } exp \right] \; \left[, \; \texttt{at}(numlist) \; \texttt{by}(varname) \; \texttt{trim} \; \texttt{per}(\#) \; \texttt{dd}(\#) \; \underline{\texttt{l}}\texttt{evel}(\#) \right.$$

$$\underline{\texttt{no}}\texttt{show} \; \texttt{jackknife} \; \texttt{smr}(groupvar \; ratevar) \; \underline{\texttt{u}}\texttt{sing}(filename) \; \underline{\texttt{t}}\texttt{itle}(string)$$

$$\left. \underline{\texttt{o}}\texttt{utput}(filename \left[,\texttt{replace} \right]) \right]$$

stptime is for use with survival-time data; see [ST] **st**. You must stset your data before using this command.

by ... : may be used with stptime; see [R] **by**.

Description

stptime calculates person-time and incidence rates. stptime also implements computation of standardized mortality/morbidity ratios (SMR), after merging the data with a suitable file of standard rates specified with the using() option.

Options

at(*numlist*) specifies time intervals at which person-time are to be computed. The intervals are specified in analysis-time t units. If at() is not specified, overall person-time and incidence rates are computed.

> If, for example, we specify at(5(5)20) and the trim option is not specified, person-time is reported for the intervals $t = (0 - 5]$, $t = (5 - 10]$, $t = (10 - 15]$, and $t = (15 - 20]$.

by(*varlist*) specifies a categorical variable by which incidence rates or SMR are to be computed.

trim specifies that observations less than or equal to the minimum or greater than the maximum value listed in at() are to be excluded from the computations.

per(*#*) defines the units used to report the rates. For example, if the analysis time is in years, specifying per(1000) results in rates per 1,000 person-years.

dd(*#*) specifies the maximum number of decimal digits to be reported for rates, ratios, and confidence intervals. This option affects only how values are displayed and not how they are calculated.

level(*#*) specifies the confidence level, in percent, for confidence intervals. The default is level(95) or as set by set level; see [U] **23.6 Specifying the width of confidence intervals**.

noshow prevents stptime from showing the key st variables. This option is rarely used since most people type stset, show or stset, noshow to reset once and for all whether they want to see these variables mentioned at the top of the output of every st command; see [ST] **stset**.

jackknife specifies that jackknife confidence intervals be produced. This is the default if pweights or iweights were specified when the dataset was stset.

smr(*groupvar ratevar*) specifies two variables in the using() dataset. The *groupvar* identifies the age-group or calendar-period variable used to match the data in memory and the using() dataset. The *ratevar* variable contains the appropriate reference rates. stptime then calculates SMRs rather than incidence rates.

using(*filename*) specifies the filename that contains a file of standard rates that is to be merged with the data so that SMR can be calculated.

title(*string*) replaces the default "person-time" label on the output table with the string specified with this option.

output(*filename*[,replace]) saves a summary dataset in *filename*. The file contains counts of failures and person-time, incidence rates (or SMRs), confidence limits, and categorical variables identifying the time intervals. This could be used for further calculations, or simply as input to the table command.

Remarks

stptime computes and tabulates the person-time and incidence rate, formed from the number of failures divided by the person-time, by different levels of one or more categorical explanatory variables specified by *varlist*. Confidence intervals for the rate are also given. By default, the confidence intervals are calculated using the quadratic approximation to the Poisson log-likelihood for the log rate parameter. However, whenever the Poisson assumption is questionable, jackknife confidence intervals can also be calculated.

stptime can also calculate and report SMRs if the data have been merged with a suitable file of reference rates.

If pweights or iweights were specified when the dataset was stset, stptime calculates jackknife confidence intervals by default.

The summary dataset can be saved to a file specified with the output() option, thus enabling further analysis or a more elaborate graphical display.

▷ Example

We begin with a simple fictitious example from Clayton and Hills' (1997, 42) book. Thirteen subjects were followed until the development of a particular disease. Here are the data for the first six subjects:

```
. use http://www.stata-press.com/data/r8/stptime
. list if _n<6
```

	id	year	fail
1.	1	19.6	1
2.	2	10.8	1
3.	3	14.1	1
4.	4	3.5	1
5.	5	4.8	1

The id variable identifies the subject, year records the time to failure in years, and fail is the failure indicator, which is one for all 30 subjects in the data. To use stptime, we must first stset the data.

```
. stset year, fail(fail) id(id)
                id:  id
     failure event:  fail != 0 & fail < .
obs. time interval:  (year[_n-1], year]
 exit on or before:  failure

        30  total obs.
         0  exclusions

        30  obs. remaining, representing
        30  subjects
        30  failures in single failure-per-subject data
     261.9  total analysis time at risk, at risk from t =         0
                              earliest observed entry t =         0
                                last observed exit t =         36.5
```

We can use `stptime` to obtain overall person-time of observation and disease incidence rate.

```
. stptime, title(person-years)
        failure _d:  fail
   analysis time _t:  year
               id:  id
```

Cohort	person-years	failures	rate	[95% Conf. Interval]
total	261.9	30	.11454754	.08009 .1638299

Note that the total 261.9 person-years reported by `stptime` matches what `stset` reported as total analysis time at risk. `stptime` computed an incidence rate of .11454754 per person-years. In epidemiology, incidence rates are often presented per 1000 person-years. We can do this by specifying `per(1000)`.

```
. stptime, title(person-years) per(1000)
        failure _d:  fail
   analysis time _t:  year
               id:  id
```

Cohort	person-years	failures	rate	[95% Conf. Interval]
total	261.9	30	114.54754	80.09001 163.8299

More interesting would be to compare incidence rates at 10 year intervals. We will specify `dd(4)` to display rates to 4 decimal places.

```
. stptime, per(1000) at(0(10)40) dd(4)
        failure _d:  fail
   analysis time _t:  year
               id:  id
```

Cohort	person-time	failures	rate	[95% Conf. Interval]
(0 - 10]	188.8000	18	95.3390	60.0676 151.3215
(10 - 20]	55.1000	10	181.4882	97.6506 337.3044
(20 - 30]	11.5000	1	86.9565	12.2490 617.3106
> 30	6.5000	1	153.8462	21.6713 1092.1648
total	261.9000	30	114.5475	80.0900 163.8299

◁

▷ Example

Using the diet data (Clayton and Hills 1997) described in Example 1 of [ST] **stsplit**, we will use
stptime to tabulate age-specific person-years and CHD incidence rates. Recall that in this dataset,
coronary heart disease (CHD) has been coded as fail = 1, 3, or 13.

We first stset the data: failure codes for CHD are specified; origin is set to date of birth, making
age analysis time; and the scale is set to 365.25, so analysis time is measured in years.

```
. use http://www.stata-press.com/data/r8/diet, clear
(Diet data with dates)
. stset dox, origin(time dob) enter(time doe) id(id) scale(365.25) fail(fail==1 3 13)

                id:  id
     failure event:  fail == 1 3 13
obs. time interval:  (dox[_n-1], dox]
 enter on or after:  time doe
 exit on or before:  failure
    t for analysis:  (time-origin)/365.25
            origin:  time dob

       337  total obs.
         0  exclusions

       337  obs. remaining, representing
       337  subjects
        46  failures in single failure-per-subject data
  4603.669  total analysis time at risk, at risk from t =         0
                                 earliest observed entry t =  30.07529
                                   last observed exit t =  69.99863
```

The incidence of CHD per 1,000 person-years can be tabulated in 10-year intervals.

```
. stptime, at(40(10)70) per(1000) trim

          failure _d:  fail == 1 3 13
    analysis time _t:  (dox-origin)/365.25
              origin:  time dob
   enter on or after:  time doe
                  id:  id
                note:  _group<=40 trimmed
```

Cohort	person-time	failures	rate	[95% Conf. Interval]	
(40 - 50]	907.00616	6	6.6151701	2.971936	14.72457
(50 - 60]	2107.0335	18	8.5428161	5.382338	13.55911
(60 - 70]	1493.3005	22	14.732467	9.700602	22.37444
total	4507.3402	46	10.205575	7.644246	13.62512

◁

The SMR for a cohort is the ratio of the total number of observed deaths to the number expected
from age-specific reference rates. This expected number can be found by multiplying the person
time in each cohort by the reference rate for that cohort. Using the smr option to define the cohort
variable and reference rate variable in the using() dataset, stptime calculates SMRs and confidence
intervals. Note that the per() option must still be specified. For example, if the reference rates are
per 100,000, then specify per(100000).

▷ Example

In smrchd.dta, we have age-specific CHD rates per 1000 person-years for a reference population. We can merge these data with our current data and use stptime to obtain SMRs and confidence intervals.

```
. stptime, smr(ageband rate) using(http://www.stata-press.com/data/r8/smrchd)
> at(40(10)70) per(1000) trim
            failure _d:  fail == 1 3 13
      analysis time _t:  (dox-origin)/365.25
                origin:  time dob
     enter on or after:  time doe
                    id:  id
                  note:  _group<=40 trimmed

                    |                observed  expected
        Cohort      |  person-time   failures  failures       SMR   [95% Conf. Interval]

       (40 - 50]    |   907.00616          6    5.62344     1.067   .4793445    2.374931
       (50 - 60]    |   2107.0335         18   18.7526      .95987   .6047571    1.523496
       (60 - 70]    |   1493.3005         22   22.8475      .96291   .6340263    1.462382

           total    |   4507.3402         46   47.2235      .97409   .7296197    1.300475
```

The stptime command can also be used to calculate person-time and incidence rates or SMRs by categories of the explanatory variable. In our diet data, the variable hienergy is coded 1 if the total energy consumption is more than 2.75 Mcals, and 0 otherwise. We want to compute the person-years and incidence rates for these two levels of hienergy.

```
. stptime, by(hienergy) per(1000)
            failure _d:  fail == 1 3 13
      analysis time _t:  (dox-origin)/365.25
                origin:  time dob
     enter on or after:  time doe
                    id:  id

      hienergy   |  person-time   failures       rate   [95% Conf. Interval]

             0   |   2059.4305         28   13.595992   9.387478    19.69123
             1   |   2544.2382         18    7.0748093  4.457431    11.2291

         total   |   4603.6687         46    9.9920309  7.484296    13.34002
```

We can also compute the incidence rate for the two levels of hienergy and the three previously defined age cohorts:

(*Continued on next page*)

```
. stptime, by(hienergy) at(40(10)70) per(1000) trim
        failure _d:  fail == 1 3 13
  analysis time _t:  (dox-origin)/365.25
           origin:  time dob
 enter on or after:  time doe
               id:  id
```

hienergy	person-time	failures	rate	[95% Conf. Interval]	
0					
(40 - 50]	346.87474	2	5.76577	1.442006	23.05407
(50 - 60]	979.33744	12	12.253182	6.958701	21.57593
> 60	699.14031	14	20.024593	11.85961	33.81091
1					
(40 - 50]	560.13142	4	7.1411813	2.680213	19.02702
(50 - 60]	1127.6961	6	5.3205824	2.390329	11.84297
> 60	794.16016	8	10.073535	5.037751	20.14314
total	4507.3402	46	10.205575	7.644246	13.62512

or compute the corresponding SMR.

```
. stptime, smr(ageband rate) using(http://www.stata-press.com/data/r8/smrchd)
> by(hienergy) at(40(10)70) per(1000) trim
        failure _d:  fail == 1 3 13
  analysis time _t:  (dox-origin)/365.25
           origin:  time dob
 enter on or after:  time doe
               id:  id
```

hienergy	person-time	observed failures	expected failures	SMR	[95% Conf. Interval]	
0						
(40 - 50]	346.87474	2	2.15062	.9299629	.2325815	3.718399
(50 - 60]	979.33744	12	8.7161	1.376762	.7818765	2.424262
> 60	699.14031	14	10.6968	1.308797	.7751381	2.209863
1						
(40 - 50]	560.13142	4	3.47281	1.151803	.4322924	3.068875
(50 - 60]	1127.6961	6	10.0365	.5978183	.2685763	1.330671
> 60	794.16016	8	12.1507	.658401	.3292648	1.316545
total	4507.3402	46	47.2235	.9740906	.7296197	1.300475

◁

Methods and Formulas

stptime is implemented as an ado-file.

References

Clayton, D. G. and M. Hills. 1993. *Statistical Models in Epidemiology*. Oxford: Oxford University Press.

Also See

Complementary:	[ST] **strate**; [ST] **stci**, [ST] **stir**, [ST] **stset**, [ST] **stsplit**
Related:	[ST] **epitab**
Background:	[ST] **st**, [ST] **survival analysis**

Title

strate — Tabulate failure rates and rate ratios

Syntax

strate [*varlist*] [if *exp*] [in *range*] [, per(#) jackknife cluster(*varname*)

  miss smr(*varname*) output(*filename*[,replace]) nolist level(#)

  graph nowhisker ciopts(*rspike_options*) *scatter_options* *twoway_options*]

stmh *varname* [*varlist*] [if *exp*] [in *range*] [, by(*varlist*) compare(*codes1,codes2*)

  miss level(#)]

stmc *varname* [*varlist*] [if *exp*] [in *range*] [, by(*varlist*) compare(*codes1,codes2*)

  miss level(#)]

strate, stmh, and stmc are for use with survival-time data; see [ST] **st**. You must stset your data before using these commands.

by ... : may be used with stmh and stmc; see [R] **by**.

Description

strate tabulates rates by one or more categorical variables declared in *varlist*. An optional summary dataset, which includes event counts and rate denominators, can be saved for further analysis or display. The combination of the commands stsplit and strate implements most, if not all, the functions of the special purpose "person-years" programs in widespread use in epidemiology. See, for example, Clayton and Hills (1993) and see [ST] **stsplit**. If your interest is solely in the calculation of person-years, see [ST] **stptime**.

stmh calculates stratified rate ratios and significance tests using a Mantel–Haenszel-type method.

stmc calculates rate ratios stratified finely by time, using the Mantel–Cox method. The corresponding significance test (the log-rank test) is also calculated.

Both stmh and stmc can be used to estimate the failure rate ratio for two categories of the explanatory variable specified by the first argument of *varlist*. Categories to be compared may be defined by specifying them with the compare() option. The remaining variables in *varlist* before the comma are categorical variables which are to be "controlled for" using stratification. Strata are defined by cross-classification of these variables.

Alternatively, stmh and stmc may be used to carry out trend tests for a metric explanatory variable. In this case, a one-step Newton approximation to the log-linear Poisson regression coefficient is computed.

Options

Options for strate

per(#) defines the units used to report the rates. For example, if the analysis time is in years, specifying per(1000) results in rates per 1,000 person-years.

jackknife specifies that jackknife confidence intervals be produced. This is the default if weights were specified when the dataset was stset.

cluster(*varname*) defines a categorical variable that indicates clusters of data to be used by the jackknife. If the jackknife option is selected and this option is not specified, the cluster variable is taken as the id variable defined in the st data. Specifying cluster() implies jackknife.

miss specifies that missing values of the explanatory variables are to be treated as extra categories. The default is to exclude such observations.

smr(*varname*) specifies a reference rate variable. strate then calculates SMRs rather than rates. This option will usually follow using stsplit to split the follow-up records by age bands and possibly calendar periods.

output(*filename*[,replace]) saves a summary dataset in *filename*. The file contains counts of failures and person-time, rates (or SMRs), confidence limits, and all the categorical variables in the *varlist*. This could be used for further calculations, or simply as input to the table command.

nolist suppresses the output. This is used only when saving results to a file specified by output().

level(#) specifies the confidence level, in percent, for confidence intervals. The default is level(95) or as set by set level; see [U] **23.6 Specifying the width of confidence intervals**.

graph produces a graph of the rate against the numerical code used for the categories of *varname*.

nowhisker omits the confidence intervals from the graph.

ciopts(*rspike_options*) affect the rendition of the confidence intervals (whiskers). See [G] **graph twoway rspike**. This option may not be combined with the nowhisker option.

scatter_options affect the rendition of the plotted points; see [G] **graph twoway scatter**.

twoway_options are any of the options documented in [G] *twoway_options* excluding by(). These include options for titling the graph (see [G] *title_options*) and options for saving the graph to disk (see [G] *saving_option*).

Options unique to stmh and stmc

by(*varlist*) specifies categorical variables by which the rate ratio is to be tabulated.

A separate rate ratio is produced for each category or combination of categories of *varlist*, and a test for unequal rate ratios (effect modification) is displayed.

In an analysis for log-linear trend the test is an approximation since the estimates are themselves based on a quadratic approximation of the log likelihood.

compare(*codes1*,*codes2*) specifies the categories of the exposure variable to be compared. The first code defines the numerator categories, and the second the denominator categories.

When compare is absent and there are only two categories, the larger is compared with the smaller; when there are more than two categories an analysis for log-linear trend is carried out.

Remarks

Remarks are presented under the headings

Tabulation of rates using strate
Stratified rate ratios using stmh
Log-linear trend test for metric explanatory variables using stmh
Controlling for age with fine strata using stmc

Tabulation of rates using strate

strate tabulates the rate, formed from the number of failures divided by the person-time, by different levels of one or more categorical explanatory variables specified by *varlist*. Confidence intervals for the rate are also given. By default, the confidence intervals are calculated using the quadratic approximation to the Poisson log-likelihood for the log rate parameter. However, whenever the Poisson assumption is questionable, jackknife confidence intervals can also be calculated. The jackknife option also allows for the case where there are multiple records for the same cluster (usually subject).

strate can also calculate and report SMRs if the data have been merged with a suitable file of reference rates.

The summary dataset can be saved to a file specified with the output() option, thus enabling further analysis or more elaborate graphical display.

If weights were specified when the dataset was stset, strate calculates jackknife confidence intervals by default.

▷ Example

Using the diet data (Clayton and Hills 1997) described in Example 1 of [ST] **stsplit**, we will use strate to tabulate age-specific CHD. Recall that in this dataset, coronary heart disease (CHD) has been coded as fail = 1, 3, or 13.

We first stset the data: failure codes for CHD are specified; origin is set to date of birth, making age analysis time; and the scale is set to 365.25, so analysis time is measured in years.

```
. use http://www.stata-press.com/data/r8/diet
(Diet data with dates)

. stset dox, origin(time doe) id(id) scale(365.25) fail(fail==1 3 13)
                id:  id
     failure event:  fail == 1 3 13
obs. time interval:  (dox[_n-1], dox]
 exit on or before:  failure
     t for analysis:  (time-origin)/365.25
            origin:  time doe
```

```
     337  total obs.
       0  exclusions

     337  obs. remaining, representing
     337  subjects
      46  failures in single failure-per-subject data
4603.669  total analysis time at risk, at risk from t =          0
                            earliest observed entry t =          0
                              last observed exit t =   20.04107
```

Now we stsplit the data into 10-year age bands.

```
. stsplit ageband=dob, at(40(10)70) trim
(26 + 0 obs. trimmed due to lower and upper bounds)
(418 observations (episodes) created)
```

stsplit added 418 observations to the dataset in memory and generated a new variable, ageband, which identifies each observation's age group.

The CHD rate per 1,000 person-years can now be tabulated against ageband:

```
. strate ageband, per(1000) graph
          failure _d:  fail == 1 3 13
    analysis time _t:  (dox-origin)/365.25
              origin:  time doe
                  id:  id
                note:  ageband<=40 trimmed
Estimated rates (per 1000) and lower/upper bounds of 95% confidence intervals
(729 records included in the analysis)
```

ageband	D	Y	Rate	Lower	Upper
40	6	0.9070	6.6152	2.9719	14.7246
50	18	2.1070	8.5428	5.3823	13.5591
60	22	1.4933	14.7325	9.7006	22.3744

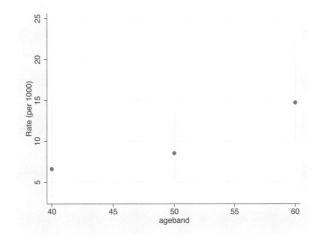

Since we specified the graph option, a plot of the estimated rate and confidence intervals was also generated.

◁

The SMR for a cohort is the ratio of the total number of observed deaths to the number expected from age-specific reference rates. This expected number can be found by first expanding on age, using stsplit, and then multiplying the person years in each age band by the reference rate for that band. merge (see [R] **merge**) can be used to add the reference rates to the dataset. Using the smr option to define the variable containing the reference rates, strate calculates SMRs and confidence intervals. Note that the per() option must still be specified. For example, if the reference rates are per 100,000, then specify per(100000). When reference rates are available by age and calendar period, stsplit must be called twice to expand on both time scales before merging the data with the reference rate file.

▷ Example

In `smrchd.dta` we have age-specific CHD rates per 1000 person-years for a reference population. We can merge these data with our current data and use `strate` to obtain SMRs and confidence intervals.

```
. merge ageband using smrchd
. strate ageband, per(1000) smr(rate)
        failure _d:  fail == 1 3 13
   analysis time _t:  (dox-origin)/365.25
            origin:  time doe
                id:  id
              note:  ageband<=40 trimmed
```

Estimated SMRs and lower/upper bounds of 95% confidence intervals
(729 records included in the analysis)

ageband	D	E	SMR	Lower	Upper
40	6	5.44	1.1025	0.4953	2.4541
50	18	16.86	1.0679	0.6728	1.6949
60	22	22.40	0.9822	0.6467	1.4916

◁

Stratified rate ratios using stmh

The `stmh` command is used for estimating rate ratios, controlled for confounding, using stratification. It can be used to estimate the ratio of the rates of failure for two categories of the explanatory variable. Categories to be compared may be defined by specifying the codes of the levels using `compare()`.

The first variable listed on the command line after `stmh` is the explanatory variable used in comparing rates, and the remaining, if any, are categorical variables, which are to be controlled for using stratification.

▷ Example

To illustrate this command, let us return to the diet data. Recall that the variable `hienergy` is coded 1 if the total energy consumption is more than 2.75 Mcals and 0 otherwise. We want to compare the rate for `hienergy`, level 1 with the rate for level 0, controlled for ageband.

To do this, we first `stset` and `stsplit` the data into age bands as before, and then we use `stmh`:

```
. use http://www.stata-pres.com/data/r8/diet
(Diet data with dates)
. stset dox, origin(time dob) enter(time doe) id(id) scale(365.25) fail(fail==1 3 13)
 (output omitted)
. stsplit ageband=dob, at(40(10)70) trim
(26 + 0 obs. trimmed due to lower and upper bounds)
(418 observations (episodes) created)
```

```
. stmh hienergy, c(1,0) by(ageband)
         failure _d:  fail == 1 3 13
   analysis time _t:  (dox-origin)/365.25
            origin:  time dob
   enter on or after:  time doe
                id:  id
              note:  ageband<=40 trimmed
Maximum likelihood estimate of the rate ratio
  comparing hienergy==1 vs. hienergy==0
  by ageband
RR estimate, and lower and upper 95% confidence limits
```

ageband	RR	Lower	Upper
40	1.24	0.23	6.76
50	0.43	0.16	1.16
60	0.50	0.21	1.20

Overall estimate controlling for ageband

RR	chi2	P>chi2	[95% Conf. Interval]	
0.534	4.36	0.0369	0.293	0.972

```
Approx test for unequal RRs (effect modification): chi2(2) =    1.19
                                                   Pr>chi2 =  0.5514
```

Note that since the RR estimates are approximate, the test for unequal rate ratios is also approximate.

We can also compare the effect of hienergy between jobs, controlling for ageband.

```
. stmh hienergy ageband, c(1,0) by(job)
         failure _d:  fail == 1 3 13
   analysis time _t:  (dox-origin)/365.25
            origin:  time dob
   enter on or after:  time doe
                id:  id
              note:  ageband<=40 trimmed
Mantel-Haenszel estimate of the rate ratio
  comparing hienergy==1 vs. hienergy==0
  controlling for ageband
  by job
RR estimate, and lower and upper 95% confidence limits
```

job	RR	Lower	Upper
0	0.42	0.13	1.33
1	0.64	0.22	1.87
2	0.51	0.21	1.26

Overall estimate controlling for ageband job

RR	chi2	P>chi2	[95% Conf. Interval]	
0.521	4.88	0.0271	0.289	0.939

```
Approx test for unequal RRs (effect modification): chi2(2) =    0.28
                                                   Pr>chi2 =  0.8695
```

◁

Log-linear trend test for metric explanatory variables using stmh

stmh may also be used to carry out trend tests for a metric explanatory variable. In addition, a one-step Newton approximation to the log-linear Poisson regression coefficient is computed.

The diet dataset contains the height for each patient recorded in the variable height. We can test for a trend of heart disease rates with height controlling for age band by typing

```
. stmh height ageband
            failure _d:  fail == 1 3 13
       analysis time _t:  (dox-origin)/365.25
               origin:  time dob
       enter on or after:  time doe
                   id:  id
                 note:  ageband<=40 trimmed
  Score test for trend of rates with height
    with an approximate estimate of the
    rate ratio for a one unit increase in height
    controlling for ageband
  RR estimate, and lower and upper 95% confidence limits
```

RR	chi2	P>chi2	[95% Conf. Interval]	
0.906	18.60	0.0000	0.866	0.948

stmh tested for trend of heart disease rates with height within age bands, and provided a rough estimate of the rate ratio for a 1 cm increase in height—this estimate is a one-step Newton approximation to the maximum likelihood estimate. It is not consistent, but it does provide a useful indication of the size of the effect.

We see that the rate ratio is significantly less than 1, so there is clear evidence for a decreasing rate with increasing height (about 9% decrease in rate per cm increase in height).

Controlling for age with fine strata using stmc

The stmc (Mantel–Cox) command is used to control for variation of rates on a time scale by breaking up time into very short intervals, or *clicks*.

Usually this approach is used only to calculate significance tests, but the rate ratio estimated remains just as useful as in the coarsely stratified analysis from stmh. The method may be viewed as an approximate form of Cox regression.

The rate ratio produced by stmc is controlled for analysis time, separately for each level of the variables specified with by(), and then combined to give a rate ratio controlled for both time and the by() variables.

▷ Example

For example, to obtain the effect of high energy controlled for age by stratifying very finely, we must first stset the data specifying the date of birth, dob, as the origin (so analysis time is age), and then we use stmc:

```
. stset dox, origin(time dob) enter(time doe) id(id) scale(365.25) fail(fail==1 3 13)
 (output omitted)

. stmc hienergy

        failure _d:  fail == 1 3 13
   analysis time _t:  (dox-origin)/365.25
            origin:  time dob
   enter on or after:  time doe
                id:  id
```

Mantel-Cox comparisons

Mantel-Haenszel estimates of the rate ratio
    comparing hienergy==1 vs. hienergy==0
    controlling for time (by clicks)
Overall Mantel-Haenszel estimate, controlling for time from dob

RR	chi2	P>chi2	[95% Conf. Interval]	
0.537	4.20	0.0403	0.293	0.982

The rate ratio of 0.537 is quite close to that obtained with stmh when controlling for age using 10-year age bands.

◁

Saved Results

stmh and stmc save in r():

Scalars
    r(RR)   overall rate ratio

Methods and Formulas

strate, stmh, and stmc are implemented as ado-files.

Nathan Mantel (1919–2002) was an American biostatistician who grew up in New York. He worked at the National Cancer Institute from 1947 to 1974 on a wide range of medical problems, and was also later affiliated with George Washington University and the American University in Washington.

William M. Haenszel (1910–1998) was an American biostatistician and epidemiologist who graduated from the University of Buffalo. He also worked at the National Cancer Institute and later at the University of Illinois.

Acknowledgments

The original versions of strate, stmh, and stmc were written by David Clayton, MRC Biostatistical Research Unit, Cambridge, and Michael Hills, London School of Hygiene and Tropical Medicine (retired).

References

Clayton, D. G. and M. Hills. 1993. *Statistical Models in Epidemiology*. Oxford: Oxford University Press.

——. 1995. ssa7: Analysis of follow-up studies. *Stata Technical Bulletin* 27: 19–26. Reprinted in *Stata Technical Bulletin Reprints*, vol. 5, pp. 219–227.

——. 1997. ssa10: Analysis of follow-up studies with Stata 5.0. *Stata Technical Bulletin* 40: 27–39. Reprinted in *Stata Technical Bulletin Reprints*, vol. 7, pp. 253–268.

Gail, M. H. 1997. A conversation with Nathan Mantel. *Statistical Science* 12: 88–97.

Hankey, B. 1997. A conversation with William M. Haenszel. *Statistical Science* 12: 108–112.

Also See

Complementary:	[ST] **stci**, [ST] **stir**, [ST] **stptime**, [ST] **stset**, [ST] **stsplit**
Related:	[ST] **epitab**
Background:	[ST] **st**, [ST] **survival analysis**

Title

> **streg** — Fit parametric survival models

Syntax

> streg [*varlist*] [if *exp*] [in *range*] [, distribution(*distname*) nohr time tr
>
> level(*#*) robust cluster(*varname*) score(*newvarlist|stub**) ancillary(*varlist*)
>
> anc2(*varlist*) strata(*varname*) frailty(gamma | invgaussian) shared(*varname*)
>
> offset(*varname*) constraints(*numlist*) noconstant nolrtest noshow nolog
>
> noheader *maximize_options*]

> stcurve [, cumhaz survival hazard range(*# #*) at(*varname*=# [*varname*=#...])
>
> [at1(*varname*=# [*varname*=#...]) [at2(*varname*=# [*varname*=#...]) [...]]]
>
> outfile(*filename*,[replace]) alpha1 unconditional kernel(*kernel*) width(*#*)
>
> *line_options twoway_options*]

where *distname* is one of exponential | weibull | gompertz | lognormal |

lnormal | loglogistic | llogistic | gamma

lognormal and lnormal are synonyms; loglogistic and llogistic are synonyms.
streg is for use with survival-time data; see [ST] **st**. You must stset your data before using streg.
by ... : may be used with streg; see [R] **by**.
stcurve may be used after streg and stcox, but is documented only here.
streg shares the features of all estimation commands; see [U] **23 Estimation and post-estimation commands**.
streg may be used with sw to perform stepwise estimation; see [R] **sw**.

Syntax for predict

> predict [*type*] *newvarname* [if *exp*] [in *range*] [, *statistic* alpha1
>
> unconditional]

where *statistic* is

median time	predicted median survival time (the default)
median lntime	predicted median ln(survival time)
mean time	predicted mean survival time
mean lntime	predicted mean ln(survival time)
hazard	predicted hazard
hr	predicted hazard ratio
xb	linear prediction $\mathbf{x}_j\boldsymbol{\beta}$
stdp	standard error of the linear prediction; SE($\mathbf{x}_j\boldsymbol{\beta}$)
surv	predicted $S(t|t_0)$
csnell	(partial) Cox–Snell residuals
mgale	(partial) martingale-like residuals
deviance	deviance residuals
csurv	predicted $S(t\,|\,$earliest t_0 for subject)
ccsnell	cumulative Cox–Snell residuals
cmgale	cumulative martingale-like residuals

All statistics are available both in and out of sample; type predict ... if e(sample) ... if wanted only for the estimation sample.

When no option is specified, the predicted median survival time is calculated for all models. The predicted hazard ratio option hr is only available for the exponential, Weibull, and Gompertz models. The mean time and mean lntime options are not available for the Gompertz model. Unconditional estimates of mean time and mean lntime are unavailable when frailty() is specified.

Description

streg performs maximum likelihood estimation for parametric regression survival-time models. These commands are appropriate for use with single- or multiple-record, single- or multiple-failure, st data. Survival models currently supported are exponential, Weibull, Gompertz, lognormal, log-logistic, and generalized gamma. Parametric frailty models and shared frailty models are also fitted using streg.

Also see [ST] **stcox** for proportional hazards models.

stcurve is used after streg to plot the cumulative hazard, survival, and hazard functions at the mean value of the covariates or at values specified by the at() options.

predict is used after streg to generate a variable containing the specified predicted values or residuals.

(Continued on next page)

Options

Options for streg

distribution(*distname*) specifies the survival model to be fitted. A specified distribution() is remembered from one estimation to the next when distribution() is not specified.

For instance, typing streg x1 x2, distribution(weibull) will fit a Weibull model. Subsequently, you do not need to specify distribution(weibull) to fit other Weibull regression models.

The currently supported distributions are listed in the syntax diagram, but new ones may have been added. Type help streg for an up-to-date list.

All Stata estimation commands, including streg, redisplay results when the command name is typed without arguments. If you wish to fit a model without any explanatory variables, type streg, distribution(*distname*)....

nohr, which may be specified at estimation or upon redisplaying results, specifies that coefficients rather than exponentiated coefficients are to be displayed or, said differently, coefficients rather than hazard ratios are displayed. This option affects only how coefficients are displayed and not how they are estimated.

This option is valid only for models with a natural proportional hazards parameterization: exponential, Weibull, and Gompertz. These three models, by default, report hazard ratios (exponentiated coefficients).

time specifies that the model is to be fitted in the accelerated failure-time metric rather than in the log relative-hazard metric. This option is only valid for the exponential and Weibull models since they have both a proportional hazards and an accelerated failure-time parameterization. Regardless of metric, the likelihood function is the same and models are equally appropriate viewed in either metric; it is just a matter of changing the interpretation.

time must be specified at estimation.

tr is appropriate only for the log-logistic, lognormal, and gamma models, or for the exponential and Weibull models when fitted in the accelerated failure-time metric. tr specifies that exponentiated coefficients are to be displayed, which have the interpretation of time ratios.

tr may be specified at estimation or upon replay.

level(*#*) specifies the confidence level, in percent, for confidence intervals. The default is level(95) or as set by set level; see [U] **23.6 Specifying the width of confidence intervals**.

robust specifies that the robust method of calculating the variance–covariance matrix is to be used instead of the conventional inverse-matrix-of-negative-second-derivatives method. If you specify robust, and if you have previously stset an id() variable, the robust calculation will be clustered on the id() variable.

We especially recommend that you specify robust if you have previously stset an id() variable because the assumption that justifies the conventional variance estimate—the independence of the observations—is presumably false.

cluster(*varname*) implies robust and specifies a variable on which clustering is to be based. This overrides the default clustering, if any. See the discussion under *Robust estimates of variance* in [ST] **stcox**; what is said there is appropriate here as well.

score(*newvarlist*) requests that new variables be created containing the score function(s). One new variable is specified in the case of an exponential model, two variables are specified for the Weibull, lognormal, Gompertz, and log-logistic models, and three new variables are specified in the case of gamma.

The first new variable will contain $\partial(\ln L_j)/\partial(\mathbf{x}_j\boldsymbol{\beta})$.

The second and third new variables, if they exist, will contain $\partial(\ln L_j)$ with respect to the second and third ancillary parameters. When an ancillary parameter is strictly positive, its logarithm is parameterized, in which case the score variable contains $\partial(\ln L_j)$ with respect to the logarithm of the ancillary parameter. See Table 1 for a list of ancillary parameters.

When frailty() is specified, an additional score variable must be specified which will contain $\partial(\ln L_j)/\partial(\ln\theta)$, where θ is the frailty parameter. The score() option is not available if shared() has been specified.

Alternatively, score(*stub**) may be specified whenever more than one score variable is to be created, in which case, the variables *stub*1, *stub*2, ..., *stub*k are created. This is especially useful if you do not know how many score variables are required for your particular model.

ancillary(*varlist*) specifies that the ancillary parameter for the Weibull, lognormal, Gompertz, and log-logistic distributions and the first ancillary parameter (sigma) of the generalized log-gamma distribution are to be estimated as a linear combination of *varlist*. This option is not available if frailty(*distname*) is specified.

Note that when an ancillary parameter is constrained to be strictly positive, it is the logarithm of the ancillary parameter that is modeled as a linear combination of *varlist*.

anc2(*varlist*) specifies that the second ancillary parameter (kappa) for the generalized log-gamma distribution is to be estimated as a linear combination of *varlist*. This option is not available if frailty(*distname*) is specified.

strata(*varname*) specifies a stratification variable. Observations with equal values of the variable are assumed to be in the same stratum. Stratified estimates (equal coefficients across strata but intercepts and ancillary parameters unique to each stratum) are then obtained. This option is not available if frailty(*distname*) is specified.

frailty(gamma | invgaussian) specifies the assumed distribution of the frailty, or heterogeneity. The estimation results, in addition to the standard parameter estimates, will contain an estimate of the variance of the frailties and a likelihood-ratio test of the null hypothesis that this variance is zero. When this null hypothesis is true, the model reduces to the model with frailty(*distname*) not specified.

A specified frailty() is remembered from one estimation to the next when distribution() is not specified. When you specify distribution(), the previously remembered specification of frailty() is forgotten.

shared(*varname*) is valid with frailty() and specifies a variable defining those groups over which the frailty is shared, analogous to random effects model for panel data where *varname* defines the panels. frailty() specified without shared() treats the frailties as occurring on the observation level.

A specified shared() is remembered from one estimation to the next when distribution() is not specified. When you specify distribution(), the previously remembered specification of shared() is forgotten.

shared() is not allowed in conjunction with distribution(gamma).

If shared() is specified without frailty(), and if there is no remembered frailty() from the previous estimation, then frailty(gamma) is assumed. This is to provide behavior analogous to stcox; see [ST] **stcox**.

offset(*varname*) specifies a variable that is to be entered directly into the linear predictor with coefficient one.

constraints(*numlist*) specifies by number the linear constraints to be applied during estimation. The default is to perform unconstrained estimation. Constraints are specified using the constraint command; see [R] **constraint**. See [R] **reg3** for the use of constraints in multiple-equation contexts.

noconstant suppresses the constant term (intercept) in the regression.

nolrtest is valid only with frailty models, in which case it suppresses the likelihood ratio test for significant frailty.

noshow prevents streg from showing the key st variables. This option is rarely used since most people type stset, show or stset, noshow to set once and for all whether they want to see these variables mentioned at the top of the output of every st command; see [ST] **stset**.

nolog prevents streg from showing the iteration log.

noheader suppresses the output header, either at estimation or upon replay.

maximize_options control the maximization process; see [R] **maximize**. You should never have to specify them, with the exception of frailty models, in which case specifying difficult may aid convergence.

Options for stcurve

cumhaz requests that the cumulative hazard function be plotted.

survival requests that the survival function be plotted.

hazard requests that the hazard function be plotted.

range(*# #*) specifies the range of the time axis to be plotted. If this option is not specified, stcurve will plot the desired curve on an interval expanding from the earliest to the latest time in the data.

at(*varname=#...*) requests that the covariates specified by *varname* be set to *#*. By default, stcurve evaluates the function by setting each covariate to its mean value. This option causes the function to be evaluated at the value of the covariates listed in at() and at the mean of all unlisted covariates.

at1(*varname=#...*), at2(*varname=#...*), ..., at10(*varname=#...*) specify that multiple curves (up to 10) are to be plotted on the same graph. at1(), at2(), ..., at10() work like the at() option. They request that the function be evaluated at the value of the covariates specified and at the mean of all unlisted covariates. at1() specifies the values of the covariates for the first curve, and at2() specifies the values of the covariates for the second curve, and so on.

outfile(*filename* [,replace]) saves in *filename*.dta the values used to plot the curve(s).

alpha1, when used after fitting a parametric frailty model, plots curves that are conditional on a frailty value of one. This is the default for shared frailty models.

unconditional, when used after fitting a parametric frailty model, plots curves that are unconditional on the frailty, i.e. the curve is "averaged" over the frailty distribution. This is the default for unshared frailty models.

kernel(*kernel*) is for use with `hazard`, and is only for use after `stcox`, since in the case of Cox regression an estimate of the hazard function is obtained by smoothing the estimated *hazard contributions*. `kernel()` specifies the kernel function for use in calculating the weighted kernel density estimate required to produce a smoothed hazard function estimator. The default kernel is Epanechinikov, yet *kernel* may be any of the kernels supported by `kdensity`; see [R] **kdensity**.

width(#) is for use with `hazard` (and only applies after `stcox`), and is used to specify the bandwidth to be used in the kernel smooth used to plot the estimated hazard function. If left unspecified, a default bandwidth is used as described in [R] **kdensity**.

line_options affect the rendition of the plotted cumulative hazard, survival, or hazard function(s); see [G] **graph twoway line**.

twoway_options are any of the options documented in [G] ***twoway_options*** (except the by() option). This includes, most importantly, options for titling the graph (see [G] ***title_options***) and options for saving the graph to disk (see [G] ***saving_option***).

Options for predict

median time calculates the predicted median survival time in analysis time units. Note that this is the prediction from time 0 conditional on constant covariates. When no options are specified with `predict`, the predicted median survival time is calculated for all models.

median lntime calculates the natural logarithm of what `median time` produces.

mean time calculates the predicted mean survival time in analysis time units. Note that this is the prediction from time 0 conditional on constant covariates. This option is not available for Gompertz regressions, and is only available for frailty models if `alpha1` is specified, in which case what you obtain is an estimate of the mean survival time conditional on a frailty effect of one.

mean lntime predicts the mean of the natural logarithm of `time`. This option is not available for Gompertz regression, and only available for frailty models if `alpha1` is specified, in which case what you obtain is an estimate of the mean log-survival time conditional on a frailty effect of one.

hazard calculates the predicted hazard.

hr calculates the hazard ratio. This option is valid only for models having a proportional hazards parameterization.

xb estimates the linear prediction from the fitted model. That is, all models can be thought of as estimating a set of parameters β_1, β_2, ..., β_k, and the linear prediction is an estimate, $\widehat{y}_j$, of $y_j = \beta_0 + \beta_1 x_{1j} + \beta_2 x_{2j} + \cdots + \beta_k x_{kj}$, often written in matrix notation as $y_j = \mathbf{x}_j\beta$.

It is important to understand that the x_{1j}, x_{2j}, ..., x_{kj} used in the calculation are obtained from the data currently in memory, and do not have to correspond to the data on the independent variables used in fitting the model.

stdp calculates the standard error of the prediction; that is, the standard error of $\widehat{y}_j$.

surv calculates each observation's predicted survivor probability $S(t|t_0)$, where t_0 is _t0, the analysis time at which each record became at risk. For multiple-record data, also see the `csurv` option below.

csnell calculates the (partial) Cox–Snell residual. If you have a single observation per subject, then `csnell` calculates the usual Cox–Snell residual. Otherwise, `csnell` calculates the additive contribution of this observation to the subject's overall Cox–Snell residual.

mgale calculates the (partial) martingale-like residual. The issues are the same as with `csnell` above.

deviance calculates the deviance residual. In the case of multiple-record data, only one value per subject is calculated, and it is placed on the last record for the subject.

csurv calculates the predicted $S(t|\text{earliest } t_0)$ for each subject in multiple-record data. This is based on calculating the conditional survivor values $S(t|t_0)$ (see option surv above) and then multiplying them together.

What you obtain from surv will differ from what you obtain from csurv only if you have multiple records for that subject, and you have delayed entry, gaps, or time-varying covariates.

ccsnell calculates the (cumulative) Cox–Snell residual in multiple-record data. This is based on calculating the partial Cox–Snell residuals (see option csnell above) and then summing them. Only one value per subject is recorded—the overall sum—and it is placed on the last record for the subject.

cmgale calculates the (cumulative) martingale-like residual in multiple-record data. This is based on calculating the partial martingale-like residuals (see option mgale above) and then summing them. Only one value per subject is recorded—the overall sum—and it is placed on the last record for the subject.

alpha1, when used after fitting a frailty model, specifies that *statistic* is to be predicted conditional on a frailty value equal to one. This is the default for shared frailty models.

unconditional, when used after fitting a frailty model, specifies that *statistic* is to be predicted unconditional on the frailty. That is, the prediction is averaged over the frailty distribution. This is the default for unshared frailty models.

Remarks

Remarks are presented under the headings

> *Introduction*
> *Distributions*
> > *Weibull and exponential models*
> > *Gompertz model*
> > *Lognormal and log-logistic models*
> > *Generalized gamma model*
> *Examples*
> *Parameterization of ancillary parameters*
> *Stratified estimation*
> *(Unshared) frailty models*
> *Shared frailty models*
> *stcurve*
> *predict*

Introduction

What follows is a brief summary of what can be done with streg and stcurve. For a complete tutorial, see Cleves, Gould, and Gutierrez (2002), who devote four chapters to this topic.

Two frequently used models for adjusting survival functions for the effects of covariates are the accelerated failure-time (AFT) model and the multiplicative or proportional hazards (PH) model. In the accelerated failure-time model, the natural logarithm of the survival time $\ln t$ is expressed as a linear function of the covariates, yielding the linear model

$$\ln t_j = \mathbf{x}_j \boldsymbol{\beta} + z_j$$

where $\mathbf{x}_j$ is a vector of covariates, $\boldsymbol{\beta}$ is a vector of regression coefficients, and z_j is the error with density $f()$. The distributional form of the error term determines the regression model. If we let $f()$ be the normal density, the lognormal regression model is obtained. Similarly, by letting $f()$ be the logistic density, the log-logistic regression is obtained. Setting $f()$ equal to the extreme-value density yields the exponential and the Weibull regression models.

The effect of the accelerated failure-time (AFT) model is to change the time scale by a factor of $\exp(-\mathbf{x}_j\boldsymbol{\beta})$. Depending on whether this factor is greater than or less than 1, time is either accelerated or decelerated (degraded). That is, if a subject who is at baseline experiences a probability of survival past time t equal to $S(t)$, then a subject with covariates $\mathbf{x}_j$ would have probability of survival past time t equal to $S()$ evaluated at the point $\exp(-\mathbf{x}_j\boldsymbol{\beta})t$ instead. Thus, accelerated failure-time does not imply a positive acceleration of time with the increase of a covariate, but instead a deceleration of time, or equivalently an increase in the expected waiting time for failure.

In the proportional hazards (PH) model, the concomitant covariates have a multiplicative effect on the hazard function

$$h(t_j) = h_0(t)g(\mathbf{x}_j)$$

for some $h_0(t)$ and $g(\mathbf{x}_j)$ is a nonnegative function of the covariates. A popular choice, and the one adopted here, is to let $g(\mathbf{x}_j) = \exp(\mathbf{x}_j\boldsymbol{\beta})$. The function $h_0(t)$ may either be left unspecified, yielding the Cox proportional hazard model (see [ST] **stcox**), or take a specific parametric form. For the `streg` command, $h_0(t)$ is assumed to be parametric. Three regression models are currently implemented as PH models: the exponential, Weibull, and Gompertz models. Note that the exponential and Weibull models are implemented as both AFT and PH models, and that the Gompertz model is only implemented in the PH metric.

The above allows for the presence of an intercept term, β_0, within $\mathbf{x}_j\boldsymbol{\beta}$. Thus, what is commonly referred to as the *baseline hazard function*, the hazard when all covariates are zero, is actually equal to $h_0(t)\exp(\beta_0)$. That is, the intercept term serves to scale the baseline hazard. Of course, specifying `noconstant` suppresses the intercept, or equivalently constrains β_0 to equal zero.

`streg` is only suitable for data that have been `stset`. By stsetting your data, you define the variables _t0, _t, and _d which serve as the trivariate response variable (t_0, t, d). Each response corresponds to a time period under observation, $(t_0, t]$, resulting in either failure $(d = 1)$ or right-censoring $(d = 0)$ at time t. As a result, `streg` is appropriate for data exhibiting delayed entry, gaps, time-varying covariates, and even multiple-failure data.

(Continued on next page)

Distributions

Six parametric survival distributions are currently supported by streg. The parameterization and ancillary parameters for each distribution are summarized in Table 1:

Table 1. Parametric survival distributions supported by streg

Distribution	Metric	Survival function	Parameterization	Ancillary parameters
Exponential	PH	$\exp(-\lambda_j l_j)$	$\lambda_j = \exp(\mathbf{x}_j \boldsymbol{\beta})$	
Exponential	AFT	$\exp(-\lambda_j t_j)$	$\lambda_j = \exp(-\mathbf{x}_j \boldsymbol{\beta})$	
Weibull	PH	$\exp(-\lambda_j t_j^p)$	$\lambda_j = \exp(\mathbf{x}_j \boldsymbol{\beta})$	p
Weibull	AFT	$\exp(-\lambda_j t_j^p)$	$\lambda_j = \exp(-p\mathbf{x}_j \boldsymbol{\beta})$	p
Gompertz	PH	$\exp\{-\lambda_j \gamma^{-1}(e^{\gamma t_j} - 1)\}$	$\lambda_j = \exp(\mathbf{x}_j \boldsymbol{\beta})$	γ
Lognormal	AFT	$1 - \Phi\left\{\frac{\ln(t_j) - \mu_j}{\sigma}\right\}$	$\mu_j = \mathbf{x}_j \boldsymbol{\beta}$	σ
Log-logistic	AFT	$\{1 + (\lambda_j t_j)^{1/\gamma}\}^{-1}$	$\lambda_j = \exp(-\mathbf{x}_j \boldsymbol{\beta})$	γ
Generalized gamma				
if $\kappa > 0$	AFT	$1 - I(\gamma, u)$	$\mu_j = \mathbf{x}_j \boldsymbol{\beta}$	σ, κ
if $\kappa = 0$	AFT	$1 - \Phi(z)$	$\mu_j = \mathbf{x}_j \boldsymbol{\beta}$	σ, κ
if $\kappa < 0$	AFT	$I(\gamma, u)$	$\mu_j = \mathbf{x}_j \boldsymbol{\beta}$	σ, κ

where PH = proportional hazard, AFT = accelerated failure time, $\Phi(z)$ is the standard normal cumulative distribution. In the case of the generalized gamma, $\gamma = |\kappa|^{-2}$, $u = \gamma \exp(|\kappa| z)$, $I(a, x)$ is the incomplete gamma function, and $z = \text{sign}(\kappa)\{\ln(t_j) - \mu_j\}/\sigma$.

Plotted in Figure 1 are example hazard functions for five of the six distributions. The exponential hazard (not separately plotted) is a special case of the Weibull hazard when the Weibull ancillary parameter $p = 1$. The generalized gamma (not plotted) is extremely flexible, and therefore can take a multitude of shapes.

(Continued on next page)

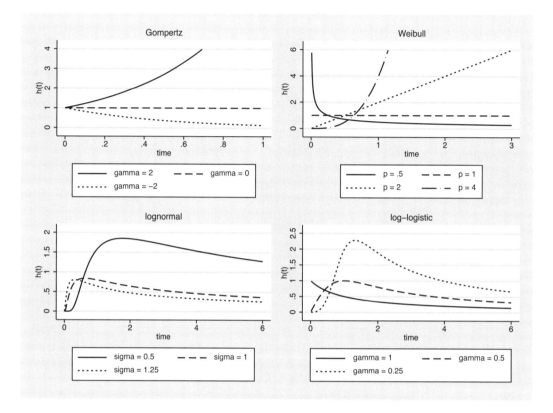

Figure 1. Example plots of hazard functions

Weibull and exponential models

The Weibull and exponential models are parameterized as both proportional hazard and accelerated failure-time models. The Weibull distribution is suitable for modeling data with monotone hazard rates that either increase or decrease exponentially with time, while the exponential distribution is suitable for modeling data with constant hazard (see Figure 1).

For the proportional hazards model, $h_0(t) = 1$ for exponential regression, and $h_0(t) = p\,t^{p-1}$ for Weibull regression, where p is the shape parameter to be estimated from the data. Some authors refer not to p but to $\sigma = 1/p$.

The accelerated failure-time model is written as

$$\ln(t_j) = \mathbf{x}_j\boldsymbol{\beta}^* + z_j$$

where z_j has an extreme-value distribution scaled by σ. Let $\boldsymbol{\beta}$ be the vector of regression coefficients derived from the proportional hazards model, so that $\boldsymbol{\beta}^* = -\sigma\boldsymbol{\beta}$. Note that this relationship only holds if the ancillary parameter p is a constant; it does not hold when the ancillary parameter is parameterized in terms of covariates.

`streg` uses, by default, for the exponential and Weibull models, the proportional-hazards metric simply because it eases comparison with those results produced by `stcox`. You can, however, specify option `time` to choose the accelerated failure-time parameterization.

The Weibull hazard and survival functions are

$$h(t) = p\lambda t^{p-1}$$

$$S(t) = \exp(-\lambda t^p)$$

where λ is parameterized as described in Table 1. If $p = 1$, then these functions reduce to those of the exponential.

Gompertz model

The Gompertz regression is parameterized only as a proportional hazards model. First described in 1825, this model has been extensively used by medical researchers and biologists modeling mortality data. The Gompertz distribution implemented is the two-parameter function as described in Lee (1992), with the following hazard and survival functions:

$$h(t) = \lambda \exp(\gamma t)$$
$$S(t) = \exp\{-\lambda\gamma^{-1}(e^{\gamma t} - 1)\}$$

The model is implemented by parameterizing $\lambda_j = \exp(\mathbf{x}_j\boldsymbol{\beta})$, implying that $h_0(t) = \exp(\gamma t)$, where γ is an ancillary parameter to be estimated from the data.

This distribution is suitable for modeling data with monotone hazard rates that either increase or decrease exponentially with time (see Figure 1).

When γ is positive, the hazard function increases with time; when γ is negative, the hazard function decreases with time; and when γ is zero, the hazard function is equal to λ for all t, which is to say, the model reduces to an exponential.

Some recent survival analysis texts, such as Klein and Moeschberger (1997), restrict γ to be strictly positive. Note that if $\gamma < 0$, then as t goes to infinity, the survivor function $S(t)$ exponentially decreases to a nonzero constant, implying that there is a nonzero probability of never failing (living forever). That is, there is always a nonzero hazard rate, yet it decreases exponentially. By restricting γ to be positive, one is assured that the survivor function always goes to zero as t tends to infinity.

Although the above argument may be desirable from a mathematical perspective, in Stata's implementation, we took the more traditional approach of not restricting γ. We did this because in survival studies subjects are not followed forever—there is a date when the study ends, and in many investigations, specifically in medical research, an exponentially-decreasing hazard rate is clinically appealing.

Lognormal and log-logistic models

The lognormal and log-logistic models are implemented in only the accelerated failure-time form. These two distributions are similar and tend to produce comparable results. In the case of the lognormal distribution, the natural logarithm of time follows a normal distribution; whereas for the log-logistic distribution, the natural logarithm of time follows a logistic distribution.

The lognormal survival and density functions are

$$S(t) = 1 - \Phi\left\{\frac{\ln(t) - \mu}{\sigma}\right\}$$

$$f(t) = \frac{1}{t\sigma\sqrt{2\pi}} \exp\left[\frac{-1}{2\sigma^2}\left\{\ln(t) - \mu\right\}^2\right]$$

where $\Phi(z)$ is the standard normal cumulative distribution function.

The lognormal regression is implemented by setting $\mu_j = \mathbf{x}_j\beta$ and treating the standard deviation σ as an ancillary parameter to be estimated from the data.

The log-logistic regression is obtained if z_j has a logistic density. The log-logistic survival and density functions are

$$S(t) = \left\{1 + (\lambda t)^{1/\gamma}\right\}^{-1}$$

$$f(t) = \frac{\lambda^{1/\gamma}t^{1/\gamma-1}}{\gamma\left\{1 + (\lambda t)^{1/\gamma}\right\}^2}$$

This model is implemented by parameterizing $\lambda_j = \exp(-\mathbf{x}_j\beta)$ and treating the scale parameter γ as an ancillary parameter to be estimated from the data.

Unlike the exponential, Weibull, and Gompertz distributions, the lognormal and the log-logistic distributions are indicated for data exhibiting nonmonotonic hazard rates, specifically initially increasing and then decreasing rates (Figure 1).

(Continued on next page)

Generalized gamma model

The generalized gamma model is implemented only in the accelerated failure-time form. The three-parameter generalized gamma survival and density functions are

$$S(t) = \begin{cases} 1 - I(\gamma, u), & \text{if } \kappa > 0 \\ 1 - \Phi(z), & \text{if } \kappa = 0 \\ I(\gamma, u), & \text{if } \kappa < 0 \end{cases}$$

$$f(t) = \begin{cases} \frac{\gamma^\gamma}{\sigma t \sqrt{\gamma} \Gamma(\gamma)} \exp(z\sqrt{\gamma} - u), & \text{if } \kappa \neq 0 \\ \frac{1}{\sigma t \sqrt{2\pi}} \exp(-z^2/2), & \text{if } \kappa = 0 \end{cases}$$

where $\gamma = |\kappa|^{-2}$, $z = \text{sign}(\kappa)\{\ln(t) - \mu\}/\sigma$, $u = \gamma \exp(|\kappa|z)$, $\Phi(z)$ is the standard normal cumulative distribution function, and $I(a, x)$ is the incomplete gamma function. See the gammap(a,x) entry in [R] **functions** to see how the incomplete gamma function is implemented in Stata.

This model is implemented by parameterizing $\mu_j = \mathbf{x}_j\boldsymbol{\beta}$ and treating the parameters κ and σ as ancillary parameters to be estimated from the data.

The hazard function of the generalized gamma distribution is extremely flexible, allowing for a large number of possible shapes, including as special cases the Weibull distribution when $\kappa = 1$, the exponential when $\kappa = 1$ and $\sigma = 1$, and the lognormal distribution when $\kappa = 0$. The generalized gamma model is, therefore, commonly used for evaluating and selecting an appropriate parametric model for the data. The Wald or likelihood-ratio test can be used to test the hypotheses that $\kappa = 1$ or that $\kappa = 0$.

Examples

▷ Example

The Weibull distribution provides a good illustration of streg, since this distribution is parameterized as both AFT and PH and serves to compare and contrast the two approaches.

You wish to analyze an experiment testing the ability of emergency generators with new-style bearings to withstand overloads. This dataset is described in [ST] **stcox**. This time, you wish to fit a Weibull model:

```
. use http://www.stata-press.com/data/r8/kva
(Generator experiment)

. streg load bearings, distribution(weibull)
        failure _d:  1 (meaning all fail)
   analysis time _t:  failtime

Fitting constant-only model:

Iteration 0:   log likelihood = -13.666193
Iteration 1:   log likelihood = -9.7427276
Iteration 2:   log likelihood = -9.4421169
Iteration 3:   log likelihood = -9.4408287
Iteration 4:   log likelihood = -9.4408286

Fitting full model:

Iteration 0:   log likelihood = -9.4408286
Iteration 1:   log likelihood =  -2.078323
```

```
Iteration 2:    log likelihood =   5.2226016
Iteration 3:    log likelihood =   5.6745808
Iteration 4:    log likelihood =   5.6934031
Iteration 5:    log likelihood =   5.6934189
Iteration 6:    log likelihood =   5.6934189

Weibull regression -- log relative-hazard form

No. of subjects =            12            Number of obs    =        12
No. of failures =            12
Time at risk    =           896
                                           LR chi2(2)       =     30.27
Log likelihood  =     5.6934189            Prob > chi2      =    0.0000
```

_t	Haz. Ratio	Std. Err.	z	P>|z|	[95% Conf. Interval]
load	1.599315	.1883807	3.99	0.000	1.269616 2.014631
bearings	.1887995	.1312109	-2.40	0.016	.0483546 .7371644
/ln_p	2.051552	.2317074	8.85	0.000	1.597414 2.505691
p	7.779969	1.802677			4.940241 12.25202
1/p	.1285352	.0297826			.0816192 .2024193

Since we did not specify otherwise, the estimation took place in the hazard metric, which is the default for distribution(weibull). The estimates are directly comparable to those produced by stcox: stcox estimated a hazard ratio of 1.526 for load and .0636 for bearings.

In this case, however, we estimated the baseline hazard function as well, assuming it is Weibull. The estimates are the full maximum-likelihood estimates. The shape parameter is fitted as $\ln p$, but streg then reports p and $1/p = \sigma$ so that you can think about the parameter however you wish.

We find that p is greater than 1, which means that the hazard of failure increases with time and, in this case, increases dramatically. After 100 hours, the bearings are over 1 million times more likely to fail per second than after 10 hours (or, to be precise, $(100/10)^{7.78-1}$). Based on our knowledge of generators, we would expect this; it is the accumulation of heat due to friction that causes bearings to expand and seize.

◁

❏ Technical Note

Regression results are often presented in a metric other than the natural regression coefficients, i.e., as hazard ratios, relative risk ratios, odds ratios, etc. When this occurs, the standard errors are calculated using the delta method.

However, the Z test and p-values given are calculated from the natural regression coefficients and their standard errors. The reason for this is that, although a test based on say a hazard ratio and its standard error would be asymptotically equivalent to that based on a regression coefficient, in real samples a hazard ratio will tend to have a more skewed distribution since it is an exponentiated regression coefficient. Also, it is more natural to think of these tests as testing whether a regression coefficient is nonzero, rather than testing whether a transformed regression coefficient is unequal to some nonzero value (one in the case of a hazard ratio).

Finally, the confidence intervals given are obtained by transforming the end points of the corresponding confidence interval for the untransformed regression coefficient. This has the advantage of ensuring that, say, strictly positive quantities such as hazard ratios have confidence intervals which do not overlap zero.

❏

▷ Example

The previous estimation took place in the proportional-hazards metric and exponentiated coefficients—hazard ratios—were reported. If we want to see the unexponentiated coefficients, we could redisplay results and specify the nohr option:

```
. streg, nohr
Weibull regression -- log relative-hazard form
No. of subjects =          12                    Number of obs   =          12
No. of failures =          12
Time at risk    =         896
                                                 LR chi2(2)      =       30.27
Log likelihood  =     5.6934189                  Prob > chi2     =      0.0000
```

_t	Coef.	Std. Err.	z	P>|z|	[95% Conf. Interval]	
load	.4695753	.1177884	3.99	0.000	.2387143	.7004363
bearings	-1.667069	.6949745	-2.40	0.016	-3.029194	-.3049443
_cons	-45.13191	10.60663	-4.26	0.000	-65.92053	-24.34329
/ln_p	2.051552	.2317074	8.85	0.000	1.597414	2.505691
p	7.779969	1.802677			4.940241	12.25202
1/p	.1285352	.0297826			.0816192	.2024193

◁

▷ Example

We could just as well have fitted this model in the accelerated failure-time metric:

```
. streg load bearings, d(weibull) time nolog
        failure _d:  1 (meaning all fail)
   analysis time _t:  failtime

Weibull regression -- accelerated failure-time form
No. of subjects =          12                    Number of obs   =          12
No. of failures =          12
Time at risk    =         896
                                                 LR chi2(2)      =       30.27
Log likelihood  =     5.6934189                  Prob > chi2     =      0.0000
```

_t	Coef.	Std. Err.	z	P>|z|	[95% Conf. Interval]	
load	-.060357	.0062214	-9.70	0.000	-.0725507	-.0481632
bearings	.2142771	.0746451	2.87	0.004	.0679753	.3605789
_cons	5.80104	.1752301	33.11	0.000	5.457595	6.144485
/ln_p	2.051552	.2317074	8.85	0.000	1.597414	2.505691
p	7.779969	1.802677			4.940241	12.25202
1/p	.1285352	.0297826			.0816192	.2024193

This is the same model as previously fitted, but presented in a different metric. Calling the previous coefficients b, these coefficients are $-\sigma b = -b/p$. For instance, in the previous example, the coefficient on load was reported as .4695753, and note that $-.4695753/7.779969 = -.06035696$. Note also that d() is a convenient shorthand for distribution().

◁

▷ Example

streg may also be applied to more complicated data. Below we have multiple records per subject on a failure that can occur repeatedly:

```
. use http://www.stata-press.com/data/r8/mfail
(streg example data)
. stdes
```

Category	total	mean	min	median	max
no. of subjects	926				
no. of records	1734	1.87257	1	2	4
(first) entry time		0	0	0	0
(final) exit time		470.6857	1	477	960
subjects with gap	6				
time on gap if gap	411	68.5	16	57.5	133
time at risk	435444	470.2419	1	477	960
failures	808	.8725702	0	1	3

In this dataset, subjects have up to 4 records—most have 2—and have up to 3 failures—most have 1—and, although you cannot tell from the above output, the data have time-varying covariates as well. There are even 6 subjects with gaps in their histories, meaning that, for a while, they went unobserved. Although we could estimate in the accelerated failure-time metric, it is easier to interpret results in the proportional-hazards metric (or the log relative-hazard metric, as it is also known):

```
. streg x1 x2, d(weibull) robust

Fitting constant-only model:

Iteration 0:   log pseudo-likelihood = -1398.2504
Iteration 1:   log pseudo-likelihood = -1382.8224
Iteration 2:   log pseudo-likelihood = -1382.7457
Iteration 3:   log pseudo-likelihood = -1382.7457

Fitting full model:

Iteration 0:   log pseudo-likelihood = -1382.7457
Iteration 1:   log pseudo-likelihood = -1328.4186
Iteration 2:   log pseudo-likelihood = -1326.4483
Iteration 3:   log pseudo-likelihood = -1326.4449
Iteration 4:   log pseudo-likelihood = -1326.4449

Weibull regression -- log relative-hazard form
```

No. of subjects	=	926	Number of obs =	1734
No. of failures	=	808		
Time at risk	=	435444		
			Wald chi2(2) =	154.45
Log pseudo-likelihood =		-1326.4449	Prob > chi2 =	0.0000

(standard errors adjusted for clustering on id)

_t	Haz. Ratio	Robust Std. Err.	z	P>|z|	[95% Conf. Interval]	
x1	2.240069	.1812848	9.97	0.000	1.911504	2.625111
x2	.3206515	.0504626	-7.23	0.000	.2355458	.436507
/ln_p	.1771265	.0310111	5.71	0.000	.1163458	.2379071
p	1.193782	.0370205			1.123384	1.268591
1/p	.8376738	.0259772			.7882759	.8901674

A one-unit change in x1 approximately doubles the hazard of failure, whereas a one-unit change in x2 cuts the hazard to one-third. We also see that these data are close to being exponentially distributed; p is nearly 1.

Above we mentioned that interpreting results in the PH metric is easier. That is not to say that regression coefficients are difficult to interpret in the AFT metric. A positive coefficient means that time is decelerated by a unit-increase in the covariate in question. This may seem awkward, but think of this instead as a unit increase in the covariate causing a delay in failure, and thus will *increase* the expected time until failure.

The difficulty that arises with the AFT metric is merely that it places an emphasis on log(time-to-failure), rather than risk (hazard) of failure. With this emphasis usually comes a desire to predict the time-to-failure, and therein lies the difficulty with complex survival data. Predicting the log(time-to-failure) using `predict` assumes the subject is at risk from time 0 until failure, and that the subject's covariate pattern is fixed over this period. With these data, such assumptions produce predictions having little to do with the test subjects, who exhibit not only time-varying covariates but multiple failures as well.

The difficulty with predicting time-to-failure with complex survival data exists regardless of the metric under which estimation took place. Instead, it is merely assumed that those who estimate in the proportional-hazards metric are used to dealing with results from Cox regression, of which predicted time-to-failure is typically not the focus.

◁

▷ Example

The multiple-failure data above are close enough to exponentially distributed that we will re-estimate using exponential regression:

```
. streg x1 x2, d(exp) robust

Iteration 0:   log pseudo-likelihood = -1398.2504
Iteration 1:   log pseudo-likelihood = -1343.6083
Iteration 2:   log pseudo-likelihood = -1341.5932
Iteration 3:   log pseudo-likelihood = -1341.5893
Iteration 4:   log pseudo-likelihood = -1341.5893

Exponential regression -- log relative-hazard form

No. of subjects     =          926              Number of obs    =       1734
No. of failures     =          808
Time at risk        =       435444
                                                 Wald chi2(2)     =     166.92
Log pseudo-likelihood =   -1341.5893             Prob > chi2      =     0.0000

                    (standard errors adjusted for clustering on id)
```

_t	Haz. Ratio	Robust Std. Err.	z	P>|z|	[95% Conf. Interval]	
x1	2.19065	.1684399	10.20	0.000	1.884186	2.54696
x2	.3037259	.0462489	-7.83	0.000	.2253552	.4093511

◁

❏ Technical Note

Note that for our "complex" survival data, we specified robust when fitting the Weibull and exponential models. This was because these data were stset with an id() variable, and given the time-varying covariates and multiple failures, it is important not to assume that the observations within each subject are independent. When we specified robust, it was implicit that we were "clustering" on the groups defined by the id() variable.

There exist situations where one has multiple observations per subject, and the multiple observations exist merely as a result of the data-organization mechanism, and are not used to record gaps, time-varying covariates, or multiple failures. Such multiple observations could be collapsed into single-observation-per-subject data with no loss of information. In these cases, we refer to the splitting of the observations to form multiple observations per subject as *noninformative*. When the episode-splitting is noninformative, the model-based (nonrobust) standard errors produced will be the same as those produced when the data are collapsed into single records per subject. Thus, for these type of data the clustering of these multiple observations that results from specifying robust is not critical.

❏

▷ Example

A reasonable question to ask is "Given that we have several possible parametric models, how can we select one?" When parametric models are nested, the likelihood-ratio or Wald tests can be used to discriminate between them. This can certainly be done in the case of Weibull versus exponential, or gamma versus Weibull or lognormal. When models are not nested, however, these tests are inappropriate and the task of discriminating between models becomes more difficult. A common approach to this problem is to use the Akaike information criterion (AIC). Akaike (1974) proposed penalizing each log likelihood to reflect the number of parameters being estimated in a particular model and then comparing them. In our case, the AIC can be defined as

$$AIC = -2(\log \text{ likelihood}) + 2(c + p + 1)$$

where c is the number of model covariates and p is the number of model-specific ancillary parameters listed in Table 1. Although the best-fitting model is the one with the largest log likelihood, the preferred model is the one with the smallest AIC value.

Using the cancer.dta distributed with Stata, let's first fit a generalized gamma model and test the hypothesis that $\kappa = 0$ (test for the appropriateness of the lognormal), and then test the hypothesis that $\kappa = 1$ (test for the appropriateness of the Weibull).

```
. use http://www.stata-press.com/data/r8/cancer, clear
(Patient Survival in Drug Trial)
. stset studytime, failure(died)
  (output omitted )
. replace drug = drug==2 | drug==3  // 0, placebo : 1, non-placebo
(48 real changes made)
```

(*Continued on next page*)

```
. streg drug age, d(gamma) nolog

        failure _d:  died
   analysis time _t:  studytime

Gamma regression -- accelerated failure-time form

No. of subjects =          48              Number of obs    =          48
No. of failures =          31
Time at risk    =         744
                                           LR chi2(2)       =       36.07
Log likelihood  =   -42.452006             Prob > chi2      =      0.0000
```

_t	Coef.	Std. Err.	z	P>|z|	[95% Conf. Interval]	
drug	1.394658	.2557198	5.45	0.000	.893456	1.895859
age	-.0780416	.0227978	-3.42	0.001	-.1227245	-.0333587
_cons	6.456091	1.238457	5.21	0.000	4.02876	8.883421
/ln_sig	-.3793632	.183707	-2.07	0.039	-.7394222	-.0193041
/kappa	.4669252	.5419478	0.86	0.389	-.595273	1.529123
sigma	.684297	.1257101			.4773897	.980881

The Wald test of the hypothesis that $\kappa = 0$ (test for the appropriateness of the lognormal) is performed and reported on the output above, $p = 0.389$, suggesting that lognormal might be an adequate model for these data.

The Wald test for $\kappa = 1$ is

```
. test [kappa]_cons = 1
 ( 1)  [kappa]_cons = 1
          chi2(  1) =     0.97
        Prob > chi2 =   0.3253
```

providing some support against rejecting the Weibull model.

We now fit the exponential, Weibull, log-logistic, and lognormal models separately. To be able to directly compare coefficients, we will ask Stata to report the exponential and Weibull models in accelerated failure-time form by specifying the `time` option. The output from fitting these models and the results from the generalized gamma model are summarized in Table 2.

Table 2. Summary of results obtained from `streg` using `cancer.dta` with drug as indicator

	Exponential	Weibull	Lognormal	Log-logistic	Generalized gamma
Age	−.088672	−.071432	−.083400	−.080329	−.078042
Drug	1.682625	1.305563	1.445838	1.420237	1.394658
Constant	7.146218	6.289679	6.580887	6.446711	6.456091
Ancillary		1.682751	0.751136	0.429276	0.684297
Kappa					0.466923
Log-likelihood	−48.397094	−42.931335	−42.800864	−43.216980	−42.452006
AIC	102.794188	93.862670	93.601728	94.433960	94.904012

We can see that the largest log likelihood was obtained for the generalized gamma model; however, the lognormal model is preferred by the AIC.

◁

Parameterization of ancillary parameters

By default, all ancillary parameters are estimated as constant quantities. For example, the ancillary parameter, p, of the Weibull distribution is assumed to be a constant not dependent on any covariates. streg's ancillary() and anc2() options allow for complete parameterization of parametric survival models. By specifying, for example,

```
. streg x1 x2, d(weibull) ancillary(x2 z1 z2)
```

both λ and the ancillary parameter, p, are parameterized in terms of covariates.

In most cases, ancillary parameters are restricted to be strictly positive, in which case it is the logarithm of the ancillary parameter that gets modeled using a linear predictor, which can assume any value on the real line.

▷ Example

Consider a dataset where we model the time until hip fracture as Weibull, for patients based on age, sex, and whether the patient wears a hip-protective device (variable protect). We believe the hazard is scaled according to sex and the presence of the device, but believe the hazards for both sexes to be of different *shapes*.

```
. use http://www.stata-press.com/data/r8/hip3, clear
(hip fracture study)

. streg protect age, d(weibull) ancillary(male) nolog

         failure _d:  fracture
   analysis time _t:  time1
                 id:  id
```

Weibull regression -- log relative-hazard form

No. of subjects =	148	Number of obs = 206
No. of failures =	37	
Time at risk =	1703	
		LR chi2(2) = 39.80
Log likelihood =	-69.323532	Prob > chi2 = 0.0000

_t	Coef.	Std. Err.	z	P>|z|	[95% Conf. Interval]
_t					
protect	-2.130058	.3567005	-5.97	0.000	-2.829178 -1.430938
age	.0939131	.0341107	2.75	0.006	.0270573 .1607689
_cons	-10.17575	2.551821	-3.99	0.000	-15.17722 -5.174269
ln_p					
male	-.4887189	.185608	-2.63	0.008	-.8525039 -.1249339
_cons	.4540139	.1157915	3.92	0.000	.2270667 .6809611

From our estimation results, we see that $\widehat{\ln(p)} = 0.454$ for females and $\widehat{\ln(p)} = 0.454 - 0.489 = -0.035$ for males. Thus, $\widehat{p} = 1.57$ for females and $\widehat{p} = 0.97$ for males. Combining this with the main equation in the model, the estimated hazards are then

$$\widehat{h}(t_j|\mathbf{x}_j) = \begin{cases} \exp\left(-10.18 - 2.13\text{protect}_j + 0.09\text{age}_j\right) 1.57 t_j^{0.57} & \text{if female} \\ \exp\left(-10.18 - 2.13\text{protect}_j + 0.09\text{age}_j\right) 0.97 t_j^{-0.03} & \text{if male} \end{cases}$$

and if we believe this model, we would say that the hazard for males given age and protect is almost constant over time.

Contrast this with what you obtain if you type,

```
. streg protect age, d(weibull) if male
. streg protect age, d(weibull) if !male
```

which is completely general, since not only will the shape parameter p differ over both sexes, but the regression coefficients as well.

◁

The anc2() option is for use only with the gamma regression model, as it contains two ancillary parameters—anc2() is used to parametrize κ.

Stratified estimation

When we type

```
. streg xvars, d(distname) strata(varname)
```

we are asking that a completely stratified model be fitted. By "completely stratified", we mean that both the model's intercept and any ancillary parameters are allowed to vary for each level of the strata variable. That is, we are constraining the coefficients on the covariates to be the same across strata, but allowing the intercept and ancillary parameters to vary.

▷ Example

We demonstrate by fitting a stratified log-normal model to the cancer data, with variable drug left in its original state: drug==1 refers to the placebo, and drug==2 and drug==3 correspond to two alternative treatments.

```
. use http://www.stata-press.com/data/r8/cancer, clear
. streg age, d(weibull) strata(drug) nolog
       failure _d:  died
   analysis time _t:  studytime
Weibull regression -- log relative-hazard form
No. of subjects =          48                    Number of obs   =          48
No. of failures =          31
Time at risk    =         744
                                                 LR chi2(3)      =       16.58
Log likelihood  =   -41.113074                   Prob > chi2     =      0.0009
```

_t	Coef.	Std. Err.	z	P>|z|	[95% Conf. Interval]	
_t						
age	.1212332	.0367538	3.30	0.001	.049197	.1932694
_Sdrug_2	-4.561178	2.339448	-1.95	0.051	-9.146411	.0240556
_Sdrug_3	-3.715737	2.595986	-1.43	0.152	-8.803776	1.372302
_cons	-10.36921	2.341022	-4.43	0.000	-14.95753	-5.780896
ln_p						
_Sdrug_2	.4872195	.332019	1.47	0.142	-.1635257	1.137965
_Sdrug_3	.2194213	.4079989	0.54	0.591	-.5802418	1.019084
_cons	.4541282	.1715663	2.65	0.008	.1178645	.7903919

◁

Completely stratified models are fit by first generating stratum specific indicator variables (dummy variables), and then adding these as independent variables in the model and as covariates in the ancillary parameter. As such, the strata() option is merely a shorthand method for generating the indicator variables from the drug categories, and then placing these indicators in *both* the main equation and the ancillary equation(s).

We associate the term *stratification* with this process by noting that the intercept term of the main equation is a component of the baseline hazard (or baseline survival) function. By allowing this intercept, as well as the ancillary shape parameter, to vary with respect to the strata, we are allowing the baseline functions to completely vary over the strata, analogous to a stratified Cox model.

▷ Example

It is possible to produce a less stratified model by combining xi (see [R] **xi**) with the ancillary() option.

```
. xi: streg age, d(weibull) ancillary(i.drug) nolog
i.drug            _Idrug_1-3          (naturally coded; _Idrug_1 omitted)

         failure _d:  died
   analysis time _t:  studytime

Weibull regression -- log relative-hazard form

No. of subjects =          48                  Number of obs    =          48
No. of failures =          31
Time at risk    =         744
                                               LR chi2(1)       =        9.61
Log likelihood  =   -44.596379                 Prob > chi2      =      0.0019
```

_t	Coef.	Std. Err.	z	P>|z|	[95% Conf. Interval]	
_t						
age	.1126419	.0362786	3.10	0.002	.0415373	.1837466
_cons	-10.95772	2.308489	-4.75	0.000	-15.48227	-6.433162
ln_p						
_Idrug_2	-.3279568	.11238	-2.92	0.004	-.5482176	-.107696
_Idrug_3	-.4775351	.1091141	-4.38	0.000	-.6913948	-.2636755
_cons	.6684086	.1327284	5.04	0.000	.4082657	.9285514

By doing this, we are restricting not only the coefficients on the covariates to be the same across "strata", but also the intercept, while allowing only the ancillary parameter to differ.

◁

By using ancillary() or strata(), one may thus consider a wide variety of models, depending on what one believes about the effect of the covariate(s) in question. For example, when fitting a Weibull PH model to the cancer data, we may choose from many models, depending on what we want to assume is the effect of the categorical variable drug. For all models considered below, it is assumed implicitly that the effect of age is proportional on the hazard function.

1. drug has no effect:

    . streg age, d(weibull)

2. Effect of drug is proportional on the hazard (scale), and the effect of age is the same for each level of drug:

    . xi: streg age i.drug, d(weibull)

3. Effect of drug is on the shape of the hazard, and the effect of age is the same for each level of drug:

    ```
    . xi: streg age, d(weibull) ancillary(i.drug)
    ```

4. Effect of drug is on both the scale and shape of the hazard, and the effect of age is the same for each level of drug:

    ```
    . streg age, d(weibull) strata(drug)
    ```

5. Effect of drug is on both the scale and shape of the hazard, and the effect of age is different for each level of drug:

    ```
    . xi: streg i.drug*age, d(weibull) strata(drug)
    ```

These models may be compared using Wald or likelihood-ratio tests when the models in question are nested (such as 3 nested within 4), or by using the AIC criterion for non-nested models.

Finally, we note that everything we said regarding the modeling of ancillary parameters and stratification applies to AFT models as well, for which interpretations may be stated in terms of the baseline survival function, i.e. the unaccelerated probability of survival past time t.

❏ Technical Note

When fitting PH models as we have done in the previous three examples, streg will, by default, display the exponentiated regression coefficients, labeled as hazard ratios. You will notice, however, that in our previous examples on using ancillary() and strata(), the regression outputs displayed the untransformed coefficients instead. The reason for this change in behavior has to do with the modeling of the ancillary parameter. When one uses one or more covariates from the main equation to model an ancillary parameter, hazard ratios (and time ratios for AFT models) lose their interpretation. streg, as a precaution, disallows the display of hazard/time ratios when either ancillary(), anc2(), or strata() is specified.

Keep this in mind when comparing results across various model specifications. For example, when comparing a stratified Weibull PH model to a standard Weibull PH model, be sure that the latter is displayed using the nohr option.

❏

(Unshared) frailty models

A frailty model is a survival model with unobservable heterogeneity, or *frailty*. At the observation level, frailty is introduced as an unobservable multiplicative effect, α, on the hazard function such that

$$h(t|\alpha) = \alpha h(t)$$

where $h(t)$ is a nonfrailty hazard function, say, the hazard function of any of the six parametric models supported by streg described earlier in this entry. The frailty, α, is a random positive quantity and for purposes of model identifiability is assumed to have mean one and variance θ.

Exploiting the relationship between the cumulative hazard function and survival function yields the expression for the survival function, given the frailty

$$S(t|\alpha) = \exp\left\{-\int_0^t h(u|\alpha)du\right\} = \exp\left\{-\alpha \int_0^t \frac{f(u)}{S(u)}du\right\} = \{S(t)\}^\alpha$$

where $S(t)$ is the survival function that corresponds to $h(t)$.

Since α is unobservable, it needs to be integrated out of $S(t|\alpha)$ in order to obtain the unconditional survival function. Let $g(\alpha)$ be the probability density function of α, in which case an estimable form of our frailty model is achieved as

$$S_\theta(t) = \int_0^\infty S(t|\alpha)g(\alpha)d\alpha = \int_0^\infty \{S(t)\}^\alpha g(\alpha)d\alpha$$

.

Given the unconditional survival function, one obtains the unconditional hazard and density in the usual way, namely

$$f_\theta(t) = -\frac{d}{dt}S_\theta(t) \qquad h_\theta(t) = \frac{f_\theta(t)}{S_\theta(t)}$$

Hence, an unshared frailty model is merely a typical parametric survival model, with the additional estimation of an overdispersion parameter, θ. In a standard survival regression, the likelihood calculations are based on $S(t)$, $h(t)$, and $f(t)$. In an unshared frailty model, the likelihood is based analogously on $S_\theta(t)$, $h_\theta(t)$, and $f_\theta(t)$.

At this stage, the only missing piece is the choice of frailty distribution $g(\alpha)$. In theory, any continuous distribution supported on the positive numbers that has expectation one and finite variance θ is allowed here. For purposes of mathematical tractability, however, we limit the choice to one of either the Gamma$(1/\theta, \theta)$ distribution or the Inverse-Gaussian distribution with parameters one and $1/\theta$, denoted as IG$(1, 1/\theta)$. The Gamma(a, b) distribution has probability density function

$$g(x) = \frac{x^{a-1}e^{-x/b}}{\Gamma(a)b^a}$$

and the IG(a, b) distribution has density

$$g(x) = \left(\frac{b}{2\pi x^3}\right)^{1/2} \exp\left\{-\frac{b}{2a}\left(\frac{x}{a} - 2 + \frac{a}{x}\right)\right\}$$

Therefore, performing the integrations described above will show that specifying `frailty(gamma)` will result in the frailty survival model (in terms of the nonfrailty survival function $S(t)$)

$$S_\theta(t) = [1 - \theta \ln\{S(t)\}]^{-1/\theta}$$

and specifying `frailty(invgaussian)` will give

$$S_\theta(t) = \exp\left\{\frac{1}{\theta}\left(1 - [1 - 2\theta \ln\{S(t)\}]^{1/2}\right)\right\}$$

Note here that regardless of the choice of frailty distribution, $\lim_{\theta \to 0} S_\theta(t) = S(t)$, and thus the frailty model reduces to $S(t)$ when there is no heterogeneity present.

When using frailty models, it is important to note the distinction between the hazard faced by the individual (subject), $\alpha h(t)$, and the "average" hazard for the population, $h_\theta(t)$. Similarly, an individual will have probability of survival past time t equal to $\{S(t)\}^\alpha$, whereas $S_\theta(t)$ will measure the proportion of the population surviving past time t. You specify $S(t)$ as before with distribution(*distname*), and the list of possible parametric forms for $S(t)$ is given in Table 1. Thus, when you specify distribution() you are specifying a model for an individual with frailty equal to one. Specifying frailty(*distname*) determines which of the two above forms for $S_\theta(t)$ is used.

The output of the estimation remains unchanged from the nonfrailty version, except for the additional estimation of θ and a likelihood-ratio test of $H_0 : \theta = 0$. For more information on frailty models, Hougaard (1986) offers an excellent introduction. For a Stata-specific overview, see Gutierrez (2002).

▷ Example

Consider as an example a survival analysis of data on women with breast cancer. Our hypothetical dataset consists of analysis times on 80 women with covariates age, smoking, and dietfat, which measures the average weekly calories from fat ($\times 10^3$) in the patient's diet over the course of the study.

```
. use http://www.stata-press.com/data/r8/bc, clear
. list in 1/12
```

	age	smoking	dietfat	t	dead
1.	30	1	4.919	14.2	0
2.	50	0	4.437	8.21	1
3.	47	0	5.85	5.64	1
4.	49	1	5.149	4.42	1
5.	52	1	4.363	2.81	1
6.	29	0	6.153	35	0
7.	49	1	3.82	4.57	1
8.	27	1	5.294	35	0
9.	47	0	6.102	3.74	1
10.	59	0	4.446	2.29	1
11.	35	0	6.203	15.3	0
12.	26	0	4.515	35	0

The data are well fit by a Weibull model for the distribution of survival time conditional on age, smoking, and dietary fat. By omitting the variable dietfat from the model, we hope to introduce unobserved heterogeneity.

```
. stset t, fail(dead)
  (output omitted )
. streg age smoking, d(weibull) frailty(gamma)

        failure _d:  dead
   analysis time _t:  t

Fitting comparison weibull model:

Fitting constant-only model:

Iteration 0:   log likelihood = -137.15363
Iteration 1:   log likelihood =  -136.3927
Iteration 2:   log likelihood = -136.01557
Iteration 3:   log likelihood = -136.01202
Iteration 4:   log likelihood = -136.01201

Fitting full model:

Iteration 0:   log likelihood = -85.933969
Iteration 1:   log likelihood =  -73.61173
Iteration 2:   log likelihood = -68.999447
Iteration 3:   log likelihood = -68.340858
Iteration 4:   log likelihood = -68.136187
Iteration 5:   log likelihood = -68.135804
Iteration 6:   log likelihood = -68.135804
```

```
Weibull regression -- log relative-hazard form
                 Gamma frailty
No. of subjects =          80              Number of obs    =          80
No. of failures =          58
Time at risk    =     1257.07
                                           LR chi2(2)       =      135.75
Log likelihood  =   -68.135804             Prob > chi2      =      0.0000
```

_t	Haz. Ratio	Std. Err.	z	P>|z|	[95% Conf. Interval]	
age	1.475948	.1379987	4.16	0.000	1.228811	1.772788
smoking	2.788548	1.457031	1.96	0.050	1.00143	7.764894
/ln_p	1.087761	.222261	4.89	0.000	.6521376	1.523385
/ln_the	.3307466	.5250758	0.63	0.529	-.698383	1.359876
p	2.967622	.6595867			1.91964	4.587727
1/p	.3369701	.0748953			.2179729	.520931
theta	1.392007	.7309092			.4973889	3.895711

```
Likelihood-ratio test of theta=0: chibar2(01) =    22.57 Prob>=chibar2 = 0.000
```

Alternatively, we could use an Inverse-Gaussian distribution to model the heterogeneity.

```
. streg age smoking, dist(weibull) frailty(invgauss) nolog
        failure _d:  dead
  analysis time _t:  t
Weibull regression -- log relative-hazard form
              Inverse-Gaussian frailty
No. of subjects =          80              Number of obs    =          80
No. of failures =          58
Time at risk    =     1257.07
                                           LR chi2(2)       =      125.44
Log likelihood  =   -73.838578             Prob > chi2      =      0.0000
```

_t	Haz. Ratio	Std. Err.	z	P>|z|	[95% Conf. Interval]	
age	1.284133	.0463256	6.93	0.000	1.196473	1.378217
smoking	2.905409	1.252785	2.47	0.013	1.247892	6.764528
/ln_p	.7173904	.1434382	5.00	0.000	.4362567	.9985241
/ln_the	.2374778	.8568064	0.28	0.782	-1.441832	1.916788
p	2.049079	.2939162			1.546906	2.714273
1/p	.4880241	.0700013			.3684228	.6464518
theta	1.268047	1.086471			.2364941	6.799082

```
Likelihood-ratio test of theta=0: chibar2(01) =    11.16 Prob>=chibar2 = 0.000
```

The results are similar with respect to the choice of frailty distribution, with the gamma frailty model producing a slightly higher likelihood. Both models show a statistically significant level of unobservable heterogeneity since the p-value for the LR test of $H_0 : \theta = 0$ is virtually zero in both cases.

❑ Technical Note

With gamma- or inverse-Gaussian-distributed frailty, hazard ratios decay over time in favor of the *frailty effect*, and thus the displayed "Haz. Ratio" in the above output is actually the hazard ratio only for $t = 0$. The degree of decay depends on θ. Should the estimated θ be close to zero, the hazard ratios do regain their usual interpretation. The rate of decay and the limiting hazard ratio differ between the gamma and inverse-Gaussian models; see Gutierrez (2002) for details.

For this reason, many researchers prefer fitting frailty models in the AFT metric, since the interpretation of regression coefficients is unchanged by the frailty—the factors in question serve to either accelerate or decelerate the survival experience. The only difference is that, with frailty models, the unconditional probability of survival is described by $S_\theta(t)$, rather than $S(t)$.

❑

❑ Technical Note

The likelihood-ratio test of $\theta = 0$ is a boundary test, and thus requires careful consideration concerning the calculation of its p-value. In particular, the null distribution of the likelihood-ratio test statistic is not the usual χ_1^2, but rather is a 50:50 mixture of a χ_0^2 (point mass at zero) and a χ_1^2, denoted as $\bar{\chi}_{01}^2$. See Gutierrez et al. (2001) for more details.

❑

To verify that the significant heterogeneity is caused by the omission of dietfat, we now refit the Weibull/Inverse-Gaussian frailty model with dietfat included.

```
. streg age smoking dietfat, d(weibull) frailty(invgauss) nolog

      failure _d:  dead
   analysis time _t:  t

Weibull regression -- log relative-hazard form
                Inverse-Gaussian frailty

No. of subjects =         80              Number of obs    =         80
No. of failures =         58
Time at risk    =    1257.07
                                          LR chi2(3)       =     246.41
Log likelihood  =    -13.352142           Prob > chi2      =     0.0000
```

_t	Haz. Ratio	Std. Err.	z	P>|z|	[95% Conf. Interval]	
age	1.74928	.0985246	9.93	0.000	1.566452	1.953447
smoking	5.203553	1.704943	5.03	0.000	2.737814	9.889993
dietfat	9.229842	2.219332	9.24	0.000	5.761311	14.78656
/ln_p	1.431742	.0978847	14.63	0.000	1.239892	1.623593
/ln_the	-14.29549	2686.786	-0.01	0.996	-5280.299	5251.708
p	4.185987	.4097441			3.45524	5.071279
1/p	.2388923	.0233839			.1971889	.2894155
theta	6.19e-07	.0016626			0	.

```
Likelihood-ratio test of theta=0: chibar2(01) =     0.00 Prob>=chibar2 = 1.000
```

Note now that the estimate of the frailty variance component θ is near zero, and the p-value of the test of $H_0 : \theta = 0$ equals one, indicating negligible heterogeneity. A regular Weibull model could be fit to these data (with dietfat included), producing almost identical estimates of the hazard ratios and ancillary parameter p, and so such an analysis is omitted here.

In addition, hazard ratios now regain their original interpretation. Thus, for example, an increase in weekly calories from fat of 1,000 would increase the risk of death by over nine-fold.

◁

Shared frailty models

A generalization of the frailty models considered in the previous section is the *shared frailty* model, where the frailty is instead assumed to be group-specific, analogous to a panel-data regression model. For observation j from the ith group, the hazard is

$$h_{ij}(t|\alpha_i) = \alpha_i h_{ij}(t)$$

for $i = 1, \ldots, n$ and $j = 1, \ldots, n_i$, where by $h_{ij}(t)$ we mean $h(t|\mathbf{x}_{ij})$, the individual hazard given covariates $\mathbf{x}_{ij}$.

Shared frailty models are appropriate when one wishes to model the frailties as being specific to groups of subjects, such as subjects within families. In this case, a shared frailty model may be used to model the degree of correlation within groups, i.e. the subjects within a group are correlated because they share the same common frailty.

▷ Example

Consider the data from a study of 38 kidney dialysis patients, as described in McGilchrist and Aisbett (1991). The study is concerned with the prevalence of infection at the catheter insertion point. Two recurrence times (in days) are measured for each patient, and each recorded time is the time from initial insertion (onset of risk) to infection or censoring.

```
. use http://www.stata-press.com/data/r8/catheter, clear
(Kidney data, McGilchrist and Aisbett, Biometrics, 1991)

. list in 1/10
```

	patient	time	infect	age	female
1.	1	16	1	28	0
2.	1	8	1	28	0
3.	2	13	0	48	1
4.	2	23	1	48	1
5.	3	22	1	32	0
6.	3	28	1	32	0
7.	4	318	1	31.5	1
8.	4	447	1	31.5	1
9.	5	30	1	10	0
10.	5	12	1	10	0

Each patient (`patient`) has two recurrence times (`time`) recorded, with each catheter insertion resulting in either infection (`infect==1`) or right-censoring (`infect==0`). Among the covariates measured are `age` and sex (`female==1` if female, `female==0` if male).

One subtlety to note concerns the use of the generic term, "subjects". In this example, the subjects are taken to be the individual catheter insertions, and not the patients themselves. This is a function of how the data were recorded—the onset of risk occurs at catheter insertion (of which there are two for each patient), and not (say) at the time of admission of the patient into the study. Thus, we have two subjects (insertions) within each group (patient).

It is reasonable to assume independence of patients, but unreasonable to assume that recurrence times within each patient are independent. One solution would be to fit a standard survival model, adjusting the standard errors of the parameter estimates to account for the possible correlation by specifying `cluster(patient)`.

Alternatively, one can model the correlation by assuming that the correlation is the result of a latent patient-level effect, or frailty. That is, rather than fitting a standard model and specifying cluster(patient), fit a frailty model and specify shared(patient). Assuming that the time-to-infection, given age and female, follows a Weibull distribution, and inverse-Gaussian distributed frailties, we get

```
. stset time, fail(infect)
 (output omitted )
. streg age female, d(weibull) frailty(invgauss) shared(patient) nolog

        failure _d:  infect
   analysis timo _t:  time

Weibull regression --
        log-relative hazard form              Number of obs      =         76
        Inverse-Gaussian shared frailty        Number of groups   =         38
Group variable: patient

No. of subjects =          76                  Obs per group: min =          2
No. of failures =          58                                  avg =          2
Time at risk    =        7424                                  max =          2

                                               LR chi2(2)         =       9.84
Log likelihood  =   -99.093527                 Prob > chi2        =     0.0073
```

_t	Haz. Ratio	Std. Err.	z	P>|z|	[95% Conf. Interval]	
age	1.006918	.013574	0.51	0.609	.9806623	1.033878
female	.2331376	.1046382	-3.24	0.001	.0967322	.5618928
/ln_p	.1900625	.1315342	1.44	0.148	-.0677398	.4478649
/ln_the	.0357272	.7745362	0.05	0.963	-1.482336	1.55379
p	1.209325	.1590676			.9345036	1.564967
1/p	.8269074	.1087666			.638991	1.070087
theta	1.036373	.8027085			.2271066	4.729362

```
Likelihood-ratio test of theta=0: chibar2(01) =     8.70 Prob>=chibar2 = 0.002
```

Contrast this with what you obtain by assuming a subject-level lognormal model:

(*Continued on next page*)

```
. streg age female, d(lnormal) frailty(invgauss) shared(patient) nolog

        failure _d:  infect
   analysis time _t:  time
```

```
Log-normal regression --
        accelerated failure-time form              Number of obs      =        76
        Inverse-Gaussian shared frailty            Number of groups   =        38
Group variable: patient

No. of subjects =          76                       Obs per group: min =         2
No. of failures =          58                                      avg =         2
Time at risk    =        7424                                      max =         2

                                                    LR chi2(2)         =     16.34
Log likelihood  =   -97.614583                      Prob > chi2        =    0.0003
```

_t	Coef.	Std. Err.	z	P>|z|	[95% Conf. Interval]	
age	-.0066762	.0099457	-0.67	0.502	-.0261694	.0128171
female	1.401719	.3334931	4.20	0.000	.7480844	2.055354
_cons	3.336709	.4972641	6.71	0.000	2.362089	4.311329
/ln_sig	.0625872	.1256185	0.50	0.618	-.1836205	.3087949
/ln_the	-1.606248	1.190775	-1.35	0.177	-3.940125	.7276282
sigma	1.064587	.1337318			.8322516	1.361783
theta	.2006389	.2389159			.0194458	2.070165

```
Likelihood-ratio test of theta=0: chibar2(01) =      1.53 Prob>=chibar2 = 0.108
```

The frailty effect is insignificant at the 10% level in the latter model, yet highly significant in the former. We thus have two possible stories to tell concerning these data: If we believe the first model, then we believe the individual hazard of infection continually rises over time (Weibull), but there is a significant frailty effect causing the population hazard to begin falling after some time. If we believe the second model, then we believe the individual hazard first rises and then declines (lognormal), meaning that if a given insertion does not become infected initially then the chances that it will become infected begin to decrease after a certain point. Since the frailty effect is insignificant, the population hazard mirrors the individual hazard in the second model.

As a result, both models view the population hazard as rising initially then falling past a certain point. The second version of our story corresponds to higher log-likelihood, yet perhaps not significantly given the limited data. Further investigation is required. One idea is to fit a more distribution-agnostic form of a frailty model, such as a piecewise exponential (Cleves, Gould, & Gutierrez, 2002, 277–280) or a Cox model with frailty; see [ST] **stcox**.

◁

Shared frailty models are also appropriate when the frailties are subject-specific, yet there exist multiple records per subject. In this case, you would share frailties across the same id() variable previously stset. When you have subject-specific frailties and uninformative episode splitting, it makes no difference whether you fit a shared or an unshared frailty model. The estimation results will be the same. Note, however, that there is a distinction between unshared and shared frailty models in the case of single-record-per-subject data with delayed entry; see Gutierrez (2002) for details.

stcurve

stcurve is used after streg to plot the fitted survival, hazard, and cumulative hazard functions. By default, stcurve computes the means of the covariates and evaluates the fitted model at each time in the data, censored or uncensored. The resulting plot is therefore the survival experience of a subject with a covariate pattern equal to the average covariate pattern in the study. It is possible to produce the plot at other values of the covariates using the at() option. You can also specify a time range using the range() option.

▷ Example

Using the cancer data with drug remapped to form an indicator of treatment, let's fit a log-logistic regression model and plot its survival curves. We can perform a log-logistic regression by issuing the following commands:

```
. use http://www.stata-press.com/data/r8/cancer
(Patient Survival in Drug Trial)
. replace drug = drug==2 | drug==3  // 0, placebo : 1, non-placebo
(48 real changes made)
. stset studytime, failure(died)
  (output omitted)
. streg age drug, d(llog) nolog

        failure _d:  died
   analysis time _t:  studytime

Log-logistic regression -- accelerated failure-time form

No. of subjects =          48                  Number of obs   =         48
No. of failures =          31
Time at risk    =         744
                                               LR chi2(2)      =      35.14
Log likelihood  =    -43.21698                 Prob > chi2     =     0.0000
```

_t	Coef.	Std. Err.	z	P>|z|	[95% Conf. Interval]	
age	-.0803289	.0221598	-3.62	0.000	-.1237614	-.0368964
drug	1.420237	.2502148	5.68	0.000	.9298251	1.910649
_cons	6.446711	1.231914	5.23	0.000	4.032204	8.861218
/ln_gam	-.8456552	.1479337	-5.72	0.000	-1.1356	-.5557105
gamma	.429276	.0635044			.3212293	.5736646

Now we wish to plot the survival and the hazard functions:

(*Continued on next page*)

```
. stcurve, survival ylabels(0 .5 1)
```

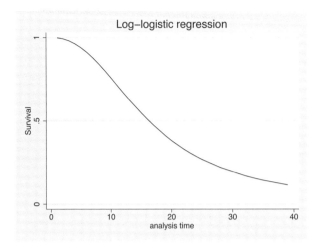

Figure 2. Log-logistic survival distribution at mean value of all covariates

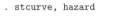

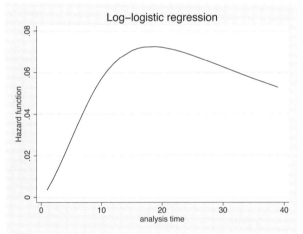

Figure 3. Log-logistic hazard distribution at mean value of all covariates

These plots show the fitted survival and hazard functions evaluated for a cancer patient of average age and receiving the average drug. Of course, the "average drug" has no meaning in this example since drug is an indicator variable. It makes more sense to plot the curves at a fixed value (level) of the drug. We can do this using the at option. For example, we may want to compare the average-age patient's survival curve under placebo (drug==0) and under treatment (drug==1).

We can plot both curves on the same graph:

```
. stcurve, surv at1(drug = 0) at2(drug = 1) ylabels(0 .5 1)
```

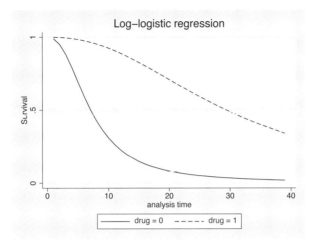

Figure 4. Log-logistic survival distribution at mean age for placebo

From the plot, we can see that based on the log-logistic model, the survival experience of an average-age patient receiving the placebo is worse than the survival experience of that same patient receiving treatment. We can also see the accelerated-failure-time feature of the log-logistic model. The survival function for treatment is a time-decelerated (stretched-out) version of the survival function for placebo.

◁

▷ Example

In our previous discussion of frailty models, much was made of the distinction between the individual hazard (or survival) function and the hazard (survival) function for the population. When significant frailty is present, the population hazard will tend to begin falling past a certain point, regardless of the shape of the individual hazard. This is due to the frailty effect—as time passes the more frail individuals will fail leaving a more homogeneous population comprised of only the most robust individuals.

The frailty effect may be demonstrated using stcurve to plot the estimated hazard (both individual and population) after fitting a frailty model. Use option alpha1 to specify the individual hazard ($\alpha = 1$) and option unconditional to specify the population hazard. Applying this to our Weibull/inverse-Gaussian shared frailty model on the kidney data,

(Continued on next page)

```
. webuse catheter, clear
(Kidney data, McGilchrist and Aisbett, Biometrics, 1991)
. stset time infect
  (output omitted)
. qui streg age female, d(weibull) frailty(invgauss) shared(patient)
. stcurve, hazard at(female = 1) alpha1
```

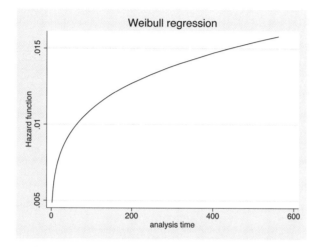

Figure 5. Individual hazard for females at mean age

Compare with

```
. stcurve, hazard at(female = 1) unconditional
```

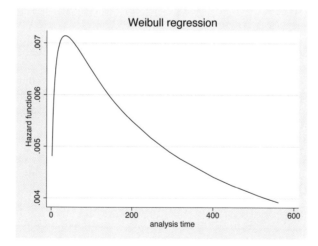

Figure 6. Population hazard for females at mean age

◁

predict

predict after streg is used to generate a variable containing predicted values or residuals.

For a more detailed discussion on residuals, read *Cox regression residuals* in the [ST] **stcox** entry. Many of the concepts and ideas presented there also apply to streg models.

Regardless of the metric used, predict can generate predicted median survival times and median log-survival times for all models, and predicted mean times and mean log-survival times where available. Predicted survival, hazard, and residuals are also available for all models. The predicted hazard ratio—emphasis on ratio—can be calculated only for models with a proportional hazards parameterization; i.e., the Weibull, exponential, and Gompertz models. It is, however, not necessary that the estimation take place in the log-hazard metric. It is possible to perform, for example, a Weibull regression specifying the time option and then to ask that hazard ratios be predicted.

After fitting a frailty model, you can use predict with the alpha1 option to generate predicted values based on $S(t)$, or with the unconditional option for predictions based on $S_\theta(t)$.

▷ Example

Let's return to the previous example of the emergency generator. Assume that we fit a proportional hazard Weibull model as before:

```
. use http://www.stata-press.com/data/r8/kva, clear
(Generator experiment)
. streg load bearings, d(weibull) nolog
        failure _d:  1 (meaning all fail)
   analysis time _t:  failtime
Weibull regression -- log relative-hazard form
No. of subjects =          12              Number of obs   =         12
No. of failures =          12
Time at risk    =         896
                                           LR chi2(2)      =      30.27
Log likelihood  =      5.6934189           Prob > chi2     =     0.0000
```

_t	Haz. Ratio	Std. Err.	z	P>|z|	[95% Conf. Interval]	
load	1.599315	.1883807	3.99	0.000	1.269616	2.014631
bearings	.1887995	.1312109	-2.40	0.016	.0483546	.7371644
/ln_p	2.051552	.2317074	8.85	0.000	1.597414	2.505691
p	7.779969	1.802677			4.940241	12.25202
1/p	.1285352	.0297826			.0816192	.2024193

Now we can predict both the (median) survival time and the log (median) survival time for each observation:

```
. predict time, time
(option median time assumed; predicted median time)
. predict lntime, lntime
(option log median time assumed; predicted median log time)
. format time lntime %9.4f
```

```
. list failtime load bearings time lntime
```

	failtime	load	bearings	time	lntime
1.	100	15	0	127.5586	4.8486
2.	140	15	1	158.0407	5.0629
3.	97	20	0	94.3292	4.5468
4.	122	20	1	116.8707	4.7611
5.	84	25	0	69.7562	4.2450
6.	100	25	1	86.4255	4.4593
7.	54	30	0	51.5845	3.9432
8.	52	30	1	63.9114	4.1575
9.	40	35	0	38.1466	3.6414
10.	55	35	1	47.2623	3.8557
11.	22	40	0	28.2093	3.3397
12.	30	40	1	34.9504	3.5539

◁

▷ Example

Using the cancer data (again with `drug` remapped into a drug treatment indicator), we can examine the various residuals that Stata produces. For a more detailed discussion on residuals, read *Cox regression residuals* in the [ST] **stcox** entry. Many of the concepts and ideas presented there also apply to `streg` models. For a more technical presentation of these residuals, see *Methods and Formulas*.

We will begin by requesting the generalized Cox–Snell residuals with the command `predict cs`, `csnell`. The `csnell` option causes `predict` to create a new variable, cs, containing the Cox–Snell residuals. If the model fits the data, then these residuals should have a standard exponential distribution with $\lambda = 1$. One way of verifying the fit is to calculate an empirical estimate of the cumulative hazard function, based, for example, on the Kaplan–Meier survival estimates or the Aalen–Nelson estimator, taking the Cox–Snell residuals as the time variable and the censoring variable as before, and plotting it against cs. If the model fits the data, then the plot should be a straight line with slope of 1.

To do this after fitting the model, we first `stset` the data, specifying cs as our new failure time variable and `died` as the failure indicator. We then use the `sts generate` command to generate the variable km containing the Kaplan–Meier survival estimates. Lastly, we generate a new variable H (cumulative hazard) and plot it against the cs. The commands are

```
. use http://www.stata-press.com/data/r8/cancer, clear
(Patient Survival in Drug Trial)

. replace drug = drug==2 | drug==3  // 0, placebo : 1, non-placebo
(48 real changes made)

. stset studytime, failure(died)
  (output omitted )
. qui streg age drug, d(exp)

. predict double cs, csnell

. qui stset cs, failure(died)

. qui sts generate km=s

. qui generate double H=-ln(km)

. line H cs cs, sort
```

We specified cs twice in the graph command so that a reference 45° line was plotted. We did this separately for each of four distributions. Results are plotted in Figure 7:

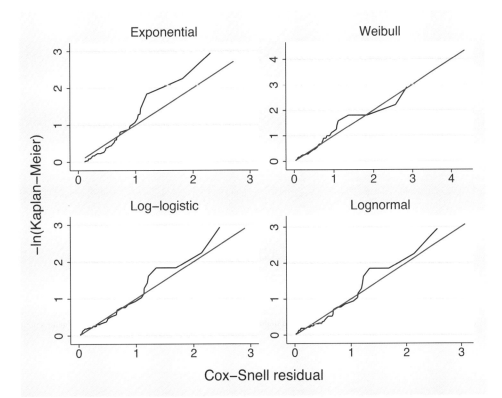

Figure 7. Cox–Snell residuals to evaluate model fit of 4 regression models

The plots indicate that the Weibull and lognormal models fit the data best and that the exponential model fits poorly. These results are consistent with our previous results based on Akaike's information criterion.

◁

▷ Example

Let's now look at the martingale-like and deviance residuals. We use the term "martingale-like" because although these residuals do not arise naturally from martingale theory in the case of parametric survival models as they do for the Cox proportional hazard model, they do share similar form. We can generate these residuals by using predict's mgale option. Martingale residuals take values between $-\infty$ and 1, and therefore are difficult to interpret. The deviance residuals are a rescaling of the martingale-like residuals so that they are symmetric about zero, and thus more like residuals obtained from linear regression. Plots of either set of residuals against survival time, ranked survival time, or observation number can be useful in identifying aberrant observations and in assessing model fit. Continuing with our modified cancer data, we plotted the deviance residual obtained after fitting a lognormal model:

```
. qui streg age drug, d(lnormal)
. predict dev, deviance
. scatter dev studytime, yline(0) m(o)
```

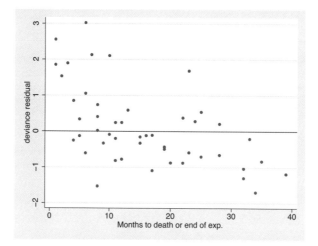

Figure 8. Deviance residuals to evaluate model fit of lognormal model

In this plot, the deviance residual is large for short survival times and then decreases with time. This pattern suggests that the lognormal model will underestimate the probability of failure for patients with short study times and will overestimate the probability of failure for patients with longer times.

◁

(Continued on next page)

Saved Results

streg saves in e():

Scalars

e(N)	number of observations		e(rc)	return code
e(N_sub)	number of subjects		e(chi2)	χ^2
e(N_fail)	number of failures		e(N_g)	number of groups
e(k)	number of parameters		e(g_max)	largest group size
e(k_eq)	number of equations		e(g_min)	smallest group size
e(k_dv)	number of dependent variables		e(g_avg)	average group size
e(rank)	rank of e(V)		e(p)	significance
e(rank0)	rank of e(V), constant-only model		e(ic)	number of iterations
e(risk)	total time at risk		e(aux_p)	ancillary parameter (weibull)
e(df_m)	model degrees of freedom		e(gamma)	ancillary parameter (gompertz,
e(ll)	log likelihood			loglogistic)
e(ll_0)	log likelihood, constant-only model		e(sigma)	ancillary parameter (gamma,
e(N_clust)	number of clusters			lnormal)
e(theta)	frailty parameter		e(kappa)	ancillary parameter (gamma)
e(chi2_c)	χ^2, comparison model		e(ll_c)	log likelihood, comparison model
e(p_c)	significance, comparison model			

Macros

e(cmd)	model or regression name		e(t0)	_t0
e(cmd2)	streg		e(frm2)	hazard or time
e(dead)	_d		e(vcetype)	covariance estimation method
e(depvar)	_t		e(user)	name of likelihood-evaluator
e(title)	title in estimation output			program
e(clustvar)	name of cluster variable		e(opt)	type of optimization
e(shared)	frailty grouping variable		e(chi2type)	Wald or LR; type of model χ^2 test
e(fr_title)	title in output identifying frailty		e(predict)	program used to implement
e(wtype)	weight type			predict
e(wexp)	weight expression		e(offset1)	offset for main equation
e(scorevars)	score variable names		e(crittype)	optimization criterion
e(stcurve)	stcurve			

Matrices

e(b)	coefficient vector		e(V)	variance–covariance matrix of the
e(ilog)	iteration log (up to 20 iterations)			estimators

Functions

e(sample)	marks estimation sample

Methods and Formulas

streg and stcurve are implemented as ado-files.

For an introduction to survival models, see Cleves, Gould, & Gutierrez (2002).

Consider for $j = 1, \ldots, n$ observations the trivariate response, (t_{0j}, t_j, d_j), representing a period of observation $(t_{0j}, t_j]$ ending in either failure ($d_j = 1$) or right-censoring ($d_j = 0$). This structure allows the analysis for a wide variety of models, and may be used to account for delayed entry, gaps, time-varying covariates, and multiple failures per subject. Regardless of the structure of the data, once stset, the data may be treated in a common manner by streg: the stset-created variable _t0 holds the t_{0j}, _t holds the t_j, and _d holds the d_j.

For a given survival function, $S(t)$, the density function is obtained as

$$f(t) = -\frac{d}{dt} S(t)$$

and the hazard function (the instantaneous rate of failure) is obtained as $h(t) = f(t)/S(t)$. The list of available forms for $S(t)$ is given in Table 1. For a set of covariates from the jth observation, $\mathbf{x}_j$, define $S_j(t) = S(t|\mathbf{x} = \mathbf{x}_j)$, and similarly define $h_j(t)$ and $f_j(t)$. For example, in a Weibull PH model, $S_j(t) = \exp\{-\exp(\mathbf{x}_j\boldsymbol{\beta})t^p\}$.

Parameter estimation

Estimation of $\boldsymbol{\beta}$ and of the ancillary parameters is done via maximum likelihood. A subject known to fail at time t_j contributes to the likelihood function the value of the density at time t_j conditional on the entry time t_{0j}, $f_j(t_j)/S_j(t_{0j})$, whereas a censored observation, only known to survive up to time t_j, contributes $S_j(t_j)/S_j(t_{0j})$, the probability of surviving beyond time t_j conditional on the entry time t_{0j}. The log-likelihood is thus given by

$$\ln L = \sum_{j=1}^{n} \{ d_j \ln f_j(t_j) + (1 - d_j) \ln S_j(t_j) - \ln S_j(t_{0j}) \}$$

Implicit in the above log-likelihood expression are the regression parameters $\boldsymbol{\beta}$ and the ancillary parameters, since both are components of the chosen $S_j(t)$ and its corresponding $f_j(t)$; see Table 1. streg reports maximum likelihood estimates of $\boldsymbol{\beta}$ and of the ancillary parameters (if any for the chosen model). The reported log-likelihood value is $\ln L_r = \ln L + T$, where $T = \sum \ln(t_j)$ is summed over uncensored observations. This adjustment is to make reported results match those of other statistical software. The intention of the adjustment is to remove the time units from $\ln L$. Whether or not the adjustment is made makes no difference to any test or result, since such tests and results depend on differences in log-likelihood functions and/or their second derivatives.

Specifying `ancillary()`, `anc2()`, or `strata()` will parameterize the ancillary parameter(s) using the linear predictor, $\mathbf{z}_j\boldsymbol{\alpha}_z$, where the covariates, $\mathbf{z}_j$, need not be distinct from $\mathbf{x}_j$. In this case, streg will report estimates of $\boldsymbol{\alpha}_z$ in addition to estimates of $\boldsymbol{\beta}$. The log-likelihood in this case is simply the log-likelihood given above, with $\mathbf{z}_j\boldsymbol{\alpha}_z$ substituted for the ancillary parameter. If the ancillary parameter is constrained to be strictly positive, then its logarithm is parameterized instead, i.e. one substitutes the linear predictor for the logarithm of the ancillary parameter in the above log-likelihood. The gamma model has two ancillary parameters, σ and κ; one parameterizes σ using `ancillary()` and κ using `anc2()`, and the linear predictors used for each may be distinct. Specifying `strata()` creates indicator variables for the strata, places these indicators in the main equation, and uses the indicators to parametrize any ancillary parameters that exist for the chosen model.

Unshared frailty models have a log-likelihood of the above form, with $S_\theta(t)$ and $f_\theta(t)$ substituted for $S(t)$ and $f(t)$, respectively. Equivalently, for gamma-distributed frailties,

$$\ln L = \sum_{j=1}^{n} \left[\theta^{-1} \ln \{ 1 - \theta \ln S_j(t_{0j}) \} - \left(\theta^{-1} + d_j \right) \ln \{ 1 - \theta \ln S_j(t_j) \} + d_j \ln h_j(t_j) \right]$$

and for inverse-Gaussian-distributed frailties,

$$\ln L = \sum_{j=1}^{n} \left[\theta^{-1} \left\{ 1 - 2\theta \ln S_j(t_{0j}) \right\}^{1/2} - \theta^{-1} \left\{ 1 - 2\theta \ln S_j(t_j) \right\}^{1/2} + \right.$$

$$\left. d_j \ln h_j(t_j) - \frac{1}{2} d_j \ln \left\{ 1 - 2\theta \ln S_j(t_j) \right\} \right]$$

In a shared frailty model, the frailty is common to a group of observations and thus to form an unconditional likelihood, the frailties need to be integrated out at the group level. The data are organized as $i = 1, \ldots, n$ groups with the ith group comprised of $j = 1, \ldots, n_i$ observations. The log-likelihood is the sum of the log-likelihood contributions for each group. Define $D_i = \sum_j d_{ij}$, the number of failures in the ith group. In the case of gamma frailties, the log-likelihood contribution for the ith group is

$$\ln L_i = \sum_{j=1}^{n_i} d_{ij} \ln h_{ij}(t_{ij}) - (1/\theta + D_i) \ln \left\{ 1 - \theta \sum_{j=1}^{n_i} \ln \frac{S_{ij}(t_{ij})}{S_{ij}(t_{0ij})} \right\} +$$

$$D_i \ln \theta + \ln \Gamma(1/\theta + D_i) - \ln \Gamma(1/\theta)$$

In the case of inverse-Gaussian frailties, define

$$C_i = \left\{ 1 - 2\theta \sum_{j=1}^{n_i} \ln \frac{S_{ij}(t_{ij})}{S_{ij}(t_{0ij})} \right\}^{-1/2}$$

The log-likelihood contribution for the ith group then becomes

$$\ln L_i = \theta^{-1}(1 - C_i^{-1}) + B(\theta C_i, D_i) + \sum_{j=1}^{n_i} d_{ij} \left\{ \ln h_{ij}(t_{ij}) + \ln C_i \right\}$$

The function $B(a, b)$ is related to the modified bessel function of the third kind (commonly known as the BesselK function; see Wolfram (1996, p. 746)). In particular,

$$B(a, b) = a^{-1} + \frac{1}{2} \left\{ \ln \left(\frac{2}{\pi} \right) - \ln a \right\} + \ln \text{BesselK} \left(\frac{1}{2} - b, a^{-1} \right)$$

For both unshared and shared frailty models, estimation of θ takes place jointly with the estimation of β and the ancillary parameters.

If the robust estimate of variance is requested, results are transformed as explained in [U] **23.14 Obtaining robust variance estimates** and, in particular, in [P] **_robust**. Note that if observations in the dataset represent repeated observations on the same subjects (that is, there are time-varying covariates), the assumption of independence of the observations is highly questionable, meaning the conventional estimate of variance is not appropriate. We strongly advise the use of the **robust** and **cluster()** options in this case. (**streg** knows to specify **cluster()** if you specify robust.) **robust** and **cluster()** do not apply in shared frailty models, where the correlation within groups is instead modeled directly.

Predictions

predict *newvar*, *options* may be used after `streg` to predict various quantities, according to the following *options*:

median time:

$$newvar_j = \{t : \widehat{S}_j(t) = 1/2\}$$

where $\widehat{S}_j(t)$ is $S_j(t)$ with the parameter estimates "plugged-in".

median lntime:

$$newvar_j = \left\{y : \widehat{S}_j(e^y) = 1/2\right\}$$

mean time:

$$newvar_j = \int_0^\infty \widehat{S}_j(t)dt$$

mean lntime:

$$newvar_j = \int_{-\infty}^\infty ye^y \widehat{f}_j(e^y)dy$$

where $\widehat{f}_j(t)$ is $f_j(t)$ with the parameter estimates plugged-in.

hazard:

$$newvar_j = \widehat{f}_j(t_j)/\widehat{S}_j(t_j)$$

hr (PH models only):

$$newvar_j = \exp(\mathbf{x}_j\widehat{\boldsymbol{\beta}})$$

xb:

$$newvar_j = \mathbf{x}_j\widehat{\boldsymbol{\beta}}$$

stdp:

$$newvar_j = \widehat{se}(\mathbf{x}_j\widehat{\boldsymbol{\beta}})$$

surv and csurv:

$$newvar_j = \widehat{S}_j(t_j)/\widehat{S}_j(t_{0j})$$

The above represents the probability of survival past time t_j given survival up until t_{0j}, and is what you obtain when you specify `surv`. If `csurv` is specified then these probabilities are multiplied (in time order) over a subject's multiple observations. What is obtained is then equal to the probability of survival past time t_j, given survival through the earliest observed t_{0j}, and given the subject's (possibly time-varying) covariate history. In single-record-per-subject data, `surv` and `csurv` are identical.

csnell and ccsnell:

$$newvar_j = -\ln \widehat{S}_j(t_j)$$

The Cox–Snell (1968) residual CS_j for observation j at time t_j is defined as $\widehat{H}_j(t_j) = -\ln \widehat{S}_j(t_j)$, the estimated cumulative hazard function obtained from the fitted model (Collett 1994, 150). Cox and Snell argued that if the correct model has been fitted to the data, these residuals are n observations from an exponential distribution with unit mean. Thus, a plot of the cumulative hazard rate of the residuals against the residuals themselves should result in a straight line of slope 1. Note that Cox–Snell residuals can never be negative, and therefore are not symmetric about zero. The option concll stores in each observation that observation's contribution to the subject's Cox–Snell residual, which we refer to as a "partial" Cox–Snell residual. The option ccsnell stores the subject's overall Cox–Snell residual in the last observation for that subject. If there is only one observation per subject, the Cox–Snell residuals stored by ccsnell and csnell are equal.

mgale and cmgale:

$$newvar_j = d_j - CS_j$$

Martingale residuals fall out naturally from martingale theory in the case of Cox proportional hazards, but their development does not carry over for parametric survival models. However, martingale-like residuals similar to those obtained in the case of Cox can be derived from the Cox–Snell residuals, $M_j = d_j - CS_j$ where CS_j are the Cox–Snell residuals as previously described.

Because martingale-like residuals are calculated from the Cox–Snell residuals, they also could be "partial" or not. Partial martingale residuals are generated with the option mgale, and overall martingale residuals are generated with the option cmgale.

Martingale residuals can be interpreted as a measurement of the difference over time between the number of deaths in the data and the expected number based on the fitted model. These residuals take values between $-\infty$ and 1 and have an expected value of zero, although like the Cox–Snell residuals, they are not symmetric about zero, making them difficult to interpret.

deviance:

$$newvar_j = \text{sign}(M_j)\left(-2\left\{M_j + d_j \ln(d_j - M_j)\right\}\right)$$

Deviance residuals are a scaling of the martingale-like residuals in an attempt to make them symmetric about zero. When the model fits the data, these residuals are symmetric about zero, and thus can be more readily used to examine the data for outliers.

predict also allows two options for use after fitting frailty models: alpha1 and unconditional. If unconditional is specified, then the above predictions are modified so as to be based on $S_\theta(t)$ and $f_\theta(t)$, rather than $S(t)$ and $f(t)$. If alpha1 is specified, then the predictions are as described above.

References

Akaike, H. 1974. A new look at the statistical model identification. *IEEE Transaction and Automatic Control* AC-19: 716–723.

Cleves, M. A. 2000. stata54: Multiple curves plotted with stcurv command. *Stata Technical Bulletin* 54: 2–4. Reprinted in *Stata Technical Bulletin Reprints*, vol. 9, pp. 7–10.

Cleves, M. A., W. W. Gould, and R. G. Gutierrez. 2002. *An Introduction to Survival Analysis Using Stata*. College Station, TX: Stata Press.

Collett, D. 1994. *Modelling Survival Data in Medical Research*. London: Chapman & Hall.

Cox, D. R. and D. Oakes. 1984. *Analysis of Survival Data*. London: Chapman & Hall.

Cox, D. R. and E. J. Snell. 1968. A general definition of residuals (with discussion). *Journal of the Royal Statistical Society B* 30: 248–275.

Crowder, M. J., A. C. Kimber, R. L. Smith, and T. J. Sweeting. 1991. *Statistical Analysis of Reliability Data*. London: Chapman & Hall.

Fisher, R. A. and L. H. C. Tippett. 1928. Limiting forms of the frequency distribution of the largest or smallest member of a sample. *Proceedings of the Cambridge Philosophical Society* 24: 180–190.

Gutierrez, R. G. 2002. Parametric frailty and shared frailty survival models. *The Stata Journal* 2: 22–44.

Gutierrez, R. G., S. L. Carter, and D. M. Drukker. 2001. On boundary-value likelihood-ratio tests. *Stata Technical Bulletin*, forthcoming.

Hosmer, D. W., Jr., and S. Lemeshow. 1999. *Applied Survival Analysis*. New York: John Wiley & Sons.

Hougaard, P. 1986. Survival models for heterogeneous populations derived from stable distributions. *Biometrika* 73: 387–396.

Kalbfleisch, J. D. and R. L. Prentice. 2002. *The Statistical Analysis of Failure Time Data*. 2d ed. New York: John Wiley & Sons.

Klein, J. P. and M. L. Moeschberger. 1997. *Survival Analysis: Techniques for Censored and Truncated data*. New York: Springer.

Lee, E. T. 1992. *Statistical Methods for Survival Data Analysis*. 2d ed. New York: John Wiley & Sons.

McGilchrist, C. A. and C. W. Aisbett. 1991. Regression with frailty in survival analysis. *Biometrics* 47: 461–466.

Peto, R. and P. Lee. 1973. Weibull distributions for continuous-carcinogenesis experiments. *Biometrics* 29: 457–470.

Pike, M. C. 1966. A method of analysis of a certain class of experiments in carcinogenesis. *Biometrics* 22: 142–161.

Schoenfeld, D. 1982. Partial residuals for the proportional hazards regression model. *Biometrika* 69: 239–241.

Scotto, M. G. and A. Tobias. 1998. sg83: Parameter estimation for the Gumbel distribution. *Stata Technical Bulletin* 43: 32–35. Reprinted in *Stata Technical Bulletin Reprints*, vol. 8, pp. 133–137.

——. 2000. sg146: Parameter estimation for the generalized extreme value distribution. *Stata Technical Bulletin* 56: 40–43.

Weibull, W. 1939. A statistical theory of the strength of materials. *Ingeniörs Vetenskaps Akademien Handlingar*, no. 151. Stockholm: Generalstabens Litografiska Anstalts Förlag.

Wolfram, S. 1996. *The Mathematica Book*. Champaign, IL: Wolfram Media.

Also See

Complementary:	[ST] **sts**, [ST] **stset**,
	[R] **adjust**, [R] **constraint**, [R] **lincom**, [R] **linktest**, [R] **lrtest**, [R] **mfx**,
	[R] **nlcom**, [R] **predict**, [R] **predictnl**, [R] **sw**, [R] **test**, [R] **testnl**, [R] **vce**
Related:	[ST] **stcox**
Background:	[U] **16.5 Accessing coefficients and standard errors**,
	[U] **23 Estimation and post-estimation commands**,
	[U] **23.14 Obtaining robust variance estimates**,
	[ST] **st**, [ST] **survival analysis**,
	[R] **maximize**

Title

> **sts** — Generate, graph, list, and test the survivor and cumulative hazard functions

Syntax

sts [g̲raph] [if *exp*] [in *range*] [, ...]

sts l̲ist [if *exp*] [in *range*] [, ...]

sts t̲est *varlist* [if *exp*] [in *range*] [, ...]

sts g̲enerate *newvar* = ... [if *exp*] [in *range*] [, ...]

sts is for use with survival-time data; see [ST] **st**. You must stset your data before using this command.

See [ST] **sts generate**, [ST] **sts graph**, [ST] **sts list**, and [ST] **sts test** for details of syntax.

Description

sts reports on and creates variables containing the estimated survivor and related functions such as the Nelson–Aalen cumulative hazard function. In the case of the survivor function, sts tests and produces Kaplan–Meier estimates or, via Cox regression, adjusted estimates.

sts graph graphs the estimated survivor, cumulative hazard, or hazard function.

sts list lists the estimated survivor and related functions.

sts test tests the equality of the survivor function across groups.

sts generate creates new variables containing the estimated survivor function, the Nelson–Aalen cumulative hazard function, or related functions.

sts is appropriate for use with single- or multiple-record, single- or multiple-failure, st data.

(Continued on next page)

239

▷ Example

Graph the Kaplan–Meier survivor function	. sts graph
	. sts graph, by(drug)
Graph the Nelson–Aalen cumulative hazard function	. sts graph, na
	. sts graph, na by(drug)
Graph the estimated hazard function	. sts graph, hazard
	. sts graph, hazard by(drug)
List the Kaplan–Meier survivor function	. sts list
	. sts list, by(drug) compare
List the Nelson–Aalen cumulative hazard function	. sts list, na
	. sts list, na by(drug) compare
Generate variable containing Kaplan–Meier survivor function	. sts gen surv = s
	. sts gen surv = s, by(drug)
Generate variable containing Nelson–Aalen cumulative hazard function	. sts gen haz = na
	. sts gen haz = na, by(drug)
Test equality of survivor functions	. sts test drug
	. sts test drug, strata(agecat)

◁

Remarks

Remarks are presented under the headings

Listing, graphing, and generating variables
Comparing survivor or cumulative hazard functions
Testing equality of survivor functions
Adjusted estimates
Counting the number lost due to censoring

sts concerns the survivor function $S(t)$, the probability of surviving to t or beyond, the cumulative hazard function, $H(t)$, and the hazard function $h(t)$. Its subcommands can list and generate variables containing $\widehat{S}(t)$ and $\widehat{H}(t)$, and test the equality of $S(t)$ over groups. In addition:

1. All subcommands share a common syntax.

2. All subcommands deal with either the Kaplan–Meier product-limit or the Nelson–Aalen estimates unless you request adjusted survival estimates.

3. If you request an adjustment, all subcommands perform the adjustment in the same way, which is described below.

The full details of each subcommand are found in the entries following this one, but each subcommand provides so many options to control exactly how the listing looks, how the graph appears, the form of the test to be performed, or what exactly is to be generated, that the simplicity of sts can be easily overlooked.

So, without getting burdened by the details of syntax, let us demonstrate the sts commands using the Stanford heart transplant data introduced in [ST] **stset**.

Listing, graphing, and generating variables

You can list the overall survivor function by typing sts list and you can graph it by typing sts graph or sts. sts assumes you mean graph when you do not type a subcommand.

Or, you can list the Nelson–Aalen cumulative hazard function by typing sts list, na and you can graph it by typing sts graph, na.

When you type sts list you are shown all the details:

```
. use http://www.stata-press.com/data/r8/stan3
(Heart transplant data)

. stset, noshow

. sts list
```

Time	Beg. Total	Fail	Net Lost	Survivor Function	Std. Error	[95% Conf. Int.]	
1	103	1	0	0.9903	0.0097	0.9331	0.9986
2	102	3	0	0.9612	0.0190	0.8998	0.9852
3	99	3	0	0.9320	0.0248	0.8627	0.9670
5	96	1	0	0.9223	0.0264	0.8507	0.9604
(output omitted)							
1586	2	0	1	0.1519	0.0493	0.0713	0.2606
1799	1	0	1	0.1519	0.0493	0.0713	0.2606

When you type sts graph, or just sts, you are shown a graph of the same result detailed by list:

```
. sts graph
```

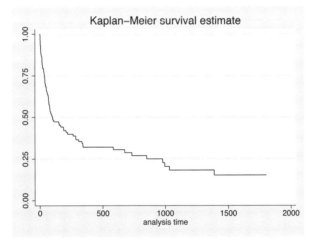

Kaplan–Meier survival estimate

sts generate is a rarely used command. Type sts generate survf = s and you will create a new variable survf containing the same survivor function that list just listed and graph just graphed:

```
. sts gen survf = s

. sort t1
```

. list t1 survf in 1/10

	t1	survf
1.	1	.99029126
2.	1	.99029126
3.	1	.99029126
4.	1	.99029126
5.	2	.96116505
6.	2	.96116505
7.	2	.96116505
8.	2	.96116505
9.	2	.96116505
10.	2	.96116505

sts generate is provided in case you want to make a calculation, listing, or graph that sts cannot already do for you.

Comparing survivor or cumulative hazard functions

sts will allow you to compare survivor or cumulative hazard functions. sts graph and sts graph, na are probably most successful at this. For example, survivor functions can be plotted using

. sts graph, by(posttran)

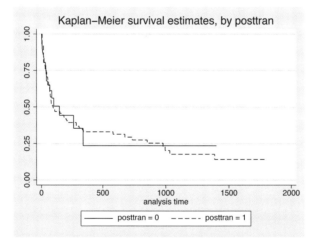

and Nelson–Aalen cumulative hazard functions can be plotted using

(*Continued on next page*)

. sts graph, na by(posttran)

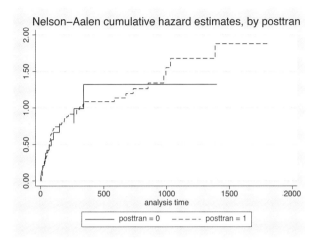

To compare survivor functions, we typed sts graph, just as previously, and then we added by(posttran) to see the survivor functions for the groups designated by posttran. In this case there are two groups, but as far as the sts command is concerned, there could have been more.

To compare cumulative hazard functions, we added na to the previous command.

You can also plot (and compare) estimated hazard functions, using sts graph, hazard. The hazard is estimated as a kernel smooth of the increments which sum to form the estimated cumulative hazard. The increments themselves do not estimate the hazard, but the smooth is weighted so that it does estimate the hazard; see [ST] **sts graph**.

Just as you can compare survivor functions graphically by typing sts graph, by(posttran) and cumulative hazard functions by typing sts graph, na by(posttran), you can obtain detailed listings by typing sts list, by(posttran) and sts list, na by(posttran), respectively. Below, we list the survivor function and also specify another option—enter—which adds a number-who-enter column:

(Continued on next page)

```
. sts list, by(posttran) enter
                    Beg.                          Survivor      Std.
        Time       Total    Fail   Lost   Enter   Function      Error     [95% Conf. Int.]

posttran=0
           0           0      0      0     103     1.0000          .           .         .
           1         103      1      3       0     0.9903       0.0097      0.9331    0.9986
           2          99      3      3       0     0.9603       0.0195      0.8976    0.9849
       (output omitted)
         427           2      0      1       0     0.2359       0.1217      0.0545    0.4882
        1400           1      0      1       0     0.2359       0.1217      0.0545    0.4882
posttran=1
           1           0      0      0       3     1.0000          .           .         .
           2           3      0      0       3     1.0000          .           .         .
           3           6      0      0       3     1.0000          .           .         .
           4           9      0      0       2     1.0000          .           .         .
           5          11      0      0       3     1.0000          .           .         .
         5.1          14      1      0       0     0.9286       0.0688      0.5908    0.9896
           6          13      0      0       1     0.9286       0.0688      0.5908    0.9896
           8          14      0      0       2     0.9286       0.0688      0.5908    0.9896
          10          16      0      0       2     0.9286       0.0688      0.5908    0.9896
       (output omitted)
        1586           2      0      1       0     0.1420       0.0546      0.0566    0.2653
        1799           1      0      1       0     0.1420       0.0546      0.0566    0.2653
```

It is easier to compare survivor or cumulative hazard functions if they are listed side-by-side and sts list has a compare option to do this:

```
. sts list, by(posttran) compare
                    Survivor Function
    posttran             0           1

    time     1        0.9903      1.0000
           225        0.4422      0.3934
           449        0.2359      0.3304
           673        0.2359      0.3139
           897        0.2359      0.2535
          1121        0.2359      0.1774
          1345        0.2359      0.1774
          1569            .       0.1420
          1793            .       0.1420
          2017            .           .
```

If we include the na option, the cumulative hazard functions will be listed:

```
. sts list, na by(posttran) compare
                    Nelson-Aalen Cum. Haz.
    posttran             0           1

    time     1        0.0097      0.0000
           225        0.7896      0.9145
           449        1.3229      1.0850
           673        1.3229      1.1350
           897        1.3229      1.3411
          1121        1.3229      1.6772
          1345        1.3229      1.6772
          1569            .       1.8772
          1793            .       1.8772
          2017            .           .
```

When you specify compare, the same detailed survivor or cumulative hazard function is calculated, but it is then evaluated at ten or so times and those evaluations listed. Above we left it to sts list to choose the comparison times, but we can specify them ourselves using the at() option:

```
. sts list, by(posttran) compare at(0 100 to 1700)
                      Survivor Function
posttran                 0            1

time       0         1.0000       1.0000
         100         0.5616       0.4814
         200         0.4422       0.4184
         300         0.3538       0.3680
         400         0.2359       0.3304
         500         0.2359       0.3304
         600         0.2359       0.3139
         700         0.2359       0.2942
         800         0.2359       0.2746
         900         0.2359       0.2535
        1000         0.2359       0.2028
        1100         0.2359       0.1774
        1200         0.2359       0.1774
        1300         0.2359       0.1774
        1400         0.2359       0.1420
        1500              .       0.1420
        1600              .       0.1420
        1700              .       0.1420
```

Testing equality of survivor functions

sts test tests equality of survivor functions:

```
. sts test posttran
Log-rank test for equality of survivor functions
                Events       Events
posttran      observed     expected

0                   30        31.20
1                   45        43.80

Total               75        75.00
              chi2(1) =        0.13
              Pr>chi2 =      0.7225
```

When you do not specify otherwise, sts test performs the log-rank test, but it can also perform the Wilcoxon test:

```
. sts test posttran, wilcoxon
Wilcoxon (Breslow) test for equality of survivor functions
                Events       Events      Sum of
posttran      observed     expected       ranks

0                   30        31.20         -85
1                   45        43.80          85

Total               75        75.00           0
              chi2(1) =        0.14
              Pr>chi2 =      0.7083
```

sts test will also perform stratified tests. This is demonstrated in [ST] **sts test**.

Adjusted estimates

All the estimates of the survivor function we have seen so far are the Kaplan–Meier product limit estimates. sts can make adjusted estimates to the survivor function. We want to illustrate this and explain how it is done.

The heart transplant dataset is not the best to demonstrate this feature because we are starting with survivor functions that are similar already, so let us switch to data on a fictional drug trial:

```
. use http://www.stata-press.com/data/r8/drug2, clear
(Patient Survival in Drug Trial)

. stset, noshow

. stdes
```

Category	total	mean	min	median	max
			— per subject —		
no. of subjects	48				
no. of records	48	1	1	1	1
(first) entry time		0	0	0	0
(final) exit time		15.5	1	12.5	39
subjects with gap	0				
time on gap if gap	0				
time at risk	744	15.5	1	12.5	39
failures	31	.6458333	0	1	1

This dataset contains 48 subjects, all observed from time 0. The st command shows us how the dataset is currently declared:

```
. st
-> stset studytime, failure(died) noshow

        failure event:  died != 0 & died < .
  obs. time interval:  (0, studytime]
   exit on or before:  failure
```

and the dataset contains variables age and drug:

```
. summarize age drug
```

Variable	Obs	Mean	Std. Dev.	Min	Max
age	48	47.125	9.492718	32	67
drug	48	.5833333	.4982238	0	1

We are comparing the outcomes of drug = 1 with that of the placebo, drug = 0. Here are the survivor curves for the two groups:

(Continued on next page)

```
. sts graph, by(drug)
```

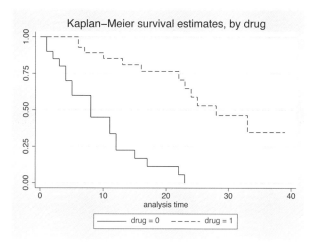

Here are the survivor curves adjusted for age (and scaled to age 50):

```
. generate age50 = age-50
. sts graph, by(drug) adjustfor(age50)
```

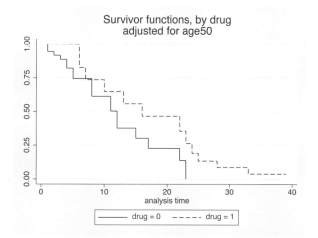

The age difference between the two samples accounts for much of the difference between the survivor functions.

When you type by(*group*) adjustfor(*vars*), sts fits a separate Cox proportional hazards model on *vars* (estimation via stcox) and retrieves the separately estimated baseline survivor functions. sts graph graphs the baseline survivor functions, sts list lists them, and sts generate saves them.

(Continued on next page)

Thus, sts list can list what sts graph plots:

```
. sts list, by(drug) adjustfor(age50) compare

        Adjusted Survivor Function
drug                    0          1

time       1       0.9463     1.0000
           5       0.7439     1.0000
           9       0.6135     0.7358
          13       0.3770     0.5588
          17       0.2282     0.4668
          21       0.2282     0.4668
          25            .     0.1342
          29            .     0.0872
          33            .     0.0388
          37            .     0.0388
          41            .          .

Survivor function adjusted for age50
```

In both the graph and the listing, it is important that we adjust for variable $age50 = age - 50$ and not just age. Adjusted survivor functions are adjusted to the adjustfor() variables and scaled to correspond to the adjustfor() variables set to 0. Here is the result of adjusting for age, which is 0 at birth:

```
. sts list, by(drug) adjustfor(age) compare

        Adjusted Survivor Function
drug                    0          1

time       1       0.9994     1.0000
           5       0.9970     1.0000
           9       0.9951     0.9995
          13       0.9903     0.9990
          17       0.9853     0.9987
          21       0.9853     0.9987
          25            .     0.9965
          29            .     0.9958
          33            .     0.9944
          37            .     0.9944
          41            .          .

Survivor function adjusted for age
```

These are equivalent to what we obtained previously but not nearly so informative because of the scaling of the survivor function. Option adjustfor(age) scales the survivor function to correspond to age = 0. age is calendar age and so the survivor function is scaled to correspond to a newborn.

There is another way sts will adjust the survivor function. Rather than specifying by(*group*) adjustfor(*vars*), we specify strata(*group*) adjustfor(*vars*):

(*Continued on next page*)

```
. sts list, strata(drug) adjustfor(age50) compare

              Adjusted Survivor Function
  drug               0            1

  time      1     0.9526       1.0000
            5     0.7668       1.0000
            9     0.6417       0.7626
           13     0.4080       0.5995
           17     0.2541       0.5139
           21     0.2541       0.5139
           25         .        0.1800
           29         .        0.1247
           33         .        0.0614
           37         .        0.0614
           41         .            .

Survivor function adjusted for age50
```

When we specify strata() instead of by(), instead of fitting separate Cox models for each stratum, a single, stratified Cox model is fitted and the stratified baseline survivor function retrieved. That is, strata() rather than by() constrains the effect of the adjustfor() variables to be the same across strata.

Counting the number lost due to censoring

sts list, in the detailed output, shows the number lost in the fourth column:

```
. sts list
```

Time	Beg. Total	Fail	Net Lost	Survivor Function	Std. Error	[95% Conf. Int.]	
1	48	2	0	0.9583	0.0288	0.8435	0.9894
2	46	1	0	0.9375	0.0349	0.8186	0.9794
3	45	1	0	0.9167	0.0399	0.7930	0.9679
(output omitted)							
8	36	3	1	0.7061	0.0661	0.5546	0.8143
9	32	0	1	0.7061	0.0661	0.5546	0.8143
10	31	1	1	0.6833	0.0678	0.5302	0.7957
(output omitted)							
39	1	0	1	0.1918	0.0791	0.0676	0.3634

and sts graph, if you specify the lost option, will show that number, too:

(Continued on next page)

`. sts graph, lost`

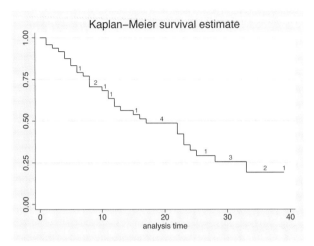

The number on the listing and on the graph is the number net lost, defined as number censored minus number who enter. With simple survival data—data with one observation per subject—net lost corresponds to lost.

With more complicated survival data—meaning delayed entry or multiple records per subject—the number net lost may surprise you. With complicated data, the vague term lost can mean many things. Sometimes subjects are lost, but mostly there are a lot of censorings followed by re-entries—a subject is censored at time 5 immediately to re-enter the data with different covariates. This is called thrashing.

There are other possibilities: A subject can be lost, but only for a while, and so re-enter the data with a gap; a subject can be censored out of one stratum to enter another. There are too many possibilities to dedicate a column in a table or a plotting symbol in a graph to each one. sts's solution is to define lost as net lost, meaning censored-minus-entered, and show that number. How we define lost does not affect the calculation of the survivor function: it merely affects a number that researchers often report.

Censored-minus-entered results in exactly what is desired for simple survival data. Since everybody enters at time 0, censored-minus-entered amounts to calculating censored − 0. The number net lost is the number censored.

In more complicated data, censored-minus-entered results in the number really lost if there are no gaps and no delayed entry. In that case, the subtraction smooths the thrashing. In an interval, 5 might be censored and 3 re-enter, so $5 - 3 = 2$ were lost.

In even more complicated data, censored-minus-entered results in something reasonable once you understand how to interpret negative numbers and are cautious in interpreting positive ones. 5 might be censored and 3 might enter (from the five? who can say?), resulting in 2 net lost; or 3 might be censored and 5 enter, resulting in −2 being lost.

sts, by default, reports net lost but will, if you specify the enter option, report the pure number censored and the pure number who enter. Sometimes you will want to do that. Earlier in this entry, we used sts list to display the survivor functions in the Stanford heart transplant data for subjects pre- and post-transplantation, and we slipped in an enter option:

```
. use http://www.stata-press.com/data/r8/stan3, clear
(Heart transplant data)
```

```
. stset, noshow

. sts list, by(posttran) enter
```

	Beg.				Survivor	Std.		
Time	Total	Fail	Lost	Enter	Function	Error	[95% Conf. Int.]	
posttran=0								
0	0	0	0	103	1.0000	.	.	.
1	103	1	3	0	0.9903	0.0097	0.9331	0.9986
2	99	3	3	0	0.9603	0.0195	0.8976	0.9849
3	93	3	3	0	0.9293	0.0258	0.8574	0.9657
(output omitted)								
427	2	0	1	0	0.2359	0.1217	0.0545	0.4882
1400	1	0	1	0	0.2359	0.1217	0.0545	0.4882
posttran=1								
1	0	0	0	3	1.0000	.	.	.
2	3	0	0	3	1.0000	.	.	.
3	6	0	0	3	1.0000	.	.	.
4	9	0	0	2	1.0000	.	.	.
5	11	0	0	3	1.0000	.	.	.
5.1	14	1	0	0	0.9286	0.0688	0.5908	0.9896
6	13	0	0	1	0.9286	0.0688	0.5908	0.9896
8	14	0	0	2	0.9286	0.0688	0.5908	0.9896
(output omitted)								
1586	2	0	1	0	0.1420	0.0546	0.0566	0.2653
1799	1	0	1	0	0.1420	0.0546	0.0566	0.2653

We did that to keep you from being shocked at negative numbers for net lost. In this complicated dataset, the value of posttran changes over time. All patients start with posttran = 0, and later some change to posttran = 1.

Thus, at time 1 in the posttran = 0 group, 3 are lost, which is to say, lost to the group but not to the experiment. Simultaneously, in the posttran = 1 group, we see that 3 enter. Had we not specified the enter option, you would not have seen that 3 enter, and you would have seen that -3 were, in net, lost:

```
. sts list, by(posttran)
```

	Beg.		Net	Survivor	Std.		
Time	Total	Fail	Lost	Function	Error	[95% Conf. Int.]	
posttran=0							
1	103	1	3	0.9903	0.0097	0.9331	0.9986
2	99	3	3	0.9603	0.0195	0.8976	0.9849
3	93	3	3	0.9293	0.0258	0.8574	0.9657
(output omitted)							
427	2	0	1	0.2359	0.1217	0.0545	0.4882
1400	1	0	1	0.2359	0.1217	0.0545	0.4882
posttran=1							
1	0	0	-3	1.0000	.	.	.
2	3	0	-3	1.0000	.	.	.
3	6	0	-3	1.0000	.	.	.
4	9	0	-2	1.0000	.	.	.
5	11	0	-3	1.0000	.	.	.
5.1	14	1	0	0.9286	0.0688	0.5908	0.9896
6	13	0	-1	0.9286	0.0688	0.5908	0.9896
8	14	0	-2	0.9286	0.0688	0.5908	0.9896
(output omitted)							
1586	2	0	1	0.1420	0.0546	0.0566	0.2653
1799	1	0	1	0.1420	0.0546	0.0566	0.2653

In this case, specifying `enter` makes the table easier to explain, but do not jump to the conclusion that specifying `enter` is always a good idea. In this same dataset, let's look at the overall survivor function, first with the `enter` option:

```
. sts list, enter
```

Time	Beg. Total	Fail	Lost	Enter	Survivor Function	Std. Error	[95% Conf. Int.]	
0	0	0	0	103	1.0000	.	.	.
1	103	1	3	3	0.9903	0.0097	0.9331	0.9986
2	102	3	3	3	0.9612	0.0190	0.8998	0.9852
3	99	3	3	3	0.9320	0.0248	0.8627	0.9670
(output omitted)								
1571	3	0	1	0	0.1519	0.0493	0.0713	0.2606
1586	2	0	1	0	0.1519	0.0493	0.0713	0.2606
1799	1	0	1	0	0.1519	0.0493	0.0713	0.2606

Note that at time 1, 3 are lost and 3 enter. There is no delayed entry in this dataset and there are no gaps, so it is the same 3 that were lost and re-entered, and no one was really lost. At time 1571, on the other hand, a patient really was lost. This is all more clearly revealed when we do not specify the `enter` option:

```
. sts list
```

Time	Beg. Total	Fail	Net Lost	Survivor Function	Std. Error	[95% Conf. Int.]	
1	103	1	0	0.9903	0.0097	0.9331	0.9986
2	102	3	0	0.9612	0.0190	0.8998	0.9852
3	99	3	0	0.9320	0.0248	0.8627	0.9670
(output omitted)							
1571	3	0	1	0.1519	0.0493	0.0713	0.2606
1586	2	0	1	0.1519	0.0493	0.0713	0.2606
1799	1	0	1	0.1519	0.0493	0.0713	0.2606

Thus, to summarize:

1. The `sts list` and `graph` commands will show the number lost or censored. `sts list` shows it on the detailed output—you specify no option to see it. `sts graph` shows the number when you specify the `lost` option.

2. By default, the number lost is the net number lost, defined as censored-minus-entered.

3. Both commands allow you to specify the `enter` option and then show the number who actually entered, and the number lost becomes the actual number censored, not censored-minus-entered.

Saved Results

`sts test` saves in `r()`:

Scalars
r(df)   degrees of freedom   r(chi2) χ^2

Methods and Formulas

sts is implemented as an ado-file.

Unless adjusted estimates are requested, sts estimates the survivor function using the Kaplan–Meier product-limit method.

When the na option is specified, sts estimates the cumulative hazard function using the Nelson–Aalen estimator.

For an introduction to the Kaplan–Meier product-limit method and the log-rank test, see Pagano and Gauvreau (2000, 495–499); for a detailed discussion, see Cox and Oakes (1984), Kalbfleisch and Prentice (2002), or Klein and Moeschberger (1997).

Let t_j, $j = 1, \ldots,$ denote the times at which failure occurs. Let n_j be the number at risk of failure just before time t_j and d_j the number of failures at time t_j. Then the nonparametric maximum-likelihood estimate of the survivor function is (Kaplan and Meier 1958)

$$\widehat{S}(t) = \prod_{j | t_j \leq t} \left(\frac{n_j - d_j}{n_j} \right)$$

(Kalbfleisch and Prentice 2002, 15).

The failure function $\widehat{F}(t)$ is defined as $1 - \widehat{S}(t)$.

The standard error reported is given by Greenwood's formula (Greenwood 1926)

$$\widehat{\text{Var}}\{\widehat{S}(t)\} = \widehat{S}^2(t) \sum_{j | t_j \leq t} \frac{d_j}{n_j(n_j - d_j)}$$

(Kalbfleisch and Prentice 2002, 17–18). These standard errors, however, are not used for confidence intervals. Instead, the asymptotic variance of $\ln[-\ln \widehat{S}(t)]$

$$\widehat{\sigma}^2(t) = \frac{\sum \frac{d_j}{n_j(n_j - d_j)}}{\left\{ \sum \ln \left(\frac{n_j - d_j}{n_j} \right) \right\}^2}$$

is used, where sums are calculated over $j | t_j \leq t$ (Kalbfleisch and Prentice 2002, 18). The confidence bounds are then $\widehat{S}(t)^{\exp(\pm z_{\alpha/2} \widehat{\sigma}(t))}$, where $z_{\alpha/2}$ is the $(1 - \alpha/2)$ quantile of the normal distribution. sts suppresses reporting the standard error and confidence bounds if the data are pweighted since these formulas are no longer appropriate.

When option adjustfor() is specified, the survivor function estimate $\widehat{S}(t)$ is the baseline survivor function estimate $\widehat{S}_0(t)$ of stcox; see [ST] **stcox**. If by() is specified, $\widehat{S}(t)$ is obtained from fitting separate Cox models on adjustfor() for each of the by() groups. If instead strata() is specified, a single Cox model on adjustfor(), stratified by strata(), is fitted.

The Nelson–Aalen estimator of the cumulative hazard rate function is due to Nelson (1972) and Aalen (1978), and is defined up to the largest observed time as

$$\widehat{H}(t) = \sum_{j | t_j \leq t} \frac{d_j}{n_j}$$

Its variance (Aalen 1978) may be estimated by

$$\widehat{\text{Var}}\{\widehat{H}(t)\} = \sum_{j|t_j \leq t} \frac{d_j}{n_j{}^2}$$

Pointwise confidence intervals are calculated using the asymptotic variance of $\ln \widehat{H}(t)$

$$\widehat{\phi}^2(t) = \frac{\widehat{\text{Var}}\{\widehat{H}(t)\}}{\{\widehat{H}(t)\}^2}$$

The confidence bounds are then $\widehat{H}(t)\exp\{\pm z_{\alpha/2}\widehat{\phi}(t)\}$. If the data are `pweighted`, these formulas are not appropriate and in that case, confidence intervals are not reported.

References

Aalen, O. O. 1978. Nonparametric inference for a family of counting processes. *Annals of Statistics* 6: 701–726.

Cleves, M. A. 1999. stata53: censored option added to sts graph command. *Stata Technical Bulletin* 50: 34–36. Reprinted in *Stata Technical Bulletin Reprints*, vol. 9, pp. 4–7.

Cox, D. R. and D. Oakes. 1984. *Analysis of Survival Data*. London: Chapman & Hall.

Greenwood, M. 1926. The natural duration of cancer. *Reports on Public Health and Medical Subjects* 33: 1–26. London: His Majesty's Stationery Office.

Kalbfleisch, J. D. and R. L. Prentice. 2002. *The Statistical Analysis of Failure Time Data*. 2d ed. New York: John Wiley & Sons.

Kaplan, E. L. and P. Meier. 1958. Nonparametric estimation from incomplete observations. *Journal of the American Statistical Association* 53: 457–481.

Klein, J. P. and M. L. Moeschberger. 1997. *Survival Analysis: Techniques for Censored and Truncated Data*. New York: Springer.

Nelson, W. 1972. Theory and applications of hazard plotting for censored failure data. *Technometrics* 14: 945–965.

Newman, S. C. 2001. *Biostatistical Methods in Epidemiology*. New York: John Wiley & Sons.

Pagano, M. and K. Gauvreau. 2000. *Principles of Biostatistics*. 2d ed. Pacific Grove, CA: Brooks/Cole.

Also See

Complementary:	[ST] **stci**, [ST] **stcox**, [ST] **sts generate**, [ST] **sts graph**, [ST] **sts list**, [ST] **sts test**, [ST] **stset**
Background:	[ST] **st**, [ST] **survival analysis**

Title

sts generate — Create survivor, hazard, and other variables

Syntax

sts <u>gen</u>erate *newvar* =

$\{$ s $|$ se(s) $|$ h $|$ se(lls) $|$ lb(s) $|$ ub(s) $|$ na $|$ se(na) $|$ lb(na) $|$ ub(na) $|$ n $|$ d $\}$

$\big[$ *newvar* = $\{\ldots\}$ $\ldots$$\big]$ $\big[$ if *exp* $\big]$ $\big[$ in *range* $\big]$ $\big[$, by(*varlist*) <u>str</u>ata(*varlist*)

<u>a</u>djustfor(*varlist*) <u>l</u>evel(*#*) $\big]$

sts generate is for use with survival-time data; see [ST] **st**. You must stset your data before using this command.

Description

sts generate creates new variables containing the estimated survivor (failure) function, the Nelson–Aalen cumulative hazard (integrated hazard) function, and related functions. See [ST] **sts** for an introduction to this command.

sts generate is appropriate for use with single- or multiple-record, single- or multiple-failure, st data.

Functions

s produces the Kaplan–Meier product-limit estimate of the survivor function $\widehat{S}(t)$ or, if adjustfor() is specified, the baseline survivor function from a Cox regression model on the adjustfor() variables.

se(s) produces the Greenwood, pointwise standard error $\widehat{se}\{\widehat{S}(t)\}$. Option adjustfor() is not allowed in this case.

h produces the estimated hazard component $\Delta H_j = H(t_j) - H(t_{j-1})$, where t_j is the current failure time and t_{j-1} is the previous one. This is mainly a utility function used to calculate the estimated cumulative hazard $H(t_j)$, yet one may estimate the hazard via a kernel smooth of the ΔH_j; see [ST] **sts graph**. It is recorded at all the points at which a failure occurs and is computed as d_j/n_j, where d_j is the number of failures occurring at time t_j and n_j is the number at risk at t_j before the occurrence of the failures.

se(lls) produces $\widehat{\sigma}(t)$, the standard error of $\ln\{-\ln\widehat{S}(t)\}$.

lb(s) produces the lower bound of the confidence interval for $\widehat{S}(t)$ based on $\ln\{-\ln\widehat{S}(t)\}$; to wit, $\widehat{S}(t)^{\exp(-z_{\alpha/2}\widehat{\sigma}(t))}$, where $z_{\alpha/2}$ is the $(1-\alpha/2)$ quantile of the standard normal distribution.

ub(s) produces the upper bound of the confidence interval for $\widehat{S}(t)$ based on $\ln\{-\ln\widehat{S}(t)\}$; to wit, $\widehat{S}(t)^{\exp(z_{\alpha/2}\widehat{\sigma}(t))}$, where $z_{\alpha/2}$ is the $(1-\alpha/2)$ quantile of the standard normal distribution.

na produces the Nelson–Aalen estimate of the cumulative hazard function. Option adjustfor() is not allowed in this case.

se(na) produces pointwise standard error for the Nelson–Aalen estimate of the cumulative hazard function, $\widehat{H}(t)$. Option adjustfor() is not allowed in this case.

lb(na) produces the lower bound of the confidence interval for $\widehat{H}(t)$ based on the log-transformed cumulative hazard function.

ub(na) produces the corresponding upper bound.

n produces n_j, the number at risk just prior to time t_j.

d produces d_j, the number failing at time t_j.

Options

by(*varlist*) produces separate survivor or cumulative hazard functions by making separate calculations for each group identified by equal values of the variables in *varlist*.

strata(*varlist*) is a subtle alternative to by(). First, you may not specify strata() unless you also specify adjustfor(), whereas you may specify by() in either case. Thus, strata() amounts to a modifier of adjustfor() and is discussed below. This option is not available for the Nelson–Aalen function.

adjustfor(*varlist*) adjusts the estimate of the survivor function to that for 0 values of the *varlist* specified. This option is not available with the Nelson–Aalen function. How sts makes the adjustment depends on whether you specify by() or strata(). This is fully discussed in [ST] **sts**, but here is a quick review:

If you specify strata(), sts performs the adjustment by estimating a stratified-on-group Cox regression model using adjustfor() as the covariates. The stratified, baseline survivor function is then retrieved.

If you specify by(), sts estimates separate Cox regression models for each group. The separately calculated baseline survivor functions are then retrieved.

Be aware that, regardless of method employed, the survivor function is adjusted to 0 values of the covariates.

level(#) specifies the confidence level, in percent, for the lb(s), ub(s), lb(na) and ub(na) functions. The default is level(95) or as set by set level; see [U] **23.6 Specifying the width of confidence intervals**.

Remarks

sts generate is a rarely used command; it gives you access to the calculations listed by sts list and graphed by sts graph.

Use of this command is demonstrated in [ST] **sts**.

Methods and Formulas

See [ST] **sts**.

References

See [ST] **sts**.

Also See

Complementary:	[ST] **sts**, [ST] **sts graph**, [ST] **sts list**, [ST] **sts test**, [ST] **stset**
Background:	[ST] **st**, [ST] **survival analysis**

Title

sts graph — Graph the survivor and the cumulative hazard functions

Syntax

sts graph [if *exp*] [in *range*] [, by(*varlist*) strata(*varlist*) adjustfor(*varlist*)

    failure gwood na cna hazard level(*#*) lost enter separate tmin(*#*) tmax(*#*)

    noorigin atrisk noshow censored(single | number | multiple) kernel(*kernel*)

    width(*#* [*#*...]) ciopts(*rline_options*) plot(*plot*) *line_options twoway_options*]

sts graph is for use with survival-time data; see [ST] **st**. You must stset your data before using this command.

Description

sts graph graphs the estimated survivor (failure) function, the Nelson–Aalen estimated cumulative (integrated) hazard function, or the estimated hazard function. See [ST] **sts** for an introduction to this command.

sts graph is appropriate for use with single- or multiple-record, single- or multiple-failure, st data.

Options

by(*varlist*) produces separate survivor, cumulative hazard, or hazard functions by making separate calculations for each group identified by equal values of the variables in *varlist*.

If you type sts graph, the single survivor function for all the data—the Kaplan–Meier estimate—is plotted. If you type sts graph, by(sex), the survivor functions for each sex—separate Kaplan–Meier estimates—are graphed (on the same axes).

Similarly, if you type sts graph, na, the Nelson–Aalen cumulative hazard function for all the data is plotted. If you type sts graph, na by(sex), the Nelson–Aalen cumulative hazard function for each sex is graphed (on the same axes).

If you type sts graph, hazard, the estimated hazard is plotted; see *Methods and Formulas*. If you type sts graph, hazard by(sex), the estimated hazard for each sex is graphed (on the same axes).

Typically, you specify a single by() variable, such as by(sex), but up to five are allowed. sts graph often restricts you to one by() variable depending on the other options specified because labeling multiple groups on a graph becomes an impossible problem.

If you have more than one by variable and find yourself in a position of needing only one, use egen to create it; see [R] **egen**. For example, to create a single variable denoting the groups defined by both sex and race, type

    . egen *newvar* = group(sex race)

strata(*varlist*) is a subtle alternative to by(). First, you may not specify strata() unless you also specify adjustfor(), whereas you may specify by() in either case. Thus, strata() amounts to a modifier of adjustfor() and is discussed below. This option is not available with na.

adjustfor(*varlist*) adjusts the estimate of the survivor function to that for 0 values of the *varlist* specified. This option is not available with na.

The two commands

```
. sts graph, by(group) adjustfor(age)
```

and

```
. sts graph, strata(group) adjustfor(age)
```

graph the survivor functions for each group. The functions graphed would be adjusted for age, meaning an attempt has been made to remove the differences in the functions due to age. How sts does this depends on whether you specify by() or strata(). This is fully discussed in [ST] **sts**, but here is a quick review:

If you specify strata(), sts performs the adjustment by fitting a stratified-on-group Cox regression model using adjustfor() as the covariates. The stratified, baseline survivor function is then retrieved.

If you specify by(), sts fits separate Cox regression models for each group. The separately calculated baseline survivor functions are then retrieved.

Be aware that, regardless of method employed, the survivor function is adjusted to 0 values of the covariates. Say you are adjusting for age and assume the ages of patients in your sample are 40 to 60. Then,

```
. sts graph, strata(group) adjustfor(age)
```

will graph results adjusted to age 0. If you want to adjust the function to age 40, you type

```
. generate age40 = age - 40
. sts graph, strata(group) adjustfor(age40)
```

failure specifies that you want $1 - S(t + 0)$, the failure function, graphed.

gwood indicates you want pointwise Greenwood confidence bands drawn around the survivor (failure) function. The default is not to include these bands. gwood is not allowed if you specify the adjustfor() option, nor is it allowed if you have pweighted data.

na specifies that the Nelson–Aalen estimate of the cumulative hazard function be plotted.

cna indicates that you want pointwise confidence bands drawn around the Nelson–Aalen cumulative hazard function. cna is not allowed with pweighted data.

hazard specifies that an estimate of the hazard function be plotted. This estimate is calculated as a weighted kernel density estimate utilizing the estimated hazard contributions, $\Delta \widehat{H}(t_j) = \widehat{H}(t_j) - \widehat{H}(t_{j-1})$. These hazard contributions are the same as those obtained by sts generate *newvar* = h.

level(*#*) specifies the confidence level, in percent, for the pointwise confidence interval around the survivor, failure, or cumulative hazard functions; see [U] **23.6 Specifying the width of confidence intervals**.

lost specifies that you wish the numbers lost shown on the plot. These numbers are shown as small numbers over the flat parts of the function.

If enter is not specified, the numbers displayed are the number censored minus the number who enter. If you do specify enter, the numbers displayed are the pure number censored. The logic underlying this is described in [ST] **sts**.

lost is not allowed with hazard.

enter specifies that the number who enter are to be shown on the graph as well as the number lost. The number who enter are shown as small numbers beneath the flat parts of the plotted function.

enter is not allowed with hazard.

separate is meaningful only with by() or strata(); it requests that each group be placed on its own graph rather than one on top of the other. Sometimes curves have to be placed on separate graphs—such as when you specify gwood or cna—because otherwise it would be too confusing.

tmin(#) specifies that the plotted curve is to be graphed only for $t \geq \#$. This option does not affect the calculation of the function, only the portion that is displayed.

tmax(#) specifies that the plotted curve is to be graphed only for $t \leq \#$. This option does not affect the calculation of the function, only the portion that is displayed.

noorigin requests that the plot of the survival (failure) curve begin at the first exit time instead of beginning at $t = 0$ (the default). This option is ignored when na, cna, or hazard is specified.

atrisk specifies that the numbers at risk at the beginning of each time interval be shown on the plot. The numbers at risk are shown as small numbers beneath the flat parts of the plotted function.

atrisk is not allowed with hazard.

noshow prevents sts graph from showing the key st variables. This option is rarely used since most people type stset, show or stset, noshow to reset once and for all whether they want to see these variables mentioned at the top of the output of every st command; see [ST] **stset**.

censored(single | number | multiple) specifies that tick marks be placed on the graph to indicate censored observations.

censored(single) places one tick at each censoring time regardless of the number of censorings at that time.

censored(number) places one tick at each censoring time and displays the number of censorings about the tick.

censored(multiple) places multiple ticks for multiple censorings at the same time. For instance, if 3 observations are censored at time 5, then 3 ticks are placed at time 5. censored(multiple) is intended for use when there are few censored observations; if there are too many, the graph can look bad and in such cases we recommend that censored(number) be used.

censored() is not allowed with hazard.

kernel(*kernel*) is for use with hazard, and is used to specify the kernel function for use in calculating the weighted kernel density estimate required to produce a smoothed hazard function estimator. The default kernel is Epanechinikov, yet *kernel* may be any of the kernels supported by kdensity; see [R] **kdensity**.

width(# [#...]) is for use with hazard, and is used to specify the bandwidth to be used in the kernel smooth used to plot the estimated hazard function. If left unspecified, a default bandwidth is used as described in [R] **kdensity**. If used with by(), then multiple bandwidths may be specified, one for each by() group. If there are more by() groups than the k bandwidths specified, then the default bandwidth is used for the $k + 1, \ldots$ remaining by() groups. If any bandwidth is specified as . (dot), then the default bandwidth is used for that group.

ciopts(*rline_options*) affect the rendition of the confidence bands; see [G] **graph twoway rline**.

plot(*plot*) provides a way to add other plots to the generated graph; see [G] *plot_option*.

line options affect the rendition of the plotted line(s); see [G] **graph twoway line**.

twoway_options are any of the options documented in [G] *twoway_options*. These include options for titling the graph (see [G] *title_options*), options for saving the graph to disk (see [G] *saving_option*), and the by() option (see [G] *by_option*).

Remarks

If you have not read [ST] **sts**, please do so.

What is important to understand about sts graph is that only one of its options — adjustfor() — modifies the calculation. All the other options merely determine how the results of the calculation are graphed.

By default, sts graph displays the Kaplan–Meier product-limit estimate of the survivor (failure) function. Specify by() if you wish to see the results of the calculation performed separately on the different groups.

Specify adjustfor() and an adjusted survival curve is calculated. Now if you specify by() or strata(), this further modifies how the adjustment is made.

Specify na or cna and sts graph displays the Nelson–Aalen estimate of the cumulative hazard function.

Specify hazard and sts graph displays the estimated hazard function.

We demonstrate many of sts graph's features in [ST] **sts**. This discussion picks up where that entry leaves off.

Including the number lost on the graph

In [ST] **sts**, we introduced a simple, one-observation-per-subject drug-trial dataset. Here is a graph of the survivor functions, by drug, including the number lost due to censoring:

(Continued on next page)

```
. use http://www.stata-press.com/data/r8/drug2
(Patient Survival in Drug Trial)

. sts graph, by(drug) lost
```

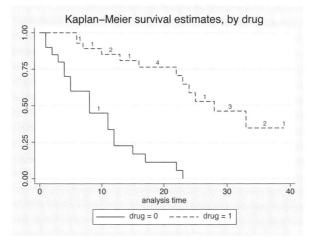

There is no late entry in these data, but pretend there is. Behind the scenes, we just modified the data so that a few subjects entered late. Here is the same graph on the modified data:

```
. use http://www.stata-press.com/data/r8/drug2b, clear

. sts graph, by(drug) lost
```

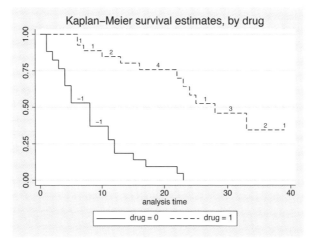

Note the negative numbers. That is because, by default, lost means censored-minus-entered. All -1 means is that 1 entered, or 2 entered and 1 was lost, etc. If we specify the enter option, we will see the censored and entered separately:

```
. sts graph, by(drug) lost enter
```

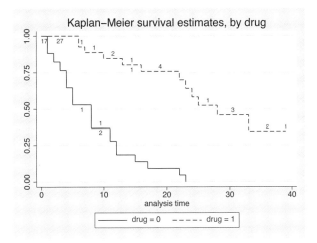

Thus, it might appear that specifying `enter` with `lost` is always a good idea. That is not true.

We have yet another version of the data—the correct data not doctored to have late entry—but in this version we have multiple records per subject. It is the same data, but where there was one record in the first dataset, sometimes there are now two because we have a covariate that is changing over time. Using this dataset, here is the graph with the number lost shown:

```
. use http://www.stata-press.com/data/r8/drug2c, clear
. sts graph, by(drug) lost
```

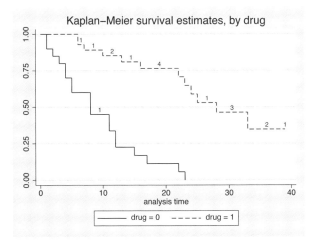

This looks just like the first graph we presented as, indeed, it should. Again we emphasize, the data are logically, if not physically, equivalent. If, however, we graph the number lost and entered, we get a graph showing a lot of activity:

```
. sts graph, by(drug) lost enter
```

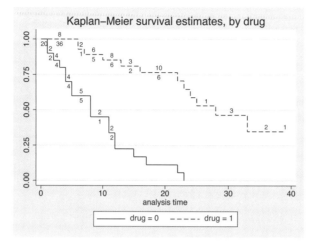

All that activity goes by the name *thrashing*—subjects are being censored to re-enter the data again, but with different covariates. This graph was better when we did not specify `enter` because the censored-minus-entered calculation smoothed out the thrashing.

Graphing the Nelson–Aalen cumulative hazard function

We can plot the Nelson–Aalen estimate of the cumulative (integrated) hazard function by specifying the `na` option. For example, using the one-observation-per-subject drug-trial dataset, here is a graph of the cumulative hazard functions, by drug:

```
. use http://www.stata-press.com/data/r8/drug2, clear
. sts graph, na by(drug)
```

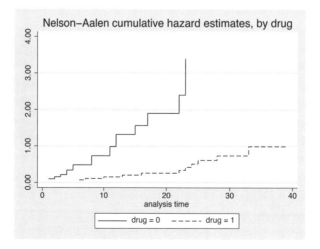

and here is a plot including the number lost due to censoring:

. sts graph, na by(drug) lost

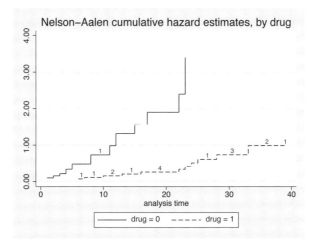

Graphing the hazard function

sts graph may also be used to plot an estimate of the hazard function. This graph is based on a weighted kernel smooth of the estimated *hazard contributions*, $\Delta \widehat{H}(t_j) = \widehat{H}(t_j) - \widehat{H}(t_{j-1})$, as obtained by sts generate *newvar* = h. As such, there are issues associated with selecting a kernel function and a bandwidth, although sts graph will use defaults if don't want to worry about this.

. sts graph, hazard by(drug)

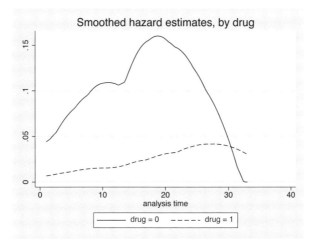

And, we can also adjust the kernel smooth and customize if we wish.

```
. sts graph, hazard by(drug) kernel(gauss) width(5 7)
> title("Comparison of hazard functions")
```

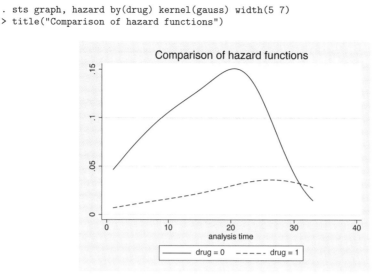

Methods and Formulas

See [ST] **sts**.

The estimated hazard is calculated as a kernel smooth of the estimated hazard contributions, $\Delta \widehat{H}(t_j) = \widehat{H}(t_j) - \widehat{H}(t_{j-1})$, using

$$\widehat{h}(t) = b^{-1} \sum_{j=1}^{D} K\left(\frac{t - t_j}{b}\right) \Delta \widehat{H}(t_j)$$

where $K()$ is the kernel function, b is the bandwidth, and the summation is over the D times at which failure occurs. (Klein and Moeschberger 1997, 153). If adjustfor() is specified, then the $\Delta \widehat{H}(t_j)$ are instead obtained from stcox, as the estimated baseline contributions from a Cox model; see [ST] **stcox** for details on how the $\Delta \widehat{H}(t_j)$ are calculated in this case.

References

See [ST] **sts**.

Also See

Complementary:	[ST] **sts**, [ST] **sts generate**, [ST] **sts list**, [ST] **sts test**, [ST] **stset**
Related:	[R] **kdensity**
Background:	[ST] **st**, [ST] **survival analysis**

Title

> **sts list** — List the survivor and the cumulative hazard functions

Syntax

sts list [if *exp*] [in *range*] [, by(*varlist*) strata(*varlist*) adjustfor(*varlist*) failure

na at(# | *numlist*) compare enter level(#) noshow]

sts list is for use with survival-time data; see [ST] **st**. You must stset your data before using this command.

Description

sts list lists the estimated survivor (failure) or the Nelson–Aalen estimated cumulative (integrated) hazard function. See [ST] **sts** for an introduction to this command.

sts list is appropriate for use with single- or multiple-record, single- or multiple-failure, st data.

Options

by(*varlist*) produces separate survivor or cumulative hazard functions by making separate calculations for each group identified by equal values of the variables in *varlist*.

If you type sts list, the single survivor function for all the data—the Kaplan–Meier estimate—is listed. If you type sts list, by(sex), the survivor function for each sex—separate Kaplan–Meier estimates—are listed.

Similarly, if you type sts list, na, the Nelson–Aalen cumulative hazard function for all the data is listed. If you type sts list, na by(sex), the Nelson–Aalen cumulative hazard function for each sex is listed.

One typically specifies a single by variable, such as by(sex), but up to five are allowed. Thus, you can specify by(sex race).

strata(*varlist*) is a subtle alternative to by(). First, you may not specify strata() unless you also specify adjustfor(), whereas you may specify by() in either case. Thus, strata() amounts to a modifier of adjustfor() and is discussed below. This option is not available with the na option.

(Continued on next page)

267

adjustfor(*varlist*) adjusts the estimate of the survivor function to that for 0 values of the *varlist* specified. This option is not available with the na option.

   The two commands

   . sts list, by(group) adjustfor(age)

and

   . sts list, strata(group) adjustfor(age)

list the survivor functions for each group. The functions listed would be adjusted for age, meaning an attempt has been made to remove the differences in the functions due to age. How sts does this depends on whether you specify by() or strata(). This is fully discussed in [ST] **sts**, but here is a quick review:

If you specify strata(), sts performs the adjustment by estimating a stratified-on-group Cox regression model using adjustfor() as the covariates. The stratified, baseline survivor function is then retrieved.

If you specify by(), sts estimates separate Cox regression models for each group. The separately calculated baseline survivor functions are then retrieved.

Be aware that, regardless of method employed, the survivor function is adjusted to 0 values of the covariates. Say you are adjusting for age and assume that the ages of patients in your sample are 40 to 60. Then

   . sts list, strata(group) adjustfor(age)

will list results adjusted to age 0. If you want to adjust the function to age 40, you type

   . generate age40 = age - 40
   . sts list, strata(group) adjustfor(age40)

failure specifies that you want $1 - S(t + 0)$, the failure function, listed.

na requests that the Nelson–Aalen estimate of the cumulative hazard function be listed.

at(# | *numlist*) specifies the time values at which the estimated survivor (failure) or cumulative hazard function is to be listed.

   The default is to list the function at all the unique time values in the data or, if functions are being compared, at about 10 times chosen over the observed interval. In any case, you can control the points chosen.

   at(5 10 20 30 50 90) would display the function at the designated times.

   at(10 20 to 100) would display the function at times 10, 20, 30, 40, . . . , 100.

   at(0 5 10 to 100 200) would display the function at times 0, 5, 10, 15, . . . , 100, and 200.

   at(20) would display the curve at (roughly) 20 equally spaced times over the interval observed in the data. We say roughly because Stata may choose to increase or decrease your number slightly if that would result in rounder values of the chosen times.

compare is specified only with by() or strata(). It states that you wish to compare the survival (failure) or cumulative hazard functions and so want them listed side-by-side rather than first one and then the next.

enter specifies that you want the table to contain the number who enter, and, correspondingly, that the number lost is to be displayed as the pure number censored rather than censored-minus-entered. The logic underlying this is explained in [ST] **sts**.

level(#) specifies the confidence level, in percent, for the Greenwood pointwise confidence interval of the survivor (failure) or for the pointwise confidence interval of the Nelson–Aalen cumulative hazard function; see [U] **23.6 Specifying the width of confidence intervals**.

noshow prevents sts list from showing the key st variables. This option is rarely used since most people type stset, show or stset, noshow to reset once and for all whether they want to see these variables mentioned at the top of the output of every st command; see [ST] **stset**.

Remarks

What is important to understand about sts list is that only one of its options—adjustfor()—modifies the calculation. All the other options merely determine how the results of the calculation are displayed.

If you do not specify adjustfor() or na, sts list displays the Kaplan–Meier product-limit estimate of the survivor (failure) function. Specify by() if you wish to see the results of the calculation performed separately on the different groups.

Specify adjustfor() and an adjusted survival curve is calculated. Now if you specify by() or strata(), this further modifies how the adjustment is made.

Specify na and sts list displays the Nelson–Aalen estimate of the cumulative hazard function.

We demonstrate many of sts list's features in [ST] **sts**. This discussion picks up where that entry leaves off.

By default, sts list will bury you in output. Using the Stanford heart transplant data introduced in [ST] **stset**, the following produces 154 lines of output.

```
. use http://www.stata-press.com/data/r8/stan3
(Heart transplant data)

. stset, noshow

. sts list, by(posttran)
```

Time	Beg. Total	Fail	Net Lost	Survivor Function	Std. Error	[95% Conf. Int.]	
posttran=0							
1	103	1	3	0.9903	0.0097	0.9331	0.9986
2	99	3	3	0.9603	0.0195	0.8976	0.9849
3	93	3	3	0.9293	0.0258	0.8574	0.9657
(output omitted)							
1400	1	0	1	0.2359	0.1217	0.0545	0.4882
posttran=1							
1	0	0	-3	1.0000	.	.	.
2	3	0	-3	1.0000	.	.	.
(output omitted)							
5.1	14	1	0	0.9286	0.0688	0.5908	0.9896
6	13	0	-1	0.9286	0.0688	0.5908	0.9896
(output omitted)							
1799	1	0	1	0.1420	0.0546	0.0566	0.2653

at() and compare() are the solutions. Here is another detailed, but more useful, view of the heart transplant data:

(Continued on next page)

```
. sts list, at(10 40 to 170) by(posttran)
```

| | Beg. | | Survivor | Std. | | |
Time	Total	Fail	Function	Error	[95% Conf. Int.]	
posttran=0						
10	74	12	0.8724	0.0346	0.7858	0.9256
40	31	11	0.6781	0.0601	0.5446	0.7801
70	17	2	0.6126	0.0704	0.4603	0.7339
100	11	1	0.5616	0.0810	0.3900	0.7022
130	10	1	0.5054	0.0903	0.3199	0.6646
160	7	1	0.4422	0.0986	0.2480	0.6204
posttran=1						
10	16	1	0.9286	0.0688	0.5908	0.9896
40	43	6	0.7391	0.0900	0.5140	0.8716
70	45	9	0.6002	0.0841	0.4172	0.7423
100	40	9	0.4814	0.0762	0.3271	0.6198
130	38	1	0.4687	0.0752	0.3174	0.6063
160	36	1	0.4561	0.0742	0.3076	0.5928

Note: Survivor function is calculated over full data and evaluated at
        indicated times; it is not calculated from aggregates shown at left.

Notice that we specified at(10 40 to 170) when that is not strictly correct; at(10 40 to 160) would make sense and so would at(10 40 to 180), but sts list is not picky.

Similar output for the Nelson–Aalen estimated cumulative hazard can be produced by specifying the na option:

```
. sts list, na at(10 40 to 170) by(posttran)
```

| | Beg. | | Nelson-Aalen | Std. | | |
Time	Total	Fail	Cum. Haz.	Error	[95% Conf. Int.]	
posttran=0						
10	74	12	0.1349	0.0391	0.0764	0.2382
40	31	11	0.3824	0.0871	0.2448	0.5976
70	17	2	0.4813	0.1124	0.3044	0.7608
100	11	1	0.5646	0.1400	0.3473	0.9178
130	10	1	0.6646	0.1720	0.4002	1.1037
160	7	1	0.7896	0.2126	0.4658	1.3385
posttran=1						
10	16	1	0.0714	0.0714	0.0101	0.5071
40	43	6	0.2929	0.1176	0.1334	0.6433
70	45	9	0.4981	0.1360	0.2916	0.8507
100	40	9	0.7155	0.1542	0.4691	1.0915
130	38	1	0.7418	0.1564	0.4908	1.1214
160	36	1	0.7689	0.1587	0.5130	1.1523

Note: Nelson-Aalen function is calculated over full data and evaluated at
        indicated times; it is not calculated from aggregates shown at left.

Here is the result of the survivor functions with the compare option:

(Continued on next page)

```
. sts list, at(10 40 to 170) by(posttran) compare
                Survivor Function
posttran               0           1

time      10      0.8724      0.9286
          40      0.6781      0.7391
          70      0.6126      0.6002
         100      0.5616      0.4814
         130      0.5054      0.4687
         160      0.4422      0.4561
```

and of the cumulative hazard functions:

```
. sts list, na at(10 40 to 170) by(posttran) compare
                Nelson-Aalen Cum. Haz.
posttran               0           1

time      10      0.1349      0.0714
          40      0.3824      0.2929
          70      0.4813      0.4981
         100      0.5646      0.7155
         130      0.6646      0.7418
         160      0.7896      0.7689
```

Methods and Formulas

See [ST] **sts**.

References

See [ST] **sts**.

Also See

Complementary:	[ST] **sts**, [ST] **sts generate**, [ST] **sts graph**, [ST] **sts test**, [ST] **stset**
Background:	[ST] **st**, [ST] **survival analysis**; [U] **14.1.8 numlist**

Title

> **sts test** — Test equality of survivor functions

Syntax

sts <u>test</u> *varlist* [if *exp*] [in *range*] [,

    [<u>log</u>rank | <u>w</u>ilcoxon | <u>tw</u>are | <u>pet</u>o | <u>f</u>h(*p q*) | <u>c</u>ox]

    <u>str</u>ata(*varlist*) <u>d</u>etail <u>tr</u>end mat(*matname₁ matname₂*) <u>not</u>itle <u>nosh</u>ow]

sts test is for use with survival-time data; see [ST] **st**. You must stset your data before using this command.

Description

sts test tests the equality of survivor functions across two or more groups. The log-rank, Wilcoxon, Tarone–Ware, Peto–Peto–Prentice, Fleming–Harrington, and "Cox" tests are provided, in both unstratified and stratified forms.

sts test also provides a test for trend.

See [ST] **sts** for an introduction to this command.

sts test is appropriate for use with single- or multiple-record, single- or multiple-failure, st data.

Options

logrank, wilcoxon, tware, peto, fh(), and cox specify which test of equality is desired. logrank is the default unless the data are pweighted, in which case, cox is the default and is the only possibility.

wilcoxon specifies the Wilcoxon–Breslow–Gehan test, tware the Tarone–Ware test, peto the Peto–Peto–Prentice test, and fh() the generalized Fleming–Harrington test. When specifying the Fleming–Harrington test, two arguments, p and q, are expected. When $p = 0$ and $q = 0$, the Fleming–Harrington test reduces to the log-rank test; and when $p = 1$ and $q = 0$, the test reduces to the Mann–Whitney–Wilcoxon test.

strata(*varlist*) requests that a stratified test be performed.

detail modifies strata(); it requests that, in addition to reporting the overall stratified test, the tests for the individual strata be reported as well. detail is not allowed with cox.

trend specifies that a test for trend of the survivor function across three or more ordered groups be performed.

mat(*matname₁ matname₂*) is not allowed with cox. The other tests are rank tests of the form $\mathbf{u}'\mathbf{V}^{-1}\mathbf{u}$. mat() requests that the vector $\mathbf{u}$ be stored in *matname₁* and that $\mathbf{V}$ be stored in *matname₂*.

notitle requests that the title printed above the test be suppressed.

noshow prevents sts test from showing the key st variables. This option is rarely used since most people type stset, show or stset, noshow to reset once and for all whether they want to see these variables mentioned at the top of the output of every st command; see [ST] **stset**.

Remarks

Remarks are presented under the headings

    The log-rank test
    The Wilcoxon (Breslow–Gehan) test
    The Tarone–Ware test
    The Peto–Peto–Prentice test
    The generalized Fleming–Harrington tests
    The "Cox" test
    The trend test

sts test tests the equality of the survivor function across groups. With the exception of "Cox", these tests are members of a family of statistical tests that are extensions to censored data of traditional nonparametric rank tests for comparing two or more distributions. A technical description of these tests can be found in the *Methods and Formulas* section of this entry. Simply, at each distinct failure time in the data, the contribution to the test statistic is obtained as a weighted standardized sum of the difference between the observed and expected number of deaths in each of the k groups. The expected number of deaths is obtained under the null hypothesis of no differences between the survival experience of the k groups.

The weights or weight function used determine the test statistic. For example, when the weight is 1 at all failure times, the log-rank test is computed, and when the weight is the number of subjects at risk of failure at each distinct failure time, the Wilcoxon–Breslow–Gehan test is computed.

Summarized in the following table are the weights used for each of the statistical tests.

Test	weight at each distinct failure time (t_i)
Log-rank	1
Wilcoxon–Breslow–Gehan	n_i
Tarone–Ware	$\sqrt{n_i}$
Peto–Peto–Prentice	$\widetilde{S}(t_i)$
Fleming–Harrington	$\widehat{S}(t_{i-1})^p \big[1 - \widehat{S}(t_{i-1})\big]^q$

where $\widehat{S}(t_i)$ is the estimated Kaplan–Meier survivor function value for the combined sample at failure time t_i, $\widetilde{S}(t_i)$ is a modified estimate of the overall survivor function described in *Methods and Formulas*, and n_i is the number of subjects in the risk pool at failure time t_i.

These tests are appropriate for testing the equality of survivor functions across two or more groups. Up to 800 groups are allowed.

The "Cox" test is related to the log-rank test but is performed as a likelihood-ratio (or, alternatively, as a Wald test) on the results from a Cox proportional-hazards regression. The log-rank test should be preferable to what we have labeled the Cox test, but with pweighted data, the log-rank test is not appropriate. In point of fact, whether one performs the log-rank or Cox test makes little substantive difference with most datasets.

sts test, trend can be used to test against the alternative hypothesis that the failure rate increases or decreases as the level of the k groups increases or decreases. This test is only appropriate when there is a natural ordering of the comparison groups; for example, when each group represents an increasing or decreasing level of a therapeutic agent.

trend is not valid when cox is specified.

The log-rank test

sts test, by default, performs the log-rank test which, to be clear, is the exponential scores test (Savage 1956, Mantel and Haenszel 1959, Mantel 1963, Mantel 1966). This test is most appropriate when the hazard functions are thought to be proportional across the groups if they are not equal.

This test statistic is constructed by giving equal weights to the contribution of each failure time to the overall test statistic.

In [ST] **sts**, we demonstrated the use of this command with the heart transplant data, a multiple-record, single-failure st dataset.

```
. use http://www.stata-press.com/data/r8/stan3
(Heart transplant data)

. sts test posttran

        failure _d:  died
   analysis time _t:  t1
                id:  id
```

Log-rank test for equality of survivor functions

posttran	Events observed	Events expected
0	30	31.20
1	45	43.80
Total	75	75.00

$$chi2(1) = 0.13$$
$$Pr>chi2 = 0.7225$$

We cannot reject the hypothesis that the survivor functions are the same.

sts test, logrank can also perform the stratified log-rank test. Say it is suggested that calendar year of acceptance also affects survival and that there are three important periods: 1967–69, 1970–72, and 1973–74. Therefore, a stratified test should be performed:

```
. stset, noshow

. generate group = 1 if year <= 69
(117 missing values generated)

. replace group=2 if year>=70 & year<=72
(78 real changes made)

. replace group=3 if year>=73
(39 real changes made)

. sts test posttran, strata(group)
```

Stratified log-rank test for equality of survivor functions

posttran	Events observed	Events expected(*)
0	30	31.51
1	45	43.49
Total	75	75.00

(*) sum over calculations within group

$$chi2(1) = 0.20$$
$$Pr>chi2 = 0.6547$$

Still finding nothing, you ask Stata to show the within-strata tests:

```
. sts test posttran, strata(group) detail
```

Stratified log-rank test for equality of survivor functions

```
-> group = 1
```

posttran	Events observed	Events expected
0	14	13.59
1	17	17.41
Total	31	31.00

```
              chi2(1) =      0.03
             Pr>chi2 =    0.8558
```

```
-> group = 2
```

posttran	Events observed	Events expected
0	13	13.63
1	20	19.37
Total	33	33.00

```
              chi2(1) =      0.09
             Pr>chi2 =    0.7663
```

```
-> group = 3
```

posttran	Events observed	Events expected
0	3	4.29
1	8	6.71
Total	11	11.00

```
              chi2(1) =      0.91
             Pr>chi2 =    0.3410
```

```
-> Total
```

posttran	Events observed	Events expected(*)
0	30	31.51
1	45	43.49
Total	75	75.00

```
(*) sum over calculations within group
              chi2(1) =      0.20
             Pr>chi2 =    0.6547
```

The Wilcoxon (Breslow–Gehan) test

sts test, wilcoxon performs the generalized Wilcoxon test of Breslow (1970) and Gehan (1965). This test is appropriate when hazard functions are thought to vary in ways other than proportionally and censoring patterns are similar across groups.

The Wilcoxon test statistic is constructed by weighting the contribution of each failure time to the overall test statistic by the number of subjects at risk. Thus, it gives heavier weights to earlier failure times when the number at risk is higher. As a result, this test is susceptible to differences in the censoring pattern of the groups.

sts test, wilcoxon works the same way as sts test, logrank:

. sts test posttran, wilcoxon

Wilcoxon (Breslow) test for equality of survivor functions

posttran	Events observed	Events expected	Sum of ranks
0	30	31.20	-85
1	45	43.80	85
Total	75	75.00	0

$$\text{chi2(1)} = 0.14$$
$$\text{Pr>chi2} = 0.7083$$

and, with the strata() option, sts test, wilcoxon performs the stratified test:

. sts test posttran, wilcoxon strata(group)

Stratified Wilcoxon (Breslow) test for equality of survivor functions

posttran	Events observed	Events expected(*)	Sum of ranks(*)
0	30	31.51	-40
1	45	43.49	40
Total	75	75.00	0

(*) sum over calculations within group

$$\text{chi2(1)} = 0.22$$
$$\text{Pr>chi2} = 0.6385$$

As with sts test, logrank, you can also specify the detail option to see the within-stratum tests.

The Tarone–Ware test

sts test, tware performs a test suggested by Tarone and Ware (1977, 383), with weights equal to the square root of the number of subjects in the risk pool at time t_i.

Like Wilcoxon's test, this test is appropriate when hazard functions are thought to vary in ways other than proportionally and censoring patterns are similar across groups. The test statistic is constructed by weighting the contribution of each failure time to the overall test statistic by the square root of the number of subjects at risk. Thus, like the Wilcoxon test, it gives heavier weights, although not as large, to earlier failure times. Although less susceptible to the failure and censoring pattern in the data than Wilcoxon's test, this could remain a problem if large differences in these patterns exist between groups.

sts test, tware works the same way as sts test, logrank:

```
. sts test posttran, tware
```

Tarone-Ware test for equality of survivor functions

posttran	Events observed	Events expected	Sum of ranks
0	30	31.20	-9.3375685
1	45	43.80	9.3375685
Total	75	75.00	0

```
             chi2(1) =      0.12
             Pr>chi2 =    0.7293
```

and, with the `strata()` option, `sts test, tware` performs the stratified test:

```
. sts test posttran, tware strata(group)
```

Stratified Tarone-Ware test for equality of survivor functions

posttran	Events observed	Events expected(*)	Sum of ranks(*)
0	30	31.51	-7.4679345
1	45	43.49	7.4679345
Total	75	75.00	0

(*) sum over calculations within group

```
             chi2(1) =      0.21
             Pr>chi2 =    0.6464
```

As with `sts test, logrank`, you can also specify the `detail` option to see the within-stratum tests.

The Peto–Peto–Prentice test

`sts test, peto` performs an alternative to the Wilcoxon test proposed by Peto and Peto (1972) and Prentice (1978). The test uses as the weight function an estimate of the overall survivor function, which is similar to that obtained using the Kaplan–Meier estimator. See *Methods and Formulas* for details.

This test is appropriate when hazard functions are thought to vary in ways other than proportionally but, unlike the Wilcoxon–Breslow–Gehan test, it is not affected by differences in censoring patterns across groups.

`sts test, peto` works the same way as `sts test, logrank`:

```
. sts test posttran, peto
```

Peto-Peto test for equality of survivor functions

posttran	Events observed	Events expected	Sum of ranks
0	30	31.20	-.86708453
1	45	43.80	.86708453
Total	75	75.00	0

```
             chi2(1) =      0.15
             Pr>chi2 =    0.6979
```

and, with the `strata()` option, `sts test, peto` performs the stratified test:

```
. sts test posttran, peto strata(group)
```

Stratified Peto-Peto test for equality of survivor functions

posttran	Events observed	Events expected(*)	Sum of ranks(*)
0	30	31.51	-.96898006
1	45	43.49	.96898006
Total	75	75.00	0

(*) sum over calculations within group

$$\text{chi2(1)} = \quad 0.19$$
$$\text{Pr>chi2} = \quad 0.6618$$

As with the previous tests, you can also specify the `detail` option to see the within-stratum tests.

The generalized Fleming–Harrington tests

`sts test, fh(p q)` performs the Fleming and Harrington (1982) class of test statistics. The weight function at each distinct failure time t is the product of the Kaplan–Meier survivor estimate at time $t-1$ raised to the p power and $1-$ the Kaplan–Meier survivor estimate at time $t-1$ raised to the q power. Thus, when specifying the Fleming and Harrington option we must specify two nonnegative arguments, p and q.

When $p > q$, the test gives more weights to earlier failures than to later ones. When $p < q$, the opposite is true, and more weight is given to later than to earlier times. When p and q are both zero, the weight is 1 at all failure times and the test reduces to the log-rank test.

`sts test, fh(p q)` works the same way as `sts test, logrank`. As we mentioned, if we specify $p = 0$ and $q = 0$ we will get the same results as the log-rank test,

```
. sts test posttran, fh(0 0)
```

Fleming-Harrington test for equality of survivor functions

posttran	Events observed	Events expected	Sum of ranks
0	30	31.20	-1.1995511
1	45	43.80	1.1995511
Total	75	75.00	0

$$\text{chi2(1)} = \quad 0.13$$
$$\text{Pr>chi2} = \quad 0.7225$$

which is what the log-rank test reported.

We could, for example, give more weight to later failures than to earlier ones.

```
. sts test posttran, fh(0 3)
```

Fleming-Harrington test for equality of survivor functions

posttran	Events observed	Events expected	Sum of ranks
0	30	31.20	-.09364975
1	45	43.80	.09364975
Total	75	75.00	0

```
              chi2(1) =     0.01
              Pr>chi2 =   0.9139
```

Similar to the previous tests, with the strata() option, sts test, fh() performs the stratified test:

```
. sts test posttran, fh(0 3) strata(group)
```

Stratified Fleming-Harrington test for equality of survivor functions

posttran	Events observed	Events expected(*)	Sum of ranks(*)
0	30	31.51	-.1623318
1	45	43.49	.1623318
Total	75	75.00	0

(*) sum over calculations within group

```
              chi2(1) =     0.05
              Pr>chi2 =   0.8266
```

As with the other tests, you can also specify the detail option to see the within-stratum tests.

The "Cox" test

The term Cox test is our own, and this test is a variation on the log-rank test using Cox regression.

One way of thinking about the log-rank test is as a Cox proportional-hazards model on indicator variables for each of the groups. The log-rank test is a test that the coefficients are zero or, if you prefer, the hazard ratios are one. The log-rank test is, in fact, a score test of that hypothesis performed on a slightly different (partial) likelihood function that handles ties more accurately.

It is generally felt that a (less precise) score test on the precise likelihood function is preferable to a (more precise) likelihood-ratio test on the approximate likelihood function used in Cox regression estimation. In our experience, it makes little difference:

```
. sts test posttran, cox
```

Cox regression-based test for equality of survival curves

posttran	Events observed	Events expected	Relative hazard
0	30	31.20	0.9401
1	45	43.80	1.0450
Total	75	75.00	1.0000

```
           LR chi2(1) =     0.13
              Pr>chi2 =   0.7222
```

By comparison, sts test, logrank also reported $\chi^2 = .13$, although the significance level was 0.7225, meaning the χ^2 values differed in the fourth digit. As mentioned by Kalbfleisch and Prentice (2002, 20), a primary advantage of the log-rank test is the ease with which it can be explained to nonstatisticians, since the test statistic is the difference between the observed and expected number of failures within groups.

Our purpose in offering sts test, cox is not to promote its use instead of the log-rank test, but to provide a test for those with sample-weighted data.

If you have sample weights (if you specified pweights when you stset the data), you cannot run the log-rank or Wilcoxon tests. Perhaps someone has worked out the generalization of these tests to sample-weighted data, but we do not know about it.

The Cox regression model, however, has been generalized to sample-weighted data, and Stata's stcox can fit models with such data. In sample-weighted data, the likelihood-ratio statistic is no longer appropriate, but the Wald test based on the robust estimator of variance is.

Thus, if we treated these data as sample-weighted data, we would obtain

```
. generate one = 1
. stset t1 [pw=one], id(id) time0(_t0) failure(died) noshow
              id:  id
   failure event:  died != 0 & died < .
obs. time interval:  (_t0, t1]
 exit on or before:  failure
          weight:  [pweight=one]

     172  total obs.
       0  exclusions

     172  obs. remaining, representing
     103  subjects
      75  failures in single failure-per-subject data
  31938.1  total analysis time at risk, at risk from t =           0
                            earliest observed entry t =           0
                               last observed exit t =        1799
. sts test posttran, cox
```

Cox regression-based test for equality of survival curves

posttran	Events observed	Events expected	Relative hazard
0	30.00	31.20	0.9401
1	45.00	43.80	1.0450
Total	75.00	75.00	1.0000

```
            Wald chi2(1) =      0.13
               Pr>chi2 =      0.7181
```

Note that sts test, cox now reports the Wald statistic which is, to two digits, 0.13 just like all the others.

The trend test

When the groups to be compared have a natural order, such as increasing or decreasing age groups, or drug dosage, we may want to test the null hypothesis of no difference in failure rate among the groups versus the alternative hypothesis that the failure rate increases or decreases as we move from one group to the next.

We illustrate this test with a dataset from a carcinogenesis experiment reprinted in Marubini and Grazia (1995, 126). Twenty-nine experimental animals were exposed to three levels (0, 1.5, 2.0) of a carcinogenic agent. The time in days to tumor formation was recorded. Here are a few of the observations:

```
. use http://www.stata-press.com/data/r8/marubini, clear
. list in 1/9
```

	time	event	group	dose
1.	67	1	2	1.5
2.	150	1	2	1.5
3.	47	1	3	2
4.	75	0	1	0
5.	58	1	3	2
6.	136	1	2	1.5
7.	58	1	3	2
8.	150	1	2	1.5
9.	43	0	2	1.5

In these data, there are two variables that indicate exposure level. The group variable is coded 1, 2, and 3, indicating a one-unit separation between exposures. The dose variable records the actual exposure dosage. To test the null hypothesis of no difference among the survival experience of the three groups versus the alternative hypothesis that the survival experience of at least one of the groups is different, it does not matter if we use group or dose.

```
. stset time, fail(event)
       failure event:  event != 0 & event < .
  obs. time interval:  (0, time]
   exit on or before:  failure

      29  total obs.
       0  exclusions

      29  obs. remaining, representing
      15  failures in single record/single failure data
    2564  total analysis time at risk, at risk from t =         0
                             earliest observed entry t =         0
                              last observed exit t =           246
. sts test group, noshow
```

Log-rank test for equality of survivor functions

group	Events observed	Events expected
1	4	6.41
2	6	6.80
3	5	1.79
Total	15	15.00

$$\text{chi2(2)} = 8.05$$
$$\text{Pr>chi2} = 0.0179$$

```
. sts test dose, noshow
```

Log-rank test for equality of survivor functions

dose	Events observed	Events expected
0	4	6.41
1.5	6	6.80
2	5	1.79
Total	15	15.00

$$chi2(2) = 8.05$$
$$Pr>chi2 = 0.0179$$

For the trend test, however, the distance between the values is important, so using group or dose will produce different results.

```
. sts test group, noshow trend
```

Log-rank test for equality of survivor functions

group	Events observed	Events expected
1	4	6.41
2	6	6.80
3	5	1.79
Total	15	15.00

$$chi2(2) = 8.05$$
$$Pr>chi2 = 0.0179$$

Test for trend of survivor functions

$$chi2(1) = 5.87$$
$$Pr>chi2 = 0.0154$$

```
. sts test dose, noshow trend
```

Log-rank test for equality of survivor functions

dose	Events observed	Events expected
0	4	6.41
1.5	6	6.80
2	5	1.79
Total	15	15.00

$$chi2(2) = 8.05$$
$$Pr>chi2 = 0.0179$$

Test for trend of survivor functions

$$chi2(1) = 3.66$$
$$Pr>chi2 = 0.0557$$

Although the above trend test was constructed using the log-rank test, any of the previously mentioned weight functions can be used. For example, a trend test on the data can be performed using the same weights as the Peto–Peto–Prentice test by specifying the option peto.

```
. sts test dose, noshow trend peto
```

Peto-Peto test for equality of survivor functions

dose	Events observed	Events expected	Sum of ranks
0	4	6.41	-1.2792221
1.5	6	6.80	-1.3150418
2	5	1.79	2.5942639
Total	15	15.00	0

```
           chi2(2) =       8.39
           Pr>chi2 =     0.0150
```

Test for trend of survivor functions
```
           chi2(1) =       2.85
           Pr>chi2 =     0.0914
```

Saved Results

sts test saves in r():

Scalars

r(df)	degrees of freedom	r(chi2)	χ^2
r(df_tr)	degrees of freedom, trend test	r(chi2_tr)	χ^2, trend test

Methods and Formulas

sts test is implemented as an ado-file.

Let $t_1 < t_2 < \cdots < t_k$ denote the ordered failure times; let d_j be the number of failures at t_j and n_j the population at risk just before t_j; and let d_{ij} and n_{ij} denote the same things for group i, $i = 1, \ldots, r$.

We are interested in testing the null hypothesis

$$H_o : \lambda_1(t) = \lambda_2(t) = \cdots = \lambda_r(t)$$

where $\lambda(t)$ is the hazard function at time t, against the alternative hypothesis that at least one of the $\lambda_i(t)$ is different for some t_j.

As described in Klein and Moeschberger (1997, 191–202), Kalbfleisch and Prentice (2002, 20–22), and Collett (1994, 45–46), if the null hypothesis is true, then the expected number of failures in group i at time t_j is $e_{ij} = n_{ij}d_j/n_j$, and the test statistic

$$\mathbf{u}' = \sum_{j=1}^{k} W(t_j)(d_{1j} - e_{1j}, \ldots, d_{rj} - e_{rj})$$

is formed. $W(t_j)$ is a positive weight function defined as zero when n_{ij} is zero. The various test statistics are obtained by selecting different weight functions $W(t_j)$. See the table in the *Remarks* section of this entry for a complete list of these weight functions. In the case of the Peto-Peto-Prentice test

$$W(t_j) = \widetilde{S}(t_j) = \prod_{\ell:t_\ell \leq t_j} \left(1 - \frac{d_\ell}{n_\ell + 1}\right)$$

The variance matrix $\mathbf{V}$ for $\mathbf{u}$ has elements

$$V_{il} = \sum_{j=1}^{k} \frac{W(t_j)^2 n_{ij} d_j (n_j - d_j)}{n_j(n_j - 1)} \left(\delta_{il} - \frac{n_{ij}}{n_j} \right)$$

where $\delta_{il} = 1$ if $i = l$ and 0 otherwise.

For the unstratified test, statistic $\mathbf{u}'\mathbf{V}^{-1}\mathbf{u}$ is distributed as χ^2 with $r - 1$ degrees of freedom.

For the stratified test, let $\mathbf{u}_s$ and $\mathbf{V}_s$ be the results of performing the above calculation separately within stratum, and define $\mathbf{u} = \sum_s \mathbf{u}_s$ and $\mathbf{V} = \sum_s \mathbf{V}_s$. The χ^2 test is given by $\mathbf{u}'\mathbf{V}^{-1}\mathbf{u}$ redefined in this way.

The "Cox" test is performed by fitting a (possibly stratified) Cox regression using stcox on $r - 1$ indicator variables, one for each of the groups with one of the indicators omitted. The χ^2 test reported is then the likelihood-ratio test (no pweights) or the Wald test (based on the robust estimate of variance); see [ST] **stcox**.

The reported relative hazards are the exponentiated coefficients from the Cox regression renormalized, and the renormalization plays no role in the calculation of the test statistic. The renormalization is chosen so that the expected-number-of-failures-within-group weighted average of the regression coefficients is 0 (meaning the hazard is 1). Let b_i, $i = 1, \ldots, r - 1$ be the estimated coefficients, and define $b_r = 0$. The constant K is then calculated by

$$K = \sum_{i=1}^{r} e_i b_i / d$$

where $e_i = \sum_j e_{ij}$ is the expected number of failures for group i, d is the total number of failures across all groups, and r is the number of groups. The reported relative hazards are $\exp(b_i - K)$.

The trend test assumes that there is natural ordering of the r groups, $r > 2$. In this case, we are interested in testing the null hypothesis

$$H_o : \lambda_1(t) = \lambda_2(t) = \ldots = \lambda_r(t)$$

against the alternative hypothesis

$$H_a : \lambda_1(t) \le \lambda_2(t) \le \ldots \le \lambda_r(t)$$

The test uses $\mathbf{u}$ as previously defined with any of the available weight functions. The test statistic is given by

$$\frac{\left(\sum_{i=1}^{r} a_i u_i \right)^2}{\mathbf{a}'\mathbf{V}\mathbf{a}}$$

where $a_1 \le a_2 \le \ldots \le a_r$ are scores defining the relationship of interest. A score is assigned to each of the comparison group, equal to the value of the grouping variable for that group. $\mathbf{a}$ is the vector of these scores.

References

Breslow, N. E. 1970. A generalized Kruskal–Wallis test for comparing k samples subject to unequal patterns of censorship. *Biometrika* 57: 579–594.

Collett, D. 1994. *Modelling Survival Data in Medical Research.* London: Chapman & Hall.

Gehan, E. A. 1965. A generalized Wilcoxon test for comparing arbitrarily singly censored data. *Biometrika* 52: 203–223.

Harrington, D. P. and T. R. Fleming. 1982. A class of rank test procedures for censored survival data. *Biometrika* 69: 553–566.

Kalbfleisch, J. D. and R. L. Prentice. 2002. *The Statistical Analysis of Failure Time Data.* 2d ed. New York: John Wiley & Sons.

Klein, J. P. and M. L. Moeschberger. 1997. *Survival Analysis: Techniques for Censored and Truncated data.* New York: Springer.

Mantel, N. 1963. Chi-squared tests with one degree of freedom: extensions of the Mantel–Haenszel procedure. *Journal of the American Statistical Association* 58: 690–700.

——. 1966. Evaluation of survival data and two new rank-order statistics arising in its consideration. *Cancer Chemotherapy Reports* 50: 163–170.

Mantel, N. and W. Haenszel. 1959. Statistical aspects of the analysis of data from retrospective studies of disease. *Journal of the National Cancer Institute* 22: 719–748. Reprinted in *Evolution of Epidemiologic Ideas*, ed. S. Greenland, 112–141. Newton Lower Falls, MA: Epidemiology Resources.

Marubini, E. and M. Grazia-Valsecchi. 1995. *Analyzing Survival Data from Clinical Trials and Observational Studies.* New York: John Wiley & Sons.

Peto, R. and J. Peto. 1972. Asymptotically efficient rank invariant test procedures (with discussion). *Journal of the Royal Statistical Society* 135: 185–206.

Prentice, R. L. 1978. Linear rank tests with right-censored data. *Biometrika* 65: 167–179.

Savage, I. R. 1956. Contributions to the theory of rank-order statistics — the two-sample case. *Annals of Mathematical Statistics* 27: 590–615.

Tarone, R. E. and J. H. Ware. 1977. On distribution-free tests for equality of survival distributions. *Biometrika* 64: 156–160.

White, I. R., S. Walker, and A. Babiker. 2002. strbee: Randomization-based efficacy estimator. *The Stata Journal* 2: 140–150.

Wilcoxon, F. 1945. Individual comparisons by ranking methods. *Biometrics* 1: 80–83.

Also See

Complementary:	[ST] **stcox**, [ST] **sts**, [ST] **sts generate**, [ST] **sts graph**, [ST] **sts list**, [ST] **stset**
Background:	[ST] **st**, [ST] **survival analysis**

Title

> **stset** — Declare data to be survival-time data

Syntax

Single-record per subject survival data

stset *timevar* [if *exp*] [*weight*] [, <u>f</u>ailure(*failvar*[==*numlist*]) time0(*varname*)

    <u>o</u>rigin(<u>t</u>ime *exp*) <u>sc</u>ale(*#*) <u>en</u>ter(<u>t</u>ime *exp*) <u>ex</u>it(<u>t</u>ime *exp*) if(*exp*) <u>nosh</u>ow]

streset [if *exp*] [*weight*] [, *same options as stset*]

st [, <u>noc</u>md <u>not</u>able]

stset, clear

Multiple-record per subject survival data

stset *timevar* [if *exp*] [*weight*] , id(*varname*) <u>f</u>ailure(*failvar*[==*numlist*]) [

    <u>o</u>rigin([*varname*==*numlist*] <u>t</u>ime *exp* | min) <u>sc</u>ale(*#*)

    <u>en</u>ter([*varname*==*numlist*] <u>t</u>ime *exp*) <u>ex</u>it(failure | [*varname*==*numlist*] <u>t</u>ime *exp*)

    time0(*varname*) if(*exp*) ever(*exp*) never(*exp*) after(*exp*) <u>befor</u>e(*exp*) <u>nosh</u>ow]

streset [if *exp*] [*weight*] [, *same options as stset*]

streset, {<u>p</u>ast | <u>f</u>uture | <u>p</u>ast <u>f</u>uture }

st [, <u>noc</u>md <u>not</u>able]

stset, clear

fweights, iweights, and pweights are allowed; see [U] **14.1.6 weight**.

Examples

. stset ftime	(*Time measured from 0, all failed*)
. stset ftime, failure(died)	(*Time measured from 0, censoring*)
. stset ftime, failure(died) id(id)	(*Time measured from 0, censoring and id*)
. stset ftime, failure(died==2,3)	(*Time measured from 0, failure codes*)
. stset ftime, failure(died) origin(time dob)	(*Time measured from dob, censoring*)

You cannot harm your data using stset, so feel free to experiment.

Description

st refers to survival-time data, which are fully described below.

stset declares the data in memory to be survival-time (st) data, informing Stata of key variables and their roles in a survival-time data analysis. When you stset your data, stset runs various data consistency checks to ensure that what you have declared makes sense. Note that if the data are weighted, you specify the weights when you stset the data, not when you issue the individual st commands.

streset changes how the st dataset is declared. In multiple record data, streset can also temporarily set the sample to include records prior to being at risk (called the past) and records after failure (called the future). In that case, typing streset without arguments resets the sample back to the analysis sample.

st displays how the dataset is currently declared.

Whenever you type stset or streset, Stata runs or reruns data consistency checks to ensure that what you are now declaring (or declared in the past) makes sense. Thus, if you have made any changes to your data or simply wish to verify how things are, you can type streset without any options.

stset, clear is for use by programmers. It causes Stata to forget the st markers, making the data no longer st data to Stata. The data remain unchanged. It is not necessary to stset, clear before doing another stset.

❑ Technical Note

Note to Stata version 5.0 users: stset was changed in Stata 6. Your old Stata 5 st datasets will work with the new st:

1. If you want to use old st datasets with the new st commands, there is nothing special you need to do. You do **not** need to re-stset your data. Just use your dataset and everything will just work.

2. There are syntax changes. If you want to use the old st syntax (perhaps you have old do-files and need to recreate an old analysis), type

    . version 5

You will then be using the old st system. When you want to use the new st system again, type

    . version 8

3. If you are using version 8 (or version 7 or 6), the stset syntax has changed from version 5. Where you previously typed 'stset *timevar failvar*', now type 'stset *timevar*, failure(*failvar*)'.

In addition, the t0() option is gone. It is literally renamed time0(), and we did that so that you would not use it unthinkingly. Where you used to specify t0(*evar*), you now want to say enter(time *evar*) or time0(*evar*), and you probably want to say the former. The new system draws a distinction among coming at risk (called origin), coming under observation (called entry), and the start date for a record (called time0).

❑

Options for use with stset and streset

id(*idvar*) specifies the subject-id variable; observations with equal, nonmissing values of *idvar* are assumed to be the same subject. *idvar* may be string or numeric. Observations for which *idvar* is missing (. or "") are ignored.

When id() is not specified, each observation is assumed to represent a different subject, and thus constitutes a single-record-per-subject survival dataset.

When you specify id(), the data are said to be multiple-record data even if it turns out there is only one record per subject. Perhaps they would better be called potentially multiple-record data.

If you specify id(), stset requires that you specify failure().

Specifying id() never hurts; we recommend it because a few st commands, such as stsplit, require that an id variable have been specified when the dataset was stset.

failure(*failvar*[==*numlist*]) specifies the failure event.

If failure() is not specified, all records are assumed to end in failure. This is allowed with single-record data only.

If failure(*failvar*) is specified, *failvar* is interpreted as an indicator variable; 0 and missing mean censored and all other values are interpreted as representing failure.

If failure(*failvar*==*numlist*) is specified, records with *failvar* taking on any of the values in *numlist* are assumed to end in failures and all other records are assumed to be censored.

origin([*varname*==*numlist*] time *exp* | min) and scale(*#*) define analysis time or, said differently, origin() defines when a subject becomes at risk. Subjects become at risk when time = origin(). All analyses are performed in terms of time since becoming at risk, called analysis time.

Let us use the terms *time*, for how time is recorded in the data, and *t*, for analysis time. Analysis time *t* is defined

$$t = \frac{time - \texttt{origin()}}{\texttt{scale()}}$$

t is time from origin in units of scale.

By default, origin(time 0) and scale(1) are assumed, meaning $t = time$. In that case, it is your responsibility to ensure that *time* in your data is measured as time since becoming at risk. Subjects are exposed at $t = time = 0$ and subsequently fail. Observations with $t = time < 0$ are ignored because information prior to becoming at risk is irrelevant.

origin() plays the substantive role of determining when the clock starts ticking. scale() plays no substantive role, but it can be handy for making *t* units more readable (such as converting days to years).

origin(time *exp*) sets the origin to *exp*. For instance, if *time* were recorded as dates such as 05jun1998 in your data and variable expdate recorded the date when subjects were exposed, you could specify origin(time expdate). If instead all subjects were exposed on 12nov1997, you could specify origin(time mdy(11,12,1997)).

origin(time *exp*) may be used with single- or multiple-record data.

origin(*varname*==*numlist*) is for use with multiple-record data; it specifies the origin indirectly. If *time* were recorded as dates in your data, variable obsdate recorded the (ending) date associated with each record, and subjects came at risk upon, say, having a certain operation— and that operation were indicated by code==217—then you could specify origin(code==217). origin(code==217) would mean, for each subject, that the origin time is the earliest time at which code==217 is observed. Records prior to that would be ignored (because $t < 0$). Subjects who never had code==217 would be ignored entirely.

origin(*varname==numlist* time *exp*) sets the origin to be the latter of the two times determined by *varname==numlist* and *exp*.

origin(min) sets origin to the earliest time observed, minus 1. This is a very odd thing to do and is described in *Example 10* under *Remarks* below.

origin() is an important concept; see the entries *Key concepts*, *Two concepts of time*, and *The substantive meaning of analysis time* under *Remarks* below.

scale() makes results more readable. If you have *time* recorded in days (such as Stata dates, which are really days since 01jan1960), specifying scale(365.25) will cause results to be reported in years.

enter([*varname==numlist*] time *exp*) specifies when a subject first comes under observation, meaning that any failures, were they to occur, would be recorded in the data.

Do not confuse enter() and origin(). origin() specifies when a subject first becomes at risk. In many datasets, becoming at risk and coming under observation are coincident. In that case, it is sufficient to specify origin().

enter(time *exp*), enter(*varname==numlist*), and enter(*varname==numlist* time *exp*) follow the same syntax as origin(). In multiple-record data, both *varname==numlist* and time *exp* are interpreted as the earliest time implied and, if both are specified, the later of the two times is used.

exit(failure | [*varname==numlist*] time *exp*) specifies the latest time under which the subject is both under observation and at risk. The emphasis is on latest; obviously subjects also exit the risk pool when their data run out.

exit(failure) is the default. When the first failure event occurs, the subject is removed from the analysis risk pool even if the subject has subsequent records in the data and even if some of those subsequent records document other failure events. Specify exit(time .) if you wish to keep all records for a subject after failure. You want to do this if you have multiple failure data.

exit(*varname==numlist*), exit(time *exp*), and exit(*varname==numlist* time *exp*) follow the same syntax as origin() and enter(). In multiple-record data, both *varname==numlist* and time *exp* are interpreted as the earliest time implied. exit differs from origin() and enter() in that, if both are specified, the earlier of the two times is used.

time0(*varname*) is rarely specified because most datasets do not contain this information. time0() should be used exclusively with multiple-record data, and even then you should consider whether origin() or enter() would be more appropriate.

time0() specifies a mechanical aspect of interpretation about the records in the dataset, namely the beginning of the period spanned by each record. See *Intermediate exit and re-entry times (gaps)* under *Remarks* below.

if(*exp*), ever(*exp*), never(*exp*), after(*exp*), and before(*exp*) select relevant records.

if(*exp*) selects records for which *exp* is true. We strongly recommend specifying this if() option rather than if *exp* following stset or streset. They differ in that if *exp* removes from consideration the data before the calculation of beginning and ending times and other quantities as well. The if() option, on the other hand, sets the restriction after all derived variables are calculated. See *The if() option versus the if exp* under *Remarks* below.

if() may be specified with single- or multiple-record data. The remaining selection options are for use with multiple-record data only.

ever(*exp*) selects only subjects for which *exp* is ever true.

never(*exp*) selects only subjects for which *exp* is never true.

after(*exp*) selects records within subject on or after the first time *exp* is true.

before(*exp*) selects records within subject before the first time *exp* is true.

show and noshow change whether the other st commands are to display the identities of the key st variables at the top of their output.

Options unique to streset

past expands the stset sample to include the entire recorded past of the relevant subjects, meaning it includes observations before becoming at risk, or excluded due to after(), etc.

future expands the stset sample to include the records on the relevant subjects after the last record that previously was included, if any, which typically means to include all observations after failure or censoring.

past future expands the stset sample to include all records on the relevant subjects.

Typing streset without arguments resets the sample back to the analysis sample. See *Past and future records* under *Remarks* for more information.

Options for st

nocmd suppresses displaying the last stset command.

notable suppresses displaying the table summarizing what has been stset.

Remarks

Remarks are presented under the headings

What is survival-time data?
Key concepts
Survival-time datasets
Using stset
Example 1: Single-record data
Example 2: Single-record data with censoring
Example 3: Multiple-record data
Example 4: Multiple-record data with multiple events
Example 5: Multiple-record data recording time rather than t
Example 6: Multiple-record data with time recorded as a date
Example 7: Multiple-record data with extraneous information
Example 8: Multiple-record data with delayed entry
Example 9: Multiple-record data with extraneous information and delayed entry
Example 10: Real data
Two concepts of time
The substantive meaning of analysis time
Setting the failure event
Setting multiple failures
First entry times
Final exit times
Intermediate exit and re-entry times (gaps)
The if() option versus the if exp
Past and future records
Using streset
Performance and multiple-record per subject datasets
Sequencing of events within t
Weights
Data warnings and errors flagged by stset
Final example: Stanford heart transplant data

What is survival-time data?

Survival-time data—what we call st data—documents spans of time ending in an event. For instance,

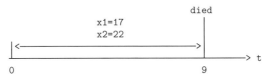

which is to say, $x1 = 17$ and $x2 = 22$ over the time span 0 to 9, and then $died = 1$. More formally, we mean $x1 = 17$ and $x2 = 22$ for $0 < t \le 9$, which we often write as $(0, 9]$. However you wish to say it, this information might be recorded by the observation

id	end	x1	x2	died
101	9	17	22	1

and we call this single-record survival data.

The data can be more complicated. For instance, we might have

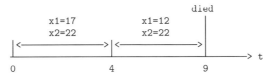

meaning

$$x1 = 17 \text{ and } x2 = 22 \text{ during } (0, 4]$$
$$x1 = 12 \text{ and } x2 = 22 \text{ during } (4, 9], \text{ and then } died = 1.$$

and this would be recorded by the data

id	begin	end	x1	x2	died
101	0	4	17	22	0
101	4	9	12	22	1

We call this multiple-record survival data.

These two formats allow recording lots of different possibilities. The last observation on a person need not be failure:

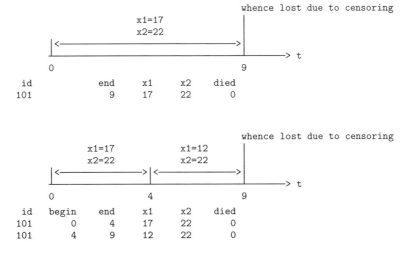

id	end	x1	x2	died
101	9	17	22	0

or

id	begin	end	x1	x2	died
101	0	4	17	22	0
101	4	9	12	22	0

Multiple-record data might have gaps,

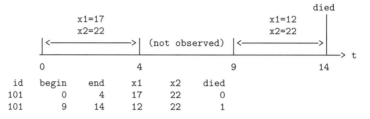

id	begin	end	x1	x2	died
101	0	4	17	22	0
101	9	14	12	22	1

or subjects might not be observed from the onset of risk:

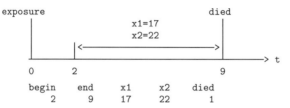

begin	end	x1	x2	died
2	9	17	22	1

and

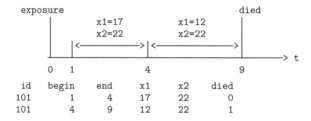

id	begin	end	x1	x2	died
101	1	4	17	22	0
101	4	9	12	22	1

The failure event might not be death but instead something that can repeat:

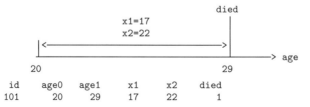

id	begin	end	x1	x2	infarc
101	0	4	17	22	1
101	4	9	12	22	0
101	9	13	10	22	1

Our data may be in different time units; rather than t where $t = 0$ corresponds to the onset of risk, we might have time recorded as age,

id	age0	age1	x1	x2	died
101	20	29	17	22	1

or time recorded as calendar dates:

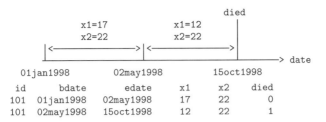

id	bdate	edate	x1	x2	died
101	01jan1998	02may1998	17	22	0
101	02may1998	15oct1998	12	22	1

Finally, you can mix these diagrams however you wish, so we might have time recorded per the calendar, unobserved periods after the onset of risk, subsequent gaps, and multiple failure events.

The st commands analyze data like these, and the first step is to tell st about your data using stset. You do not change your data to fit some predefined mold; you describe your data using stset and then the rest of the st commands just do the right thing.

Before we turn to using stset, let us describe one more style of recording time-to-event data because it is common and is inappropriate for use with st. It is inappropriate, but it is easy to convert to the survival-time form. It is called snapshot data. In snapshot data you do not know spans of time, but you have information recorded at various points in time:

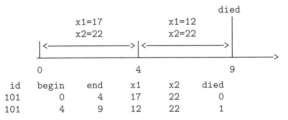

id	t	x1	x2	died
101	0	17	22	0
101	4	12	22	0
101	9	.	.	1

In this snapshot dataset all we know are the values of x1 and x2 at $t = 0$ and $t = 4$, and we know that the subject died at $t = 9$. Snapshot data can be converted to survival-time data if we are willing to assume that x1 and x2 remained constant between times:

id	begin	end	x1	x2	died
101	0	4	17	22	0
101	4	9	12	22	1

The snapspan command makes this conversion. If you have snapshot data, first see [ST] **snapspan** to convert it to survival-time data and then use stset to tell st about the converted data, but see *Example 10: Real data* below, first.

Key concepts

time, or better, *time units*, is how time is recorded in your data. It might be numbers such as 0, 1, 2, ..., with *time* = 0 corresponding to some exposure event, or it might be subject's age, or it might be calendar time, or it might be recorded in some other way.

events are things that happen at an instant in time, such as being exposed to an environmental hazard, being diagnosed as myopic, becoming employed, being promoted, becoming unemployed, having a heart attack, and dying.

failure event is the *event* indicating failure as it is defined for the purpose of analysis. This can be a single or compound event. The *failure* event might be when variable dead is 1 or it might be when variable diag is any of 115, 121, or 133.

at risk means the subject is at risk of the *failure event* occurring. For instance, if the *failure event* is becoming unemployed, a person must be employed. The subject is not *at risk* prior to being employed. Once employed, the subject becomes *at risk* and, once again, the subject is no longer *at risk* once the *failure event* occurs. If subjects become *at risk* upon the occurrence of some *event*, it is called the exposure event. Gaining employment is the exposure event in our example.

origin is the *time* when the subject became at risk. If *time* is recorded as numbers such as 0, 1, 2, ..., with *time* = 0 corresponding to the exposure event, then *origin* = 0. Alternatively, *origin* might be the age of the subject when diagnosed or the date when the subject was exposed. Regardless, *origin* is expressed in *time units*.

scale is just a fixed number, typically 1, used in mapping *time* to analysis time *t*.

t, or *analysis time*, is $(time - origin)/scale$, which is to say, time since onset of being at risk measured in *scale* units.

$t = 0$ corresponds to the onset of risk and *scale* just provides a way to make the units of *t* more readable. You might have *time* recorded in days from 01jan1960 and want *t* recorded in years, in which case *scale* would be 365.25.

time is how time is recorded in your data and *t* is how time is reported in the analysis.

under observation means that, should the *failure event* occur, it would be observed and recorded in the data. Sometimes subjects are *under observation* only after they are *at risk*. This would be the case, for instance, if subjects enrolled in a study after being diagnosed with cancer and, in order to enroll in the study, subjects were required to be diagnosed with cancer.

Being *under observation* does not mean the subject is necessarily *at risk*. A subject may come *under observation* prior to being *at risk* and, in fact, a subject *under observation* may never come to be *at risk*.

entry time and *exit time* mark when a subject is first and last *under observation*. The emphasis here is on the words first and last; *entry time* and *exit time* do not record observational gaps, if any; there is only one *entry time* and one *exit time* per subject.

entry time and *exit time* might be expressed as *times* (meaning recorded in *time units*) or they might correspond to the occurrence of some *event* (such as enrolling in the study).

Often the *entry time* corresponds to $t = 0$ or, since $t = (time - origin)/scale$, $time = origin$, or, substituting true meanings, the onset of risk.

Often the *exit time* corresponds to when the *failure event* occurs or, failing that, the end of data for the subject.

delayed entry means that *entry time* corresponds to $t > 0$; the subject became at risk but it was some time afterwards that the subject was under observation.

id refers to a subject identification variable; equal values of *id* indicate that the records are on the same subject. An *id* variable is required for multiple-record data and is optional, but recommended, with single-record data.

time0 refers to the beginning *time* (meaning recorded in *time units*) of a record. Some datasets have this variable but most do not. If the dataset does not contain the beginning time for each record, then subsequent records are assumed to begin where previous records ended. A *time0* variable may be created for these datasets using the snapspan command; see [ST] **snapspan**. Do not confuse *time0*—a mechanical aspect of datasets—with *entry time*—a substantive aspect of analysis.

gaps refer to gaps in observation between *entry time* and *exit time*. During a *gap* a subject is not *under observation*. *gaps* can arise only if the data contain a *time0* variable, because otherwise, subsequent records beginning when previous records end precludes there being *gaps* in the data. Note that *gaps* are distinct from *delayed entry*.

past history is a term we use to mean information recorded in the data prior to the subject being both *at risk* and *under observation*. In complex datasets there can be such observations. Say the dataset contains histories on subjects from birth to death. You might tell st that a subject becomes at risk once diagnosed with a particular kind of cancer. The *past history* on the subject would then refer to records prior to being diagnosed.

The word *history* is often dropped and the term simply becomes *past*. For instance, one might want to know whether the subject smoked in the *past*.

future history is a term we use to mean information recorded in the data after the subject is no longer *at risk*. Perhaps the failure event is not so serious as to preclude the possibility of data after failure.

The word *history* is often dropped and the term simply becomes *future*. Perhaps the *failure event* is cardiac infarction and you want to know whether the subject died soon in the *future* so that you can exclude them.

Survival-time datasets

The key concept of the st system is that observations (records) document a span of time. The span might be explicitly indicated, such as

```
begin    end    x1    x2
    3      9    17    22          <- spans (3,9]
```

or it might be implied that the record begins at 0,

```
         end    x1    x2
           9    17    22          <- spans (0,9]
```

or it might be implied because there are multiple records per subject:

```
id    end    x1    x2
 1      4    17    22             <- spans (0,4]
 1      9    12    22             <- spans (4,9]
```

Records may have an event indicator:

```
begin    end    x1    x2    died
    3      9    17    22       1   <- spans (3,9], died at t=9
```

```
         end    x1    x2    died
           9    17    22       1   <- spans (0,9], died at t=9
```

```
id    end    x1    x2    died
 1      4    17    22       0      <- spans (0,4],
 1      9    12    22       1      <- spans (4,9], died at t=9
```

The first two examples are called single-record survival-time data because there is one record per subject.

The final example is called multiple-record survival-time data. There are two records for subject with id = 1.

Either way, understand that survival-time data document time spans; characteristics are assumed to remain constant over the span; and the event is assumed to occur at the end of the span.

Using stset

Once you have `stset` your data, you can use the other st commands.

If you `save` your data after `stset`ting, you will not have to re-`stset` in the future; Stata will remember.

`stset` declares things about your data. It does not change the data, although it does add a few variables to your dataset.

This means you can re-`stset` your data as often as you wish. In fact, the `streset` command encourages this. Using complicated datasets often requires typing long `stset` commands, such as

    . stset date, fail(event==27 28) origin(event==15) enter(event==22)

Later, you might want to try `fail(event==27)`. You could retype the `stset` command, making the substitution, or you could type

    . streset, fail(event==27)

`streset` takes what you type, merges it with what you have previously declared with `stset`, and performs the combined `stset` command.

Example 1: Single-record data

Generators are run until they fail. Here is some of your dataset:

    . use http://www.stata-press.com/data/r8/kva
    (Generator experiment)
    . list in 1/3

	failtime	load	bearings
1.	100	15	0
2.	140	15	1
3.	97	20	0

The `stset` command for this dataset is

    . stset failtime
          failure event: (assumed to fail at time=failtime)
    obs. time interval: (0, failtime]
     exit on or before: failure

        12 total obs.
         0 exclusions

        12 obs. remaining, representing
        12 failures in single record/single failure data
       896 total analysis time at risk, at risk from t =   0
                              earliest observed entry t =   0
                                 last observed exit t =   140

When you type 'stset *timevar*', *timevar* is assumed to be the time of failure. More generally, you will learn, *timevar* is the time of failure or censoring.

Example 2: Single-record data with censoring

Generators are run until they fail, but during the experiment, the room flooded so some were run only until the flood. Here are some of your data:

```
. use http://www.stata-press.com/data/r8/kva2, clear
(Generator experiment)
. list in 1/4
```

	failtime	load	bearings	failed
1.	100	15	0	1
2.	140	15	1	0
3.	97	20	0	1
4.	122	20	1	1

In this case the second generator did not fail at time 140; the experiment was merely discontinued at that point. The stset command for this dataset is

```
. stset failtime, failure(failed)

       failure event:  failed != 0 & failed < .
  obs. time interval:  (0, failtime]
   exit on or before:  failure
```

```
     12  total obs.
      0  exclusions

     12  obs. remaining, representing
     11  failures in single record/single failure data
    896  total analysis time at risk, at risk from t =          0
                            earliest observed entry t =          0
                              last observed exit t =          140
```

When you type 'stset *timevar*, failure(*failvar*)', *timevar* is interpreted as the time of failure or censoring, which is determined by the value of *failvar*. *failvar* = 0 and *failvar* = . (missing) indicate censorings and all other values indicate failure.

Example 3: Multiple-record data

Assume that we are analyzing survival time of patients with a particular kind of cancer. In this dataset, the characteristics of patients vary over time, perhaps because new readings were taken or perhaps because the drug therapy was changed. Some of the data are

```
. list, separator(0)
```

	patid	t	died	x1	x2
1.	90	100	0	1	0
2.	90	150	1	0	0
3.	91	50	1	1	1
4.	92	100	0	0	0
5.	92	150	0	0	1
6.	92	190	0	0	0
7.	93	100	0	0	0
	(output omitted)				

Note that there are two records for patient 90 and that died is 0 in the first record, but 1 in the second. The interpretation of these two records is

Interval (0,100]:	x1 = 1 and x2 = 0
Interval (100,150]:	x1 = 0 and x2 = 0
at t = 150:	the patient died

Similarly, here is how you interpret the other records:

Patient 91:

Interval (0, 50]:	x1 = 1 and x2 = 1
at t = 50:	the patient died

Patient 92:

Interval (0,100]:	x1 = 0 and x2 = 0
Interval (100,150]:	x1 = 0 and x2 = 1
Interval (150,190]:	x1 = 0 and x2 = 0
at t = 190:	the patient was lost due to censoring

Look again at patient 92's data:

patid	t	died	x1	x2
92	100	0	0	0
92	150	0	0	1
92	190	0	0	0

Note that died = 0 for the first event. Mechanically, this removes the subject from the data at t = 100—the patient is treated as censored. The next record, however, adds the patient back into the data (at t = 100) with new characteristics.

The stset command for this dataset is

```
. stset t, id(patid) failure(died)

                id:  patid
     failure event:  died != 0 & died < .
 obs. time interval:  (t[_n-1], t]
  exit on or before:  failure

        126  total obs.
          0  exclusions

        126  obs. remaining, representing
         40  subjects
         26  failures in single failure-per-subject data
       2989  total analysis time at risk, at risk from t =          0
                              earliest observed entry t =          0
                                last observed exit t =        139
```

When you have multiple-record data, you specify stset's id(*idvar*) option. When you type 'stset *timevar*, id(*idvar*) failure(*failvar*)', *timevar* denotes the end of the period (just as it does in single-record data). The first record within *idvar* is assumed to begin at time 0, and subsequent records are assumed to begin where the previous record left off. *failvar* should contain 0 on all but, possibly, the last record within *idvar* unless your data contain multiple failures (in which case you must specify the exit() option; see *Setting multiple failures* below).

Example 4: Multiple-record data with multiple events

You have the following data on hospital patients admitted to a particular ward:

patid	day	sex	x1	x2	code
101	5	1	10	10	177
101	13	1	20	8	286
101	21	1	16	11	208
101	24	1	11	17	401
102	8	0	20	19	204
102	18	0	19	1	401
103	*etc.*				

Variable code records various actions and codes 401 and 402 indicate discharge, 401 being discharged alive and 402 being discharged dead. You stset this dataset by typing

```
. stset day, id(patid) fail(code==402)
                id:  patid
     failure event:  code == 402
obs. time interval:  (day[_n-1], day]
 exit on or before:  failure
```

```
    243  total obs.
      0  exclusions

    243  obs. remaining, representing
     40  subjects
     15  failures in single failure-per-subject data
   1486  total analysis time at risk, at risk from t =         0
                            earliest observed entry t =         0
                                last observed exit t =         62
```

When you specify failure(*eventvar*==*#*), the failure event is as specified. You may include a list of numbers following the equal signs. If failure were codes 402 or 403, you could specify failure(code == 402 403). If failure were codes 402, 403, 404, 405, 406, 407, and 409, you could specify failure(code == 402/407 409).

Example 5: Multiple-record data recording time rather than t

More reasonably, the hospital data in the above example would not contain days since admission, it would contain admission and current dates. Below adday contains the day of admission and curdate the ending date of the record, both recorded as number of days since the ward opened:

patid	adday	curday	sex	x1	x2	code
101	287	292	1	10	10	177
101	.	300	1	20	8	286
101	.	308	1	16	11	208
101	.	311	1	11	17	401
102	289	297	0	20	19	204
102	.	307	0	19	1	401
103	*etc.*					

This is the same dataset as shown in Example 4. Previously the first record on patient 101 recorded 5 days after admission. In this dataset, $292 - 287 = 5$. You would stset this dataset by typing

```
. stset curday, id(patid) fail(code==402) origin(time adday)
                id:  patid
     failure event:  code == 402
 obs. time interval:  (curday[_n-1], curday]
  exit on or before:  failure
    t for analysis:  (time-origin)
            origin:  time adday
```
_____

```
   243  total obs.
     0  exclusions
```
_____

```
   243  obs. remaining, representing
    40  subjects
    15  failures in single failure-per-subject data
  1486  total analysis time at risk, at risk from t =          0
                         earliest observed entry t =          0
                            last observed exit t =         62
```

origin() sets when a subject becomes at risk. It does this by defining *analysis time*.

When you specify 'stset *timevar*, ... origin(time *originvar*)', analysis time is defined as $t = (timevar - originvar)/scale()$. In analysis-time units, subjects become at risk at $t = 0$. See *Two concepts of time* and *The substantive meaning of analysis time* below.

Example 6: Multiple-record data with time recorded as a date

Even more reasonably, dates would not be recorded as integers 428, 433, and 453, meaning the number of days since the ward opened. The dates would be recorded as dates:

patid	addate	curdate	sex	x1	x2	code
101	18aug1998	23aug1998	1	10	10	177
101	.	31aug1998	1	20	8	286
101	.	08sep1998	1	16	11	208
101	.	11sep1998	1	11	17	401
102	20aug1998	28aug1998	0	20	19	204
102	.	07sep1998	0	19	1	401
103	*etc.*					

That, in fact, changes nothing. You still type what you previously typed:

```
. stset curdate, id(patid) fail(code==402) origin(time addate)
```

Stata dates are in fact integers—they are the number of days since 01jan1960—and it is merely Stata's %d display format that makes them display as dates.

Example 7: Multiple-record data with extraneous information

Perhaps we wish to study the outcome following a certain operation, said operation being indicated by code 286. Subjects become at risk when and if the operation is performed. In that case, we do not type

```
. stset curdate, id(patid) fail(code==402) origin(time addate)
```

We instead type

```
. stset curdate, id(patid) fail(code==402) origin(code==286)
```

The result of this would be to set analysis time t to

$$t = \texttt{curdate} - (\text{the value of } \texttt{curdate} \text{ when } \texttt{code==286})$$

Let us work through this for the first patient:

patid	addate	curdate	sex	x1	x2	code
101	18aug1998	23aug1998	1	10	10	177
101	.	31aug1998	1	20	8	286
101	.	08sep1998	1	16	11	208
101	.	11sep1998	1	11	17	401

The event 286 occurred on 31aug1998, and thus the values of t for the four records are

$$t_1 = curdate_1 - 31\text{aug}1998 = 23\text{aug}1998 - 31\text{aug}1998 = -8$$
$$t_2 = curdate_2 - 31\text{aug}1998 = 31\text{aug}1998 - 31\text{aug}1998 = 0$$
$$t_3 = curdate_3 - 31\text{aug}1998 = 08\text{sep}1998 - 31\text{aug}1998 = 8$$
$$t_4 = curdate_4 - 31\text{aug}1998 = 11\text{sep}1998 - 31\text{aug}1998 = 11$$

Information prior to $t = 0$ is not relevant because the subject is not yet at risk. Thus, the relevant data on this subject are

t in $(0, 8]$ sex $= 1$, x1 $= 16$, x2 $= 11$
t in $(8, 11]$ sex $= 1$, x1 $= 11$, x2 $= 17$ and the subject is censored (code $\neq 402$)

That is precisely the logic stset went through. For your information, stset quietly creates the variables

_st	1 if the record is to be used, 0 if ignored
_t0	analysis time when record begins
_t	analysis time when record ends
_d	1 if failure, 0 if censored

You can examine these variables after issuing the stset command:

```
. list _st _t0 _t _d
```

	_st	_t0	_t	_d
1.	0	.	.	.
2.	0	.	.	.
203.	1	0	8	0
204.	1	8	11	0

Results are just as we anticipated. Do not let the observation numbers bother you; stset sorts the data in a way it finds convenient. Feel free to resort the data; if any of the st commands need the data in a different order, they will sort it themselves.

There are two ways of specifying origin():

origin(time *timevar*)   or   origin(time *exp*)
origin(*eventvar*== *numlist*)

In the first syntax—which is denoted by typing the word time—you directly specify when a subject becomes at risk. In the second syntax—which is denoted by typing a variable name and equal signs—you specify the same thing indirectly. The subject becomes at risk when the specified event occurs (which may be never).

Information prior to origin() is ignored. That information composes what we call the past history.

Example 8: Multiple record data with delayed entry

In another analysis, we want to use the above data to analyze all patients; not just those undergoing a particular operation. In this analysis, subjects become at risk when they enter the ward. For this analysis, however, we need information from a particular test, and that information is available only if the test is administered to the patient. Even if the test is administered, some amount of time passes before that. Assume that when the test is administered, code==152 is inserted into the patient's hospital record.

To summarize, we want origin(time addate), but patients do not enter our sample until code==152. The way to stset this is

```
. stset curdate, id(patid) fail(code==402) origin(time addate) enter(code==152)
```

Patient 107 has code 152:

patid	addate	curdate	sex	x1	x2	code
107	22aug1998	25aug1998	1	9	13	274
107	.	28aug1998	1	19	19	152
107	.	30aug1998	1	18	12	239
107	.	07sep1998	1	12	11	401

In terms of analysis time, $t = 0$ corresponds to 22aug1998. The test was not administered, however, until 6 days later. The analysis times for these records are

$$t_1 = curdate_1 - 22aug1998 = 25aug1998 - 22aug1998 = 3$$

$$t_2 = curdate_2 - 22aug1998 = 28aug1998 - 22aug1998 = 6$$

$$t_3 = curdate_3 - 22aug1998 = 30aug1998 - 22aug1998 = 8$$

$$t_4 = curdate_4 - 22aug1998 = 07sep1998 - 22aug1998 = 16$$

and the data we want in our sample are

t in $(6, 8]$ $sex = 1$, $x1 = 18$, $x2 = 12$
t in $(8, 16]$ $sex = 1$, $x1 = 12$, $x2 = 11$ and patient was censored (code $\neq$ 402)

The above stset command produced that:

```
. list _st _t0 _t _d
```

	_st	_t0	_t	_d
1.	0	.	.	.
2.	0	.	.	.
39.	1	6	8	0
40.	1	8	16	0

Example 9: Multiple-record data with extraneous information and delayed entry

Options origin() and enter() can be combined. For instance, you want to analyze patients receiving a particular operation (time at risk begins upon code == 286, but patients may not enter the sample prior to a test being administered, denoted by code == 152). You type

```
. stset curdate, id(patid) fail(code==402) origin(code==286) enter(code==152)
```

If you typed the above, it would not matter whether the test was performed before or after the operation.

A patient who had the test and then the operation would enter at analysis time $t = 0$.

A patient who had the operation and then the test would enter at analysis time $t > 0$, the analysis time being when the test was performed.

If you wanted to require that the operation be performed after the test, you could type

```
. stset curdate, id(patid) fail(code==402) origin(code==286) after(code==152)
```

Admittedly this can be confusing. The way to proceed is find a complicated case in your data and then list _st _t0 _t _d for that case after st setting your data.

Example 10: Real data

All of our hospital-ward examples are artificial in one sense: it is unlikely the data would have come to us in survival-time form:

patid	addate	curdate	sex	x1	x2	code
101	18aug1998	23aug1998	1	10	10	177
101	.	31aug1998	1	20	8	286
101	.	08sep1998	1	16	11	208
101	.	11sep1998	1	11	17	401
102	20aug1998	28aug1998	0	20	19	204
102	.	07sep1998	0	19	1	401
103	etc.					

Rather, what we would have received would have been a snapshot dataset:

patid	date	sex	x1	x2	code
101	18aug1998	1	10	10	22
101	23aug1998	.	20	8	177
101	31aug1998	.	16	11	286
101	08sep1998	.	11	17	208
101	11sep1998	.	.	.	401
102	20aug1998	0	20	19	22
102	28aug1998	.	19	1	204
102	07sep1998	.	.	.	401
103	etc.				

In a snapshot dataset, we have a time (date in this case) and values of the variables as of that instant.

This dataset can be converted to the appropriate form by typing

```
. snapspan patid date code
```

The result would be

patid	date	sex	x1	x2	code
101	18aug1998	.	.	.	22
101	23aug1998	1	10	10	177
101	31aug1998	.	20	8	286
101	08sep1998	.	16	11	208
101	11sep1998	.	11	17	401
102	20aug1998	.	.	.	22
102	28aug1998	0	20	19	204
102	07sep1998	.	19	1	401

This is virtually the same dataset with which we have been working. It differs in two ways:

1. The variable sex is not filled in for all the observations because it was not filled in on the original form. The hospital wrote down the sex on admission and then never bothered to document it again.

2. We have no admission date (addate) variable. Instead, we have an extra first record for each patient with code = 22 (22 is the code the hospital uses for admissions).

The first problem is easily fixed and the second, it turns out, is not a problem because we can vary what we type when we stset the data.

First, let's fix the problem with variable sex. There are two ways to proceed. One would be simply to fill in the variable ourselves:

```
. by patid (date), sort: replace sex = sex[_n-1] if sex>=.
```

Alternatively, we could perform a phony stset that is good enough to set all the data, and then use stfill to fill in the variable for us. Let us begin with the phony stset:

```
. stset date, id(patid) origin(min) fail(code==-1)
                    id:  patid
         failure event:  code == -1
    obs. time interval:  (date[_n-1], date]
     exit on or before:  failure
        t for analysis:  (time-origin)
                origin:  min

      283  total obs.
        0  exclusions

      283  obs. remaining, representing
       40  subjects
        0  failures in single failure-per-subject data
     2224  total analysis time at risk, at risk from t =          0
                             earliest observed entry t =          0
                                last observed exit t =           89
```

Typing stset date, id(patid) origin(min) fail(code == -1) does not produce anything we would want to use for analysis. This is a trick to get the dataset temporarily stset so that we can use some st data-management commands on it.

The first part of the trick was to specify origin(min). This defines analysis time as $t = 0$, corresponding to the minimum observed value of time variable, minus 1. The time variable is date in this case. Why the minimum minus 1? Because st is bound and determined to ignore observations for which analysis time $t < 0$. origin(min) provides a phony definition of t that ensures $t > 0$ for all observations.

The second part of the trick was to specify fail(code == -1), and you might have to vary what you type. We just wanted to choose an event that we know never happens, thus ensuring that no observations are ignored following failure.

Now that we have the dataset stset, we can use the other st commands. Do not use the st analysis commands unless you want ridiculous results, but one of the st data-management commands is just what we want:

```
. stfill sex, forward
        failure _d:  code == -1
   analysis time _t:  (date-origin)
           origin:  min
               id:  patid
replace missing values with previously observed values:
           sex:  203 real changes made
```

Problem one solved.

The second problem concerns the lack of an admission date variable. That is not really a problem because we have a new first observation with code $= 22$ recording the date of admission, so every place we previously coded origin(time addate), we substitute origin(code == 22).

Problem two solved.

We also solved the big problem—converting a snapshot dataset into a survival-time dataset; see [ST] **snapspan**.

Two concepts of time

The st system has two concepts of time. The first is *time* in italics, which corresponds to how time is recorded in your data. The second is analysis time, which we write as t. Substantively, analysis time is time at risk. stset defines analysis time in terms of *time* via

$$t = \frac{time - \text{origin}()}{\text{scale}()}$$

t and time can be the same thing and, by default they are because, by default, origin() is 0 and scale() is 1.

All the st analysis commands work with analysis time t.

By default, if you do not specify the origin() and scale() options, your time variables are expected to be the analysis time variables. This means *time* $= 0$ corresponds to when subjects became at risk and that means, among other things, that observations for which *time* < 0 are ignored because survival analysis concerns the analysis of persons who are at risk, and no one is at risk prior to $t = 0$.

origin plays the substantive role of determining when the clock starts ticking. If you do not specify origin(), origin(time 0) is assumed, meaning that $t = time$ and that persons are at risk from $t = time = 0$.

In many datasets, *time* and t will differ. Time might be calendar time and t the length of time since some event, such as being born, being exposed to some risk factor, etc. origin() sets when $t = 0$. scale() merely sets a constant that makes t more readable.

The syntax for the origin() option makes it look more complicated than it really is:

$$\text{origin}\left(\left[varname == numlist\right] \text{ time } exp \mid \text{min}\right)$$

This says that there are four different ways to specify origin():

> origin(time *exp*)
> origin(*varname* == *numlist*)
> origin(*varname* == *numlist* time *exp*)
> origin(min)

The first syntax can be used with single- or multiple-record data. It states that the origin is given by *exp*, which can be a constant for all observations, a variable (and hence varying subject by subject), or even an expression composed of variables and constants. Perhaps origin is a fixed date or a date recorded in the data when the subject was exposed or when the subject turned 18.

The second and third syntaxes are for use with multiple-record data. The second states that origin corresponds to the (earliest) time when the designated event occurred. Perhaps origin is when an operation was performed. The third syntax calculates origin both ways and then selects the later one.

The fourth syntax does something odd; it sets the origin to the minimum time observed, minus 1. This is not useful for analysis but is sometimes useful for playing data-management tricks; see *Example 10: Real data* above.

Let's start with the first syntax. Pretend that you had the data

```
faildate      x1      x2
28dec1997     12      22
12nov1997     15      22
03feb1998     55      22
```

and that all the observations came at risk on the same date, 01nov1997. You could type

```
. stset faildate, origin(time mdy(11,1,1997))
```

Remember that `stset` secretly adds _t0 and _t to your dataset and that they contain the time span for each record, documented in analysis-time units. After typing `stset`, you can `list` the results:

```
. list faildate x1 x2 _t0 _t
```

	faildate	x1	x2	_t0	_t
1.	28dec1997	12	22	0	57
2.	12nov1997	15	22	0	11
3.	03feb1998	55	22	0	94

Record 1 reflects the period $(0, 57]$ in analysis-time units, which are, in this case, days. `stset` calculated the 57 from $28\text{dec}1997 - 01\text{nov}1997 = 13{,}876 - 13{,}819 = 57$. (Remember that dates such as 28dec1997 are really just integers containing the number of days from 01jan1960, and it is Stata's %d display format that makes them display nicely. 28dec1997 is really the number 13,876.)

As another example, we might have data recording exposure and failure dates:

```
expdate      faildate       x1      x2
07may1998    22jun1998      12      22
02feb1998    11may1998      11      17
```

The way to `stset` this dataset is

```
. stset faildate, origin(time expdate)
```

and the result, in analysis units, is

```
. list expdate faildate x1 x2 _t0 _t
```

	expdate	faildate	x1	x2	_t0	_t
1.	07may1998	22jun1998	12	22	0	46
2.	02feb1998	11may1998	11	17	0	98

There is nothing magic about dates. Our original data could just as well have been

expdate	faildate	x1	x2
32	78	12	22
12	110	11	17

and the final result would still be the same because $78 - 32 = 46$ and $110 - 12 = 98$.

Specifying an expression can sometimes be useful. Suppose that your dataset has the variable date recording the date of event and variable age recording the subject's age as of date. You want to make $t = 0$ correspond to when the subject turned 18. You could type $origin\big(time\ date-int((age-18)*365.25)\big)$.

$origin(varname == numlist)$ is for use with multiple-record data. It states when each subject became at risk indirectly; the subject became at risk at the earliest time that *varname* takes on any of the enumerated values. Pretend you had

patid	date	x1	x2	event
101	12nov1997	15	22	127
101	28dec1997	12	22	155
101	03feb1998	55	22	133
101	05mar1998	14	22	127
101	09apr1998	12	22	133
101	03jun1998	13	22	101
102	22nov1997	.	.	.

and assume event $= 155$ represents the onset of exposure. You might stset this dataset by typing

```
. stset date, id(patid) origin(event==155) ...
```

If you did that, the information for patient 101 prior to 28dec1997 would be ignored in subsequent analysis. The prior information would not be removed from the dataset; it would just be ignored. Probably something similar would happen for patient 102 or, if patient 102 has no record with event $= 155$, all the records on the patient would be ignored.

In terms of analysis time, $t = 0$ would correspond to when event 155 occurred. Here are the results in analysis-time units:

patid	date	x1	x2	event	_t0	_t
101	12nov1997	15	22	127	.	.
101	28dec1997	12	22	155	.	.
101	03feb1998	55	22	133	0	37
101	05mar1998	14	22	127	37	67
101	09apr1998	12	22	133	67	102
101	03jun1998	13	22	101	102	157
102	22nov1997	.	.	.	.	.

Note that patient 101's second record is excluded from the analysis. That is not a mistake. Records document durations, date reflects the end of the time period, and events occur at the end of time periods. Thus, event 155 occurred at the instant date $= $ 28dec1997 and the relevant first record for the patient is $(28dec1997, 03feb1998]$ in time units, which is $(0, 37]$ in t units.

The substantive meaning of analysis time

In specifying $origin()$, you must ask yourself whether two subjects with identical characteristics face the same risk of failure. The answer is that they face the same risk when they have the same value of $t = (time - origin())/scale()$ or, equivalently, when the same amount of time has elapsed from $origin()$.

Say we have the following data on smokers who have died:

```
        ddate     x1     x2   reason
     11mar1984     23     11        2
     15may1994     21      9        1
     22nov1993     22     13        2
     etc.
```

We wish to analyze death due to reason==2. However, typing

```
. stset ddate, fail(reason==2)
```

would probably not be adequate. We would be saying that smokers were at risk of death from 01jan1960. Would it matter? If we planned on doing anything parametric, it would, because parametric hazard functions, except for the exponential, are functions of analysis time and the location of 0 makes a difference.

Even if we were thinking of performing nonparametric analysis, there would probably be difficulties. We would be asserting that two "identical" persons (identical, presumably, in terms of x1 and x2) face the same risk on the same calendar date. Does the risk of death due to smoking really change as the calendar changes?

It would be more reasonable to assume that the risk changes with how long you have been smoking and that our data would likely include that date. We would type

```
. stset ddate, fail(reason==2) origin(time smdate)
```

if smdate were the name of the date-started-smoking variable. We would now be saying that the risk is equal when the number of days smoked is the same. We might prefer to see t in years,

```
. stset ddate, fail(reason==2) origin(time smdate) scale(365.25)
```

but that would make no substantive difference.

Consider single-record data on firms which went bankrupt:

```
      incorp    bankrupt     x1     x2   btype
   22jan1983   11mar1984     23     11       2
   17may1992   15may1994     21      9       1
   03nov1991   22nov1993     22     13       2
   etc.
```

Say we wish to examine the risk of a particular kind of bankruptcy, btype == 2, among firms that become bankrupt. Typing

```
. stset bankrupt, fail(btype==2)
```

would be more reasonable than it was in the smoking example. It would not be reasonable if we were thinking of performing any sort of parametric analysis, of course, because then location of $t = 0$ matters, but it might be reasonable for semiparametric analysis. We would be asserting that two "identical" firms (identical with respect to the characteristics we model) have the same risk of bankruptcy when the calendar dates are the same. We would be asserting that the overall state of the economy matters.

Alternatively, it might be reasonable to measure time from the date of incorporation:

```
. stset bankrupt, fail(btype==2) origin(time incorp)
```

Understand that the choice of origin() is a substantive decision.

Setting the failure event

You set the failure event using the `failure()` option.

In single-record data, if `failure()` is not specified, every record is assumed to end in a failure. For instance, with

```
      failtime     load   bearings
1.        100       15          0
2.        140       15          1
etc.
```

you would type `stset failtime` and the first observation would be assumed to fail at time = 100, the second at time = 140, and so on.

`failure`(*varname*) specifies that a failure occurs whenever *varname* is not zero and is not missing. For instance, with

```
      failtime     load   bearings   burnout
1.        100       15          0         1
2.        140       15          1         0
3.         97       20          0         1
4.        122       20          1         0
5.         84       25          0         1
6.        100       25          1         1
etc.
```

you might type `stset failtime, failure(burnout)`. Observations 1, 3, 5, and 6 would be assumed to fail at times 100, 97, 84, and 100, respectively; observations 2 and 4 would be assumed to be censored at times 140 and 122.

Similarly, were the data

```
      failtime     load   bearings   burnout
1.        100       15          0         1
2.        140       15          1         0
3.         97       20          0         2
4.        122       20          1         .
5.         84       25          0         2
6.        100       25          1         3
etc.
```

the result would be the same. Nonzero, nonmissing values of the failure variable are assumed to represent failures. (Perhaps `burnout` contains a code on how the burn out occurred.)

`failure`(*varname* == *numlist*) specifies that a failure occurs whenever *varname* takes on any of the values of *numlist*. In the above example, specifying

```
. stset failtime, failure(burnout==1 2)
```

would treat observation 6 as censored.

```
. stset failtime, failure(burnout==1 2 .)
```

would also treat observation 4 as a failure.

```
. stset failtime, failure(burnout==1/3 6 .)
```

would treat `burnout==1`, `burnout==2`, `burnout==3`, `burnout==6`, and `burnout==.` as representing failures, and all other values as representing censorings. (Perhaps we want to examine "failure due to meltdown", and these are the codes that represent the various kinds of meltdown.)

`failure()` is treated the same way in both single- and multiple-record data. Consider

	patno	t	x1	x2	died
1.	1	4	23	11	1
2.	2	5	21	9	0
3.	2	8	22	13	1
4.	3	7	20	5	0
5.	3	9	22	5	0
6.	3	11	21	5	0
7.	4	...			

Typing

    . stset t, id(patno) failure(died)

would treat

patno==1	as dying	at t==4
patno==2	as dying	at t==8
patno==3	as being censored	at t==11

Note that intervening records on the same subject are marked as "censored". Technically, they are not really censored if you think about it carefully; they are simply marked as not failing. Look at the data for subject 3:

patno	t	x1	x2	died
3	9	22	5	0
3	11	21	5	0

The subject is not censored at $t = 9$ because there are more data on the subject; it is merely the case that the subject did not die at that time. At $t = 9$, x1 changed from 22 to 21. The subject is really censored at $t = 11$ because the subject did not die and there are no more records on the subject.

Typing `stset t, id(patno) failure(died)` would mark the same persons as dying and the same persons as censored as in the previous case. If `died` contained not 0 and 1, but 0 and nonzero, nonmissing codes for the reason for death are

	patno	t	x1	x2	died
1.	1	4	23	11	103
2.	2	5	21	9	0
3.	2	8	22	13	207
4.	3	7	20	5	0
5.	3	9	22	5	0
6.	3	11	21	5	0
7.	4	...			

Typing

    . stset t, id(patno) failure(died)

or

    . stset t, id(patno) failure(died==103 207)

would yield the same results; subjects 1 and 2 would be treated as dying and subject 3 as censored.

Typing

    . stset t, id(patno) failure(died==207)

would treat subject 2 as dying and subjects 1 and 3 as censored. Thus, when you specify the values for the code, it is not necessary that the code variable ever contain 0. In

	patno	t	x1	x2	died
1.	1	4	23	11	103
2.	2	5	21	9	13
3.	2	8	22	13	207
4.	3	7	20	5	11
5.	3	9	22	5	12
6.	3	11	21	5	12
7.	4	...			

Typing

```
. stset t, id(patno) failure(died==207)
```

treats patient 2 as dying, and 1 and 3 as censored. Typing

```
. stset t, id(patno) failure(died==103 207)
```

treats patients 1 and 2 as dying and 3 as censored.

Setting multiple failures

In multiple-record data, records after the first failure event are ignored unless you specify the exit() option. Consider the following data:

	patno	t	x1	x2	code
1.	1	4	21	7	14
2.	1	5	21	7	11
3.	1	8	22	7	22
4.	1	7	20	7	17
5.	1	9	22	7	22
6.	1	11	21	7	29
7.	2	...			

Perhaps code 22 represents the event of interest—say the event "visited the doctor". Were you to type stset t, failure(code == 22), the result would be as if the data contained

	patno	t	x1	x2	code
1.	1	4	21	7	14
2.	1	5	21	7	11
3.	1	8	22	7	22

Records after the first occurrence of the failure event are ignored. If you do not want this, you must specify the exit() option. Probably you would want to specify exit(time .) in this case, meaning subjects are not to exit the risk group until their data run out. Alternatively, perhaps code 142 means "entered the nursing home" and, once that event happens, you no longer want them in the risk group. In that case you would code exit(code == 142); see *Final exit times* below.

First entry times

Do not confuse enter() with origin(). origin() specifies when a subject first becomes at risk. enter() specifies when a subject first comes under observation. In most datasets, becoming at risk and coming under observation are coincident. In that case, it is sufficient to specify origin() alone, although you could specify both options.

The issue here has to do with persons who enter the data after they have been at risk of failure. Say we are studying deaths due to exposure to substance X, and we know the date at which a person was first exposed to the substance. We are willing to assume that persons are at risk from the date of exposure forward. A person arrives at our door who was exposed 15 years ago. Can we add this person to our data? The statistical issue is labeled *left-truncation* and the problem is that, had the person died before arriving at our door, we would never have known about her. We can add her to our data, but we must be careful to treat her subsequent survival time as conditional on having already survived 15 years.

Say we are examining visits to the widget repair facility, "failure" being defined as a visit (so failures can be repeated). The risk begins once a person buys a widget. We have a woman who bought a widget three years ago and she has no records on when she has visited the facility in the last three years. Can we add her to our data? Yes, as long as we are careful to treat her subsequent behavior as already being three years after she first became at risk.

The jargon for this is "under observation". All this means is that failures, were they to occur, would be observed. Before being under observation, failures, were they to occur, would not be observed.

If `enter()` is not specified, it is assumed that subjects are under observation at the time they enter the risk group, which is to say, as specified by `origin()`, 0 if `origin()` is not specified, or possibly `time0()`. To be precise, subject i is assumed to first enter the analysis risk pool at

$$time_i = \max\big(\text{earliest } \texttt{time0()} \text{ for } i, \texttt{enter()}, \texttt{origin()}\big)$$

For example, say we have multiple-record data recording "came at risk" (`mycode == 1`), "enrolled in our study" (`mycode == 2`), and "failed due to risk" (`mycode == 3`). We `stset` this dataset by typing

        . stset time, id(id) origin(mycode==1) enter(mycode==2) failure(mycode==3)

The above `stset` correctly handles the came at risk/came under observation problem regardless of the order of events 1, 2, and 3. For instance, if the subject comes under observation before he or she becomes at risk, the subject will be treated as entering the analysis risk pool at the time he or she came at risk.

Say we have the same data in single-record format: variable `riskdate` documents becoming at risk and variable `enr_date` the date of enrollment in our study. We would `stset` this dataset by typing

        . stset time, origin(time riskdate) enter(time enr_date) failure(mycode==3)

As a final example, let's return to the multiple-record way of recording our data and pretend that we started enrolling people in our study on 12jan1998 but that, up until 16feb1998, we do not trust that our records are complete (we had start-up problems). We would `stset` that dataset by typing

        . stset time, origin(mycode==1) enter(mycode==2 time mdy(2,16,1998)) fail(mycode==3)

`enter(`*varname*`==`*numlist* `time` *exp*`)` is interpreted as

$$\max(\text{time of earliest event in } numlist, exp)$$

Thus, persons having `mycode == 2` occurring before 16feb1998 are assumed to be under observation from 16feb1998, and those having `mycode == 2` thereafter are assumed to be under observation from the time of `mycode == 2`.

Final exit times

`exit()` specifies the latest time under which the subject is both under observation and at risk of the failure event. The emphasis is on latest; obviously subjects also exit the data when their data run out.

When you type

        . stset ..., ... failure(outcome==1/3 5) ...

the result is as if you had typed

```
. stset ..., ... failure(outcome==1/3 5) exit(failure) ...
```

which in turn is the same as

```
. stset ..., ... failure(outcome==1/3 5) exit(outcome==1/3 5) ...
```

When is a person to be removed from the analysis risk pool? When their data end, of course, and in addition, when the event 1, 2, 3, or 5 first occurs. How are they to be removed? According to their status at that time. If the event is 1, 2, 3, or 5 at that instant, then they exit as a failure. If the event is something else, they exit as censored.

Perhaps events 1, 2, 3, and 5 represent death due to heart disease, and that is what we are studying. Pretend outcome == 99 represents death for some other reason. Obviously, once the person dies, they are no longer at risk of dying from heart disease, so we would want to specify

```
. stset ..., ... failure(outcome==1/3 5) exit(outcome==1/3 5 99) ...
```

When we explicitly specify exit(), it is our responsibility to list all the reasons a person is to be removed other than simply running out of data. In this case, it would have been a mistake to specify just exit(99), because that would have left persons in the analysis risk pool who died for reasons 1, 2, 3, and 5. We would have treated those people as if they were still at risk of dying.

In fact, it probably would not have mattered had we specified exit(99) because, once a person is dead, he or she is unlikely to have any subsequent records anyway. By that logic, we did not even have to specify exit(99) because death is death and there should be no records following it.

For other kinds of events, however, exit() becomes important. Let us assume the failure event is diagnosed with heart disease. A person may surely have records following diagnosis, but even so,

```
. stset ..., ... failure(outcome==22) ...
```

would be adequate because, by not specifying exit(), we are accepting the default that exit() is equivalent to failure(). Once outcome 22 occurs, subsequent records on the subject will be ignored—they comprise the future history of the subject.

Say, however, that we wish to treat as censored persons diagnosed with kidney disease. We would type

```
. stset ..., ... failure(outcome==22) exit(outcome==22 29) ...
```

assuming outcome = 29 is "diagnosed with kidney disease". It is now of great importance that we specified exit(outcome==22 29) and not just exit(outcome==29) because, had we omitted code 22, persons would have remained in the analysis risk pool even after the failure event, i.e., being diagnosed with heart disease.

If, in addition, our data were untrustworthy after 22nov1998 (perhaps not all the data have been entered yet), we would type

```
. stset ..., ... failure(outcome==22) exit(outcome==22 29 time mdy(11,22,1998)) ...
```

If you type exit(*varname*==*numlist* time *exp*), the exit time is taken to be

$$\min(\text{time of earliest event in } numlist, exp)$$

For some analyses, repeated failures are possible. If you have repeated failure data, you specify the exit() option and include whatever reasons, if any, that would cause the person to be removed. If there are no such reasons and you wish to retain all observations for the person, you type

```
. stset ..., ... exit(time .) ...
```

exit(time .) specifies that the maximum time a person can be in the risk pool is infinite; thus, they will not be removed until their data run out.

Intermediate exit and re-entry times (gaps)

Gaps arise when a subject is temporarily not under observation. The statistical importance of gaps is that, if failure is death and had the person died, then they would not have been around to be found again. The solution to this is to remove the person from the risk pool during the observational gap.

In order even to know you have gaps, your data must provide starting and ending times for each record. Most datasets provide only ending times, and that precludes the possibility of gaps.

time0() is how you specifying the beginning times of records. Understand that time0() specifies a mechanical aspect of interpretation about the records in the dataset, namely the beginning of the period spanned by each record. Do not confuse time0() with origin(), which specifies the substantive issue of when a subject became at risk, and do not confuse time0() with enter(), which specifies when a subject first comes under observation.

time0() merely identifies the beginning of the time span covered by each record. Pretend that we had two records on a subject, the first covering the span (40,49] and the second (49,57]:

```
                |<—— record 1 ——>|<- record 2 ->|
     _____|_____|_____|_____> time
               40               49             57
```

A time0() variable would contain

40 in record 1
49 in record 2

and not, for instance, 40 and 40. Note that a time0() variable varies record-by-record for a subject.

Most datasets merely provide an end-of-record time value, *timevar*, which you specify by typing stset *timevar*, ... When you have multiple records per subject and you do not specify a time0() variable, stset assumes that the records begin where the previous one left off.

The if() option versus the if exp

Both the if *exp* and if(*exp*) option select records for which *exp* is true. We strongly recommend specifying the if() option in preference to the if *exp*. They differ in that if *exp* removes data from consideration before the calculation of beginning and ending times, and other quantities as well. The if() option, on the other hand, sets the restriction after all derived variables are calculated. To appreciate this difference, consider the following multiple-record data:

patno	t	x1	x2	code
3	7	20	5	14
3	9	22	5	23
3	11	21	5	29

Consider the difference in results between typing

```
. stset t if x1!=22, failure(code==14)
```

and

```
. stset t,  if(x1!=22) failure(code==14)
```

The first would remove record 2 from consideration at the outset. In constructing beginning and ending times, stset and streset would see

patno	t	x1	x2	code
3	7	20	5	14
3	11	21	5	29

and would construct the result

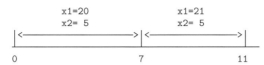

In the second case, the result would be

The latter result is correct and the former incorrect, because x1 = 21 is not true in the interval $(7, 9)$.

The only reason to specify if *exp* is to ignore errors in the data—observations that would confuse stset and streset should they see them—without actually dropping the offending observations from the dataset.

You specify the if() option to ignore information in the data that are not themselves errors. Specifying if() yields the same result as specifying if *exp* on the subsequent st commands after the dataset has been stset.

Past and future records

Consider the hospital-ward data that we have seen before:

patid	addate	curdate	sex	x1	x2	code
101	18aug1998	23aug1998	1	10	10	177
101	.	31aug1998	1	20	8	286
101	.	08sep1998	1	16	11	208
101	.	11sep1998	1	11	17	401
etc.						

Let us imagine that you stset this dataset such that you selected the middle two records. Perhaps you typed

```
. stset curdate, id(patid) origin(time addate) enter(code==286) failure(code==208)
```

The first record for the subject, since it was not selected, is called a *past history record*. If there were more early records that were not selected, they would all be called past history records.

The last record for the subject, since it was not selected, is called a *future history record*. If there were more later records that were not selected, they would all be called future history records.

Were you to type

```
. streset, past
```

the first three records for this subject would be selected.

Were you to type

```
. streset, future
```

the last three records for this subject would be selected.

Were you to type

```
. streset, past future
```

all four records for this subject would be selected.

No matter which, were you then to type

. streset

the original two records would be selected and things would be back just as they were before.

After typing `streset, past`, or `streset, future`, or `streset, past future`, you would not want to use any analysis commands. `streset` did some strange things, especially with the analysis time variable, in order to include the extra records. It would be the wrong sample, anyway.

You might, however, want to use certain data management commands on the data; especially those for creating new variables.

Typically, it is `streset, past` that is of greater interest. Past records—records prior to being at risk or excluded for other reasons—are not supposed to play a role in survival analysis. `stset` makes sure they do not. But it is sometimes reasonable to ask questions about them such as, was the subject ever on the drug cisplatin? Or has the subject ever been married? Or did the subject ever have a heart attack?

To answer questions like that, you sometimes want to dig into the past. Typing `streset, past` makes that easy and, once the past is set, the data can be used with `stgen` and a few other st commands. You might well type the following:

```
. stset curdate, id(patid) origin(addate) enter(code==286) failure(code==208)
. streset, past
. stgen attack = ever(code==177)
. streset
. stcox attack ...
```

Do not be concerned about doing something inappropriate while having the past or future set; st will not let you:

```
. stset curdate, id(patid) origin(time addate) enter(code==286) failure(code==208)
(output omitted)
. streset, past
(output omitted)
. stcox x1
you last "streset, past"
you must type "streset" to restore the analysis sample
r(119);
```

Using streset

`streset` is a useful tool for gently modifying what you have previously `stset`. Rather than typing the whole `stset` command, you can type `streset` followed by just what has changed.

For instance, you might

```
. stset curdate, id(patid) origin(time addate) enter(code==286) failure(code==208)
```

and then later want to restrict the analysis to subjects who ever have x1>20. You could retype the whole `stset` command and add `ever(x1>20)`, but it would be easier to type

```
. streset, ever(x1>20)
```

If later you decide you want to remove the restriction, type

```
. streset, ever(.)
```

That is the general rule for resetting options to the default: type '.' as the option's argument.

Be careful using streset because you can make subtle mistakes. In another analysis with another dataset, consider the following:

```
. stset date, fail(code==2) origin(code==1107)
. ...
. streset date, fail(code==9) origin(code==1422) after(code==1423)
. ...
. streset, fail(code==2) origin(code==1107)
```

If, in the last step, the user is trying to get back to the results of the first stset, the user fails. The last streset is equivalent to

```
. stset date, fail(code==9) origin(code==1107) after(code==1423)
```

streset() remembers the previously specified options and uses them if you do not override them. Note that both stset and streset display the current command line. Make sure that you verify that the command is as you intended.

Performance and multiple-record per subject datasets

stset and streset do not drop data; they simply mark data to be excluded from consideration. Some survival-time datasets can be very large, yet the relevant subsamples small. In such cases, you can reduce memory requirements and speed execution by dropping the irrelevant observations.

stset and streset mark the relevant observations by creating a variable named _st (it is always named this). The variable contains 1 and 0; _st = 1 marks the relevant observations and _st = 0 marks the irrelevant ones. If you type

```
. drop if _st==0
```

or equivalently

```
. keep if _st==1
```

or equivalently

```
. keep if _st
```

you will drop the irrelevant observations. All st commands produce the same results whether you do this or not. Be careful, however, if you are planning future stsets or stresets. Observations that are irrelevant right now might be relevant later.

One solution to this conundrum is to keep only those observations that are relevant after setting the entire history:

```
. stset date, fail(code==9) origin(code==1422) after(code==1423)
. streset, past future
. keep if _st
. streset
```

Final note: you may drop the irrelevant observations as marked by _st = 0, but do not drop the _st variable itself. The other st commands expect to find variable _st.

Sequencing of events within t

Consider the following bit of data:

etime	failtime	fail
0	5	1
0	5	0
5	7	1

Note all the different events happening at time 5: the first observation fails, the second is censored, and the third enters.

What does it mean for something to happen at time 5? In particular, is it at least potentially possible for the second observation to have failed at time 5, i.e., was it in the risk group when the first observation failed? How about the third observation? Was it in the risk group and could it have potentially failed at time 5?

Stata sequences events within a time as follows:

first,	at time	t	the failures occur
then,	at time	$t + 0$	the censorings are removed from the risk group
finally,	at time	$t + 0 + 0$	the new entries are added to the risk group

Thus, to answer the questions:

Could the second observation have potentially failed at time 5? Yes.

Could the third observation have potentially failed at time 5? No, because it was not yet in the risk group.

By this logic, the following makes no sense:

etime	failtime	fail
5	5	1

because this would mark a subject as failing before being at risk. It would make no difference if fail were 0—the subject would then be marked as being censored too soon. Either way, stset would flag this as an error. If you had a subject who entered and immediately exited, you would code this as

etime	failtime	fail
4.99	5	1

Weights

stset allows you to specify fweights, pweights, and iweights.

fweights are Stata's frequency or replication weights. Consider the data

failtime	load	bearings	count
100	15	0	3
140	15	1	2
97	20	0	1

and the stset command

```
. stset failtime [fw=count]

     failure event:  (assumed to fail at time=failtime)
 obs. time interval:  (0, failtime]
 exit on or before:  failure
             weight:  [fweight=count]
```

```
     3  total obs.
     0  exclusions
```

```
     3  physical obs. remaining, equal to
     6  weighted obs., representing
     6  failures in single record/single failure data
   677  total analysis time at risk, at risk from t =           0
                            earliest observed entry t =           0
                                last observed exit t =         140
```

This combination is equivalent to the expanded data

failtime	load	bearings
100	15	0
100	15	0
100	15	0
140	15	1
140	15	1
97	20	0

and

```
. stset failtime
```

So much for fweights.

pweights are Stata's sampling weights—the inverse of the probability the subject was chosen from the population. pweights are typically integers, but they do not have to be. For instance, you might have

time0	time	died	sex	reps
0	300	1	0	1.50
0	250	0	1	4.50
30	147	1	0	2.25

Here, reps is how many patients each observation represents in the underlying population—perhaps when multiplied by 10. The stset command for this data is

```
. stset time [pw=reps], origin(time time0) failure(died)
```

In terms of variance calculations, the scale of the pweights does not matter. reps in the three observations shown could just as well be 3, 9, and 4.5. Nevertheless, the scale of the pweights is used when you ask for counts. For instance, stsum would report the person-time at risk as

$$(300 - 0) 1.5 + (250 - 0) 4.5 + (147 - 30) 2.25 = 1,838.25$$

for the three observations shown. stsum would count that $1.5 + 2.25 = 3.75$ persons died, and so the incidence rate for these three observations would be $3.75/1,838.25 = .0020$. The incidence rate is thus unaffected by the scale of the weights. Similarly, the coefficients and confidence intervals reported by, for instance, streg, dist(exponential) would be unaffected. The 95% confidence interval for the incidence rate would be $[.0003, .0132]$, regardless of the scale of the weights.

(Note that were these three observations examined unweighted, the incidence rate would be .0030 and the 95% confidence interval would be $[.0007, .0120]$.)

Finally, stset allows you to set iweights, Stata's "importance" weights, but we recommend you do not. This is provided for those who wish to create special effects by manipulating standard formulas. The st commands treat iweights just as they would fweights, although they do not require the weights be integers, and push their way through conventional variance calculations. Thus, results—counts, rates, and variances—depend on the scale of these weights.

Data warnings and errors flagged by stset

When you stset your data, stset runs various checks to verify that what you are setting makes sense. stset refuses to set the data only if, in multiple-record, weighted data, weights are not constant within id. Otherwise, stset merely warns you about any inconsistencies it identifies.

Although stset will set the data, it will mark out records that it cannot understand; for instance,

```
. stset curdate, origin(time addate) failure(code==402) id(patid)

                id:  patid
     failure event:  code == 402
 obs. time interval:  (curdate[_n-1], curdate]
  exit on or before:  failure
     t for analysis:  (time-origin)
             origin:  time addate
```

243	total obs.	
1	event time missing (curdate>=.)	PROBABLE ERROR
4	multiple records at same instant (curdate[_n-1]==curdate)	PROBABLE ERROR
238	obs. remaining, representing	
40	subjects	
15	failures in single failure-per-subject data	
1478	total analysis time at risk, at risk from t =	0
	earliest observed entry t =	0
	last observed exit t =	62

It is your responsibility to ensure that the final result, after exclusions, is correct.

The warnings stset might issue include

```
    ignored because patid missing
    event time missing                           PROBABLE ERROR
    entry time missing                           PROBABLE ERROR
    entry on or after exit (etime>t)             PROBABLE ERROR
    obs. end on or before enter()
    obs. end on or before origin()
    multiple records at same instant (t[_n-1]==t)  PROBABLE ERROR
    overlapping records (t[_n-1]>entry time)     PROBABLE ERROR
    weights invalid                              PROBABLE ERROR
```

stset sets _st = 0 when observations are excluded for whatever reason. Thus, observations with any of the above problems can be found among the _st = 0 observations.

Final example: Stanford heart transplant data

In the examples above, we have shown you how Stata wants survival-time data recorded. To summarize:

1. Each subject's history is represented by one or more observations in the dataset.

2. Each observation documents a span of time. The observation must contain when the span ends (exit time) and may optionally contain when the span begins (entry time). If the entry time is not recorded, it is assumed to be 0, or in multiple-record data, the exit time of the subject's previous observation if there is one. Previous here means previous after the data are temporally ordered on exit times within subject. The physical order of the observations in your dataset does not matter.

3. Each observation documents an outcome associated with the exit time. Unless otherwise specified with failure():, 0 and missing mean censored and nonzero means failed.

4. Each observation contains other variables (called covariates) that are assumed to be constant over the span of time recorded by the observation.

Data rarely arrive in this neatly organized form. For instance, Kalbfleisch and Prentice (2002, 4–5) present heart-transplant survival data from Stanford (Crowley and Hu 1977). These data can be converted into the correct st format in at least two ways. Here we will describe the process using the standard Stata commands. A second, shorter, method using the st commands is described as an example in the stsplit entry.

```
. use http://www.stata-press.com/data/r8/stan2
(Heart transplant data)

. describe

Contains data from http://www.stata-press.com/data/r8/stan2.dta
  obs:           103                          Heart transplant data
 vars:             5                          26 Sep 2002 10:29
 size:         1,442 (99.9% of memory free)

              storage  display    value
variable name   type   format     label      variable label

id              int    %8.0g                 Patient Identifier
died            byte   %8.0g                 Survival Status (1=dead)
stime           float  %8.0g                 Survival Time (Days)
transplant      byte   %8.0g                 Heart Transplant
wait            int    %8.0g                 Waiting Time

Sorted by:
```

The data are from 103 patients selected as transplantation candidates. There is one record on each patient and the important variables, from an st-command perspective, are

id	the patient's id number
transplant	whether the patient received a transplant
wait	when (after acceptance) the patient received the transplant
stime	when (after acceptance) the patient died or was censored
died	the patient's status at stime

To better understand, let us show you two records from this dataset:

```
. list id transplant wait stime died if id==44 | id==16

        id  transp~t  wait  stime  died

33.     44         0     0     40     1
34.     16         1    20     43     1
```

Patient 44 never did receive a new heart; he or she died 40 days after acceptance while still on the waiting list. Patient 16 did receive a new heart—20 days after acceptance—yet died 43 days after acceptance.

Our goal is to turn this into st data that contains the histories of each of these patients. That is, we want records that appear as

```
id        t1     died  posttran
16        20      0       0
16        43      1       1
44        40      1       0
```

or, even more explicitly, as

```
id        t0       t1     died  posttran
16         0       20       0       0
16        20       43       1       1
44         0       40       1       0
```

The new variable `posttran` would be 0 before transplantation and 1 afterwards.

Patient 44 would have one record in this new dataset, recording that he or she died at time 40 and that `posttran` was 0 over the entire interval.

Patient 16, however, would have two records. Patient 16's first record would document the duration $(0, 20]$, during which `posttran` was 0, and 16's second record would document the duration $(20, 43]$, during which `posttran` was 1.

Our goal is to take the first dataset and convert it into the second, which we can then `stset`. We make the transformation using Stata's other data management commands. One way we could do this is

```
. expand 2 if transplant
(69 observations created)

. by id, sort: gen byte posttran = (_n==2)

. by id: gen t1 = stime if _n==_N
(69 missing values generated)

. by id: replace t1 = wait if _n==1 & transplant
(69 real changes made)

. by id: replace died=0 if _n==1 & transplant
(45 real changes made)
```

`expand 2 if transplant` duplicated the observations for patients who had `transplant` $\neq$ 0. Considering our two sample patients, we would now have the following data:

```
id  transp~t     wait     stime     died
44         0        0        40        1
16         1       20        43        1
16         1       20        43        1
```

We would have one observation for patient 44, and two identical observations for patient 16.

We then `by id, sort: gen posttran = (_n==2)`. This resulted in

```
id  transp~t     wait     stime     died  posttran
16         1       20        43        1        0
16         1       20        43        1        1
44         0        0        40        1        0
```

This type of trickiness is discussed in [U] **16.7 Explicit subscripting**. Statements like `_n==2` produce values 1 (meaning true) and 0 (meaning false), so new variable `posttran` will contain 1 or 0 depending on whether `_n` is or is not 2. `_n` is the observation counter, and combined with `by id:`, becomes the observation-within-id counter. Thus, we set `posttran` to 1 on second records, but to 0 on all first records.

Finally, we produce the exit-time variable. Final exit time is just stime, and that is handled by the command by id: gen t1 = stime if _n==_N. _n is the observation-within-id counter and _N is the total number of observations within id, so we just set the last observation on each patient to stime. Now we have

```
id  transp~t      wait      stime     died  posttran        t1
16         1        20         43        1         0         .
16         1        20         43        1         1        43
44         0         0         40        1         0        40
```

All that is left to do is to fill in t1 with the value from wait on the interim records, which is to say, replace t1=wait if it is an interim record.

There are lots of ways we could identify the interim records. In the output above, we did it by

```
. by id: replace t1 = wait if _n==1 & transplant
```

which is to say, if the record is a first record of a person who did receive a transplant. More easily, but with more trickery, we could have just said

```
. replace t1=wait if t1>=.
```

because the only values of t1 left to be filled in are the missing ones. Another alternative would be

```
. by id: replace t1 = wait if _n==1 & _N==2
```

which would identify the first record of two-record pairs. There are lots of alternatives, but they would all produce the same thing:

```
id  transp~t      wait      stime     died  posttran        t1
16         1        20         43        1         0        20
16         1        20         43        1         1        43
44         0         0         40        1         0        40
```

There is one more thing we must do, which is reset died to contain 0 on the interim records:

```
. by id: replace died=0 if _n==1 & transplant
```

The result is

```
id  transp~t      wait      stime     died  posttran        t1
16         1        20         43        0         0        20
16         1        20         43        1         1        43
44         0         0         40        1         0        40
```

We now have the desired result and are ready to stset our data:

```
. stset t1, failure(died) id(id)
                 id:  id
      failure event:  died != 0 & died < .
 obs. time interval:  (t1[_n-1], t1]
  exit on or before:  failure

       172  total obs.
         2  multiple records at same instant              PROBABLE ERROR
           (t1[_n-1]==t1)

       170  obs. remaining, representing
       102  subjects
        74  failures in single failure-per-subject data
     31933  total analysis time at risk, at risk from t =          0
                            earliest observed entry t =          0
                               last observed exit t =       1799
```

Well, something went wrong. Two records were excluded. There is a small enough amount of data here that we could just list the dataset and look for the problem, but let's pretend otherwise. We want to find the records which, within patient, are marked as exiting at the same time:

```
. bys id: gen problem = t1==t1[_n-1]
. sort id died
. list id if problem
```

	id
61.	38

```
. list id transplant wait stime died posttran t1 if id==38
```

	id	transp~t	wait	stime	died	posttran	t1
60.	38	1	5	5	0	0	5
61.	38	1	5	5	1	1	5

There is no typographical error in these data—we checked that variables transplant, wait, and stime contain what the original source published. What those variables say is that patient 38 waited 5 days for a heart transplant, received one on the fifth day, and then died on the fifth day, too.

That makes perfect sense, but not to Stata. Remember that Stata orders events within t as failures, followed by censorings, followed by entries. Reading t1, Stata went for this literal interpretation: patient 38 was censored at time 5 with posttran $= 0$, then, at time 5, patient 38 died and then, at time 5, patient 38 re-entered the data, but this time with posttran $= 1$. That made no sense to Stata.

Stata's sequencing of events may surprise you but, trust us, there are good reasons for it and, really, the ordering convention does not matter. To fix this problem, we just have to put a little time between the implied entry at time 5 and the subsequent death:

```
. replace t1 = 5.1 in 61
(1 real change made)
. list id transplant wait stime died posttran t1 if id==38
```

	id	transp~t	wait	stime	died	posttran	t1
60.	38	1	5	5	0	0	5
61.	38	1	5	5	1	1	5.1

Now the data make sense both to us and to Stata: Up until time 5, the patient had posttran $= 0$, then at time 5 the value of posttran changed to 1, and then at time 5.1, the patient died.

(Continued on next page)

```
. global S_FN

. stset t1, id(id) failure(died)
                 id:  id
      failure event:  died != 0 & died < .
 obs. time interval:  (t1[_n-1], t1]
 exit on or before:  failure
```

```
         172  total obs.
           0  exclusions

         172  obs. remaining, representing
         103  subjects
          75  failures in single failure-per-subject data
     31938.1  total analysis time at risk, at risk from t =         0
                             earliest observed entry t =            0
                              last observed exit t =             1799
```

This dataset is now ready for use with all the other st commands. Here is an illustration:

```
. use http://www.stata-press.com/data/r8/stan3, clear
(Heart transplant data)

. stset, noshow

. stsum, by(posttran)
```

posttran	time at risk	incidence rate	no. of subjects	Survival time		
				25%	50%	75%
0	5936	.0050539	103	36	149	340
1	26002.1	.0017306	69	39	96	979
total	31938.1	.0023483	103	36	100	979

```
. stcox age posttran surgery year

Iteration 0:   log likelihood = -298.31514
Iteration 1:   log likelihood =  -289.7344
Iteration 2:   log likelihood = -289.53498
Iteration 3:   log likelihood = -289.53378
Iteration 4:   log likelihood = -289.53378
Refining estimates:
Iteration 0:   log likelihood = -289.53378

Cox regression -- Breslow method for ties

No. of subjects =         103                  Number of obs   =        172
No. of failures =          75
Time at risk    =     31938.1

                                               LR chi2(4)      =      17.56
Log likelihood  =    -289.53378                Prob > chi2     =     0.0015
```

_t _d	Haz. Ratio	Std. Err.	z	P>|z|	[95% Conf. Interval]	
age	1.030224	.0143201	2.14	0.032	1.002536	1.058677
posttran	.9787243	.3032597	-0.07	0.945	.5332291	1.796416
surgery	.3738278	.163204	-2.25	0.024	.1588759	.8796
year	.8873107	.059808	-1.77	0.076	.7775022	1.012628

References

Cleves, M. A., W. W. Gould, and R. G. Gutierrez. 2002. *An Introduction to Survival Analysis Using Stata.* College Station, TX: Stata Press.

Cleves, M. A. 1999. ssa13: Analysis of multiple failure-time data with Stata. *Stata Technical Bulletin* 49: 30–39. Reprinted in *Stata Technical Bulletin Reprints*, vol. 9, pp. 338–349.

Crowley, J. and M. Hu. 1977. Covariance analysis of heart transplant data. *Journal of the American Statistical Association* 72: 27–36.

Hills, M. and B. L. De Stavola. 2002. *A Short Introduction to Stata for Biostatistics.* London: Timberlake Consultants Press.

Kalbfleisch, J. D. and R. L. Prentice. 2002. *The Statistical Analysis of Failure Time Data.* 2d ed. New York: John Wiley & Sons.

Also See

Complementary: [ST] **snapspan**, [ST] **stdes**

Background: [ST] **st**, [ST] **survival analysis**

Title

> **stsplit** — Split and join time-span records

Syntax

stsplit, syntax one

```
stsplit newvarname [if exp] , {at(numlist) | every(#)} [ trim nopreserve ]
```

stsplit, syntax two

```
stsplit newvarname [if exp] , after(spec) {at(numlist) | every(#)}
```

```
[ trim nopreserve ]
```

where

$$spec = \{\texttt{time} \,|\, \texttt{t} \,|\, \texttt{\_t}\} = \{exp \,|\, \texttt{asis}(exp) \,|\, \texttt{min}(exp)\}$$

stsplit, syntax three

```
stsplit [if exp] , at(failures) [ strata(varlist) riskset(newvar) nopreserve ]
```

Syntax for stjoin

```
stjoin [ , censored(numlist)]
```

stsplit and stjoin are for use with survival-time data; see [ST] **st**. You must stset your dataset using the id() option before using these commands; see [ST] **stset**.

Description

stsplit with option at(*numlist*) or every(#) splits episodes into two or more episodes at the implied time points since being at risk (syntax one) or after a time point specified via after() (syntax two). Each resulting record contains the follow-up on one subject through one time band. Expansion on multiple time scales may be obtained by repeatedly using stsplit. *newvarname* specifies the name of the variable to be created containing the observation's category. It records the time interval to which each new observation belongs. It is bottom coded.

stsplit, at(failures) (syntax three) performs episode splitting at the failure times (per stratum).

stjoin performs the reverse operation, namely joining episodes back together when that can be done without a loss of information.

Options

Options for stsplit, syntax one

at (*numlist*) or every (*#*) are not optional. They specify the analysis times at which the records are to be split.

at (5(5)20) splits records at $t = 5$, $t = 10$, $t = 15$, and $t = 20$.

If at ([...] max) is specified, max is replaced by a suitably large value. For instance, if we wish to split records every five analysis time units from time zero to the largest follow-up time in our data, we could find out what the largest time value is by typing summarize _t and then explicitly typing it into the at() option, or we could just specify at (0(5)max).

every (*#*) is a shorthand for at (*#*(*#*)max), i.e., episodes are split at each positive multiple of *#*.

trim specifies that observations less than the minimum or greater than the maximum value listed in at() are to be excluded from subsequent analysis. Such observations are not dropped from the data; trim merely sets their value of variable _st to 0 so that they will not be used, and yet are still retrievable the next time the dataset is stset.

nopreserve is intended for use by programmers. It speeds the transformation by not saving the original data, which can be restored should things go wrong or if you press *Break*. Programmers often specify this option when they have already preserved the original data. nopreserve changes nothing about the transformation that is made.

Options for stsplit, syntax two

at (*numlist*) or every (*#*) are not optional. They specify the analysis times at which the records are to be split.

at (5(5)20) splits the records at t, corresponding to 5, 10, 15, and 20 analysis time units after the time expression given by *spec* is evaluated.

If at ([...] max) is specified, max is replaced by a suitably large value. For more details on max, see the explanation for at() in the above section.

every (*#*) is shorthand for at (*#*(*#*)max), i.e., episodes are split at each positive multiple of *#*.

after (*spec*) specifies the reference time for at() or every(). Syntax one above can be thought of as corresponding to after (*time of onset of risk*), although you cannot really type this. You could type, however, after (time= birthdate) or after (time= marrydate).

spec has syntax

$$\{ \texttt{time} \,|\, \texttt{t} \,|\, \texttt{\_t} \} = \{ exp \,|\, \texttt{asis}(exp) \,|\, \texttt{min}(exp) \}$$

where

time specifies that the expression is to be evaluated in the same time units as *timevar* in stset timevar, ... This is the default.

t and _t specify that the expression is to be evaluated in units of "analysis time". t and _t are synonyms; it makes no difference whether you specify one or the other.

exp specifies the reference time. In the case of multi-episode data, *exp* should be constant within subject id.

min(*exp*) specifies that in the case of multi-episode data, the minimum of *exp* is taken within id.

asis (*exp*) specifies that in the case of multi-episode data, *exp* is allowed to vary within id.

trim specifies that observations less than the minimum or greater than the maximum value listed in at() are to be excluded from subsequent analysis. Such observations are not dropped from the data; trim merely sets their value of variable _st to 0 so that they are retrievable the next time the dataset is stset.

nopreserve is intended for use by programmers. See the description under syntax one.

Options for stsplit, syntax three

strata(*varlist*) specifies up to 5 strata variables. Observations with equal values of the variables are assumed to be in the same stratum. strata() restrict episode splitting to failures that occur within the stratum, and memory requirements are reduced when strata are specified.

riskset(*newvar*) specifies the name for a new variable recording the unique riskset in which an episode occurs, and missing otherwise.

nopreserve is intended for use by programmers. See the description under syntax one.

Option for stjoin

censored(*numlist*) specifies values of the failure variable, *failvar*, from
stset, failure(*failvar*=...), that indicate "no event" (censoring).

If you are using stjoin to rejoin records after stsplit, you do not need to specify censored(). Just do not forget to drop the variable created by stsplit before typing stjoin. See Example 4 below.

Neither do you need to specify censored() if, when you stset your dataset, you specified failure(*failvar*) and not failure(*failvar*=...). In that case, stjoin knows that *failvar* = 0 and *failvar* = . (missing) correspond to no event. Two records can be joined if they are contiguous and record the same data and the first record has *failvar* = 0 or *failvar* = ., meaning no event at that time.

You may need to specify censored(), and you probably do if, when you stset the dataset, you specified failure(*failvar*=...). If stjoin is to join records, it needs to know what events do not count, which is to say, which events can be discarded. If the only such event is *failvar* = ., then you do not need to specify censored().

Remarks

Remarks are presented under the headings

> What stsplit does and why
> Using stsplit to split at designated times
> Time versus analysis time
> Splitting data on recorded ages
> Example 1: Splitting on age
> Example 2: Splitting on age and time-in-study
> Example 3: Explanatory variables that change with time
> Using stsplit to split at failure times
> Example 4: Splitting on failure times to test the proportional hazards assumption
> Example 5: Cox versus conditional logistic regression
> Example 6: Joining data split with stsplit

What stsplit does and why

stsplit splits records into two or more records based on analysis time or based on a variable that depends on analysis time such as age. The intention is to start with something like

id	_t0	_t	x1	x2	_d
1	0	18	12	11	1

and produce

id	_t0	_t	x1	x2	_d	tcat
1	0	5	12	11	0	0
1	5	10	12	11	0	5
1	10	18	12	11	1	10

or

id	_t0	_t	x1	x2	_d	agecat
1	0	7	12	11	0	30
1	7	17	12	11	0	40
1	17	18	12	11	1	50

The above alternatives record the same underlying data: subject 1 had $x1 = 12$ and $x2 = 11$ during $0 < t \leq 18$ and, at $t = 18$, the subject failed.

The difference between them is that the first alternative breaks out the analysis time periods 0–5, 5–10, and 10–20 (although subject 1 failed before $t = 20$). The second alternative breaks out age 30–40, 40–50, and 50–60. You cannot tell from what is presented above, but at $t = 0$, subject 1 was 33 years old.

In our example, that the subject started with a single record is not important. The original data on the subject might have been

id	_t0	_t	x1	x2	_d
1	0	14	12	11	0
1	14	18	12	9	1

and then we would have obtained

id	_t0	_t	x1	x2	_d	tcat
1	0	5	12	11	0	0
1	5	10	12	11	0	5
1	10	14	12	11	0	10
1	14	18	12	9	1	10

or

id	_t0	_t	x1	x2	_d	agecat
1	0	7	12	11	0	30
1	7	14	12	11	0	40
1	14	17	12	9	0	40
1	17	18	12	9	1	50

In addition, we could just as easily have produced records with analysis time or age recorded in single-year categories. That is, we could start with

id	_t0	_t	x1	x2	_d
1	0	14	12	11	0
1	14	18	12	9	1

and produce

id	_t0	_t	x1	x2	_d	tcat
1	0	1	12	11	0	0
1	1	2	12	11	0	1
1	2	3	12	11	0	2
...						

or

id	_t0	_t	x1	x2	_d	agecat
1	0	1	12	11	0	30
1	1	2	12	11	0	31
1	2	3	12	11	0	32
...						

Moreover, we can even do this splitting on more than one variable. Let's go back and start with

id	_t0	_t	x1	x2	_d
1	0	18	12	11	1

Let's split it into the analysis-time intervals 0–5, 5–10, and 10–20, *and* let's split it into 10-year age intervals 30–40, 40–50, and 50–60. The result would be

id	_t0	_t	x1	x2	_d	tcat	agecat
1	0	5	12	11	0	0	30
1	5	7	12	11	0	5	30
1	7	10	12	11	0	5	40
1	10	17	12	11	0	10	40
1	17	18	12	11	1	10	50

Why would we want to do any of this?

We might want to split on a time-dependent variable such as age if we want to estimate a Cox proportional hazards model and include current age among the regressors (although we could instead use stcox's tvc() option), or if we want to make tables by age groups (see, for instance, [ST] **strate**).

Using stsplit to split at designated times

stsplit's syntax to split at designated times is, ignoring other options,

$$\text{stsplit } \textit{newvarname } \big[\text{if } \textit{exp}\big] \text{, at}(\textit{numlist})$$

$$\text{stsplit } \textit{newvarname } \big[\text{if } \textit{exp}\big] \text{, at}(\textit{numlist}) \text{ after}(\textit{spec})$$

at() specifies the analysis times at which records are to be split. Typing at(5 10 15) splits records at the indicated analysis times, and separates records into the four intervals 0–5, 5–10, 10–15, and 15+.

In the first syntax, the splitting is done on analysis time t. In the second syntax, the splitting is done on 5, 10, and 15 analysis time units after the time given by after(*spec*).

In either case, stsplit also creates *newvarname* containing the interval to which each observation belongs. In this case, *newvarname* would contain 0, 5, 10, and 15; 0 if the observation occurred in the interval 0–5, 5 if the observation occurred in the interval 5–10, and so on. To be precise,

category	precise meaning	*newvarname* value
0–5	$(-\infty, 5]$	0
5–10	$(5, 10]$	5
10–15	$(10, 15]$	10
15+	$(15, \infty)$	15

If any of the at() numbers are negative (which would be allowed only by specifying the after() option and would still be odd), then the first category is labeled one less than the minimum value specified by at().

Consider the data

```
id     yr0    yr1  yrborn   x1  event
 1    1990   1995    1960    5     52
 2    1993   1997    1964    3     47
```

In these data, subjects became at risk in yr0. The failure event of interest is event = 47, so we stset our dataset by typing

```
. stset yr1, id(id) origin(time yr0) failure(event==47)
(output omitted)
```

and that results in

```
id   _t0   _t    yr0    yr1  yrborn   x1  event   _d
 1     0    5   1990   1995    1960    5     52    0
 2     0    4   1993   1997    1964    3     47    1
```

In the jargon of st, variables _t0 and _t record the span of each record in analysis time (t) units. Variables yr0 and yr1 also record the time span, but in time units. Variable _d records 1 if failure and 0 otherwise.

Typing stsplit cat, at(2 4 6 8) would split the records based on analysis time:

```
. stsplit cat, at(2 4 6 8)
(3 observations (episodes) created)
. order id _t0 _t yr0 yr1 yrborn x1 event _d cat
. list id-cat
```

	id	_t0	_t	yr0	yr1	yrborn	x1	event	_d	cat
1.	1	0	2	1990	1992	1960	5	.	0	0
2.	1	2	4	1990	1994	1960	5	.	0	2
3.	1	4	5	1990	1995	1960	5	52	0	4
4.	2	0	2	1993	1995	1964	3	.	0	0
5.	2	2	4	1993	1997	1964	3	47	1	2

The first record, which represented the analysis time span $(0, 5]$, was split into three records: $(0, 2]$, $(2, 4]$, and $(4, 5]$. The yrborn and x1 values from the single record were duplicated in $(0, 2]$, $(2, 4]$, and $(4, 5]$. The original event variable was changed to missing at $t = 2$ and $t = 4$ because we do not know the value of event; all we know is that event is 52 at $t = 5$. The _d variable was correspondingly set to 0 for $t = 2$ and $t = 4$ because we do know, at least, that the subject did not fail.

stsplit also keeps your original time variables up to date in case you want to streset or re-stset your dataset. Note that yr1 was updated, too.

Now let's go back to our original dataset after it was `stset` but before we split it,

```
id  _t0   _t    yr0     yr1   yrborn    x1  event    _d
 1    0    5    1990    1995    1960      5     52     0
 2    0    4    1993    1997    1964      3     47     1
```

and consider splitting on age. Note that, in 1990, subject 1 is age $1990 - \text{yrborn} = 1990 - 1960 = 30$, and subject 2 is 29. If we type

    . stsplit acat, at(30 32 34) after(time=yrborn)

we will split the data according to

```
        age <= 30   (called acat=0)
 30 < age <= 32      (called acat=30)
 32 < age <= 34      (called acat=32)
 34 < age            (called acat=34)
```

The result would be

```
id  _t0   _t    yr0     yr1   yrborn    x1  event    _d   acat
 1    0    2    1990    1992    1960      5     .      0     30
 1    2    4    1990    1994    1960      5     .      0     32
 1    4    5    1990    1995    1960      5     52     0     34
 2    0    1    1993    1994    1964      3     .      0      0
 2    1    3    1993    1996    1964      3     .      0     30
 2    3    4    1993    1997    1964      3     47     1     32
```

The original record on subject 1 corresponding to $(0,5]$ was split into $(0,2]$, $(2,4]$, and $(4,5]$ because those are the t values at which age becomes 32 and 34.

You can `stsplit` the data more than once. Now having these data, were we to type

    . stsplit cat, at(2 4 6 8)

the result would be

```
id  _t0   _t    yr0     yr1   yrborn    x1  event    _d   acat   cat
 1    0    2    1990    1992    1960      5     .      0     30     0
 1    2    4    1990    1994    1960      5     .      0     32     2
 1    4    5    1990    1995    1960      5     52     0     34     4
 2    0    1    1993    1994    1964      3     .      0      0     0
 2    1    2    1993    1995    1964      3     .      0     30     0
 2    2    3    1993    1996    1964      3     .      0     30     2
 2    3    4    1993    1997    1964      3     47     1     32     2
```

Whether we typed

    . stsplit acat, at(30 32 34) after(time=yrborn)

    . stsplit cat, at(2 4 6 8)

or

    . stsplit cat, at(2 4 6 8)

    . stsplit acat, at(30 32 34) after(time=yrborn)

would make no difference.

Time versus analysis time

Be careful using the `after()` option, if, when you `stset` your dataset, you specified `stset`'s `scale()` option. We say be careful, but actually we mean be appreciative, because `stsplit` will do just what you would expect if you did not think too hard.

When splitting on a time-dependent variable, `at()` is still specified in analysis time units, which is to say, the units of time/`scale()`.

For instance, if your original data recorded time as Stata dates, i.e., number of days since 1960,

```
id     date0      date1   birthdate   x1   event
 1   14apr1993  27mar1995  12jul1959    5     52
...
```

and you previously `stset` your dataset by typing

```
. stset date1, id(id) origin(time date0) scale(365.25) ...
```

and you now wanted to split on the age implied by `birthdate`, you would specify the split points in *years* since birth:

```
. stsplit agecat, at(20(5)60) after(time=birth)
```

`at(20(5)60)` means at the ages, measured in years, of 20, 25, ..., 60.

When you `stset` your dataset, you basically told st how you recorded times (you recorded them as dates), and how to map such times (dates) into analysis time. That was implied by what you typed and all of st remembers that. The way to read

```
. stsplit agecat, at(20(5)60) after(time=birth)
```

is, "please, `stsplit`, split the data on 20, 25, ..., 60 *analysis time units* after `birthdate` for each subject".

Splitting data on recorded ages

Be careful. Consider the data

```
id    yr0    yr1   age   x1   event
 1   1980   1996    30    5     52
...
```

When was `age` = 30 recorded—1980 or 1996? Put aside that question because things are about to get worse. Say you `stset` this dataset so that yr0 is the `origin()`

```
id   _t0   _t    yr0    yr1   age   x1   event
 1     0   16   1980   1996    30    5     52
...
```

and then split on analysis time by typing `stsplit cat, at(5(5)20)`. The result would be

```
id   _t0   _t    yr0    yr1   age   x1   event
 1     0    5   1980   1985    30    5      .
 1     5   10   1980   1990    30    5      .
 1    10   15   1980   1995    30    5      .
 1    15   16   1980   1996    30    5     52
```

Regardless of the answer to the question on when age was measured, `age` is most certainly not 30 in the newly created records, although you might argue that age at baseline was 30 and that is what you wanted anyway.

The only truly safe way to deal with ages is to convert them back to birth dates at the outset. In this case we would, early on, type

    . gen bdate = yr1 - age   (if age was measured at yr1)

    or

    . gen bdate = yr0 - age   (if age was measured at yr0)

In fact, stsplit tries to protect you from making age errors. Pretend you did not do as we just recommended. Say age was measured at yr1, and say you type, knowing that stsplit wants a date,

    . stsplit acat, at(20(5)50) after(time= yr1-age)

on these already stsplit data. stsplit will issue the error message "after() should be constant within id". To use the earliest date, you need to type

    . stsplit acat, at(20(5)50) after(time= min(yr1-age))

Nevertheless, be aware that when you stsplit data, if you have recorded ages in your data, and if the records were not already split to control for the range of those ages, then age values, just like all the other variables, are carried forward and no longer reflect the age of the newly created record.

Example 1: Splitting on age

Consider the data from a heart disease and diet survey. The data arose from a study described more fully in Morris, Marr, and Clayton (1977) and analyzed in Clayton and Hills (1993). (Their results differ slightly from ours because the dataset has been updated.)

    . use http://www.stata-press.com/data/r8/diet, clear
    (Diet data with dates)
    . describe
    Contains data from http://www.stata-press.com/data/r8/diet.dta
      obs: 337   Diet data with dates
      vars: 11   3 Nov 2002 10:11
      size: 17,861 (99.8% of memory free)

    ----
                  storage   display   value
    variable name   type   format   label   variable label
    ----
    id   float   %9.0g   Subject identity number
    fail   int   %8.0g   Outcome (CHD = 1 3 13)
    job   int   %8.0g   Occupation
    month   byte   %8.0g   month of survey
    energy   float   %9.0g   Total energy (1000kcals/day)
    height   float   %9.0g   Height (cm)
    weight   float   %9.0g   Weight (kg)
    hienergy   float   %9.0g   Indicator for high energy
    doe   double   %dDmCY   Date of entry
    dox   double   %dDmCY   Date of exit
    dob   double   %dDmCY   Date of birth

    Sorted by: id

In this dataset, the outcome variable, fail, has been coded as 0, 1, 3, 5, 12, 13, 14, and 15. Codes 1, 3, and 13 indicated coronary heart disease (CHD), other nonzero values code other events such as cancer, and 0 is used to mean "no event" at the end of the study.

The variable hienergy is coded 1 if the total energy consumption is more than 2.75 Mcals and 0 otherwise.

We would like to expand the data using age as the time scale with 10-year age bands. We do this by first stsetting the dataset, specifying the date of birth as the origin.

```
. stset dox, failure(fail) origin(time dob) enter(time doe) scale(365.25) id(id)

                id:  id
     failure event:  fail != 0 & fail < .
obs. time interval:  (dox[_n-1], dox]
 enter on or after:  time doe
  exit on or before:  failure
     t for analysis:  (time-origin)/365.25
             origin:  time dob

      337  total obs.
        0  exclusions

      337  obs. remaining, representing
      337  subjects
       80  failures in single failure-per-subject data
 4603.669  total analysis time at risk, at risk from t =         0
                             earliest observed entry t =   30.07529
                              last observed exit t =   69.99863
```

Note that the origin is set to date of birth, making time-since-birth analysis time, and the scale is set to 365.25, so that time-since-birth is measured in years.

Let's list a few records and verify that the analysis-time variables _t0 and _t are indeed recorded as we expect:

```
. list id dob doe dox fail _t0 _t if id==1 | id==34
```

	id	dob	doe	dox	fail	_t0	_t
1.	1	04Jan2015	16Aug2064	01Dec2076	0	49.615332	61.908282
34.	34	12Jun1999	16Apr2059	31Dec2066	3	59.843943	67.55373

We see that patient 1 was 49.6 years old at time of entry into our study and left at age 61.9. Patient 34 entered the study at age 59.8 and exited the study with CHD at age 67.6.

Now we can split the data by age:

```
. stsplit ageband, at(40(10)70)
(418 observations (episodes) created)
```

stsplit added 418 observations to the dataset in memory and generated a new variable, ageband, which identifies each observation's age group.

```
. list id _t0 _t ageband fail height if id==1 | id==34
```

	id	_t0	_t	ageband	fail	height
1.	1	49.615332	50	40	.	175.387
2.	1	50	60	50	.	175.387
3.	1	60	61.908282	60	0	175.387
61.	34	59.843943	60	50	.	177.8
62.	34	60	67.55373	60	3	177.8

Note that the single record for subject with id = 1 has expanded to three records. The first refers to the age band 40–49, coded 40, and the subject spends _t − _t0 = .384668 years in this band. The second refers to the age band 50–59, coded 50, and the subject spends 10 years in this band, and so on. The follow-up in each of the three bands is censored (fail = .). The single record for the subject with id = 34 is expanded to two age bands; the follow-up for the first band was censored (fail = .) and the follow-up for the second band ended in CHD (fail = 3).

The values for variables which do not change with time, such as height, are simply repeated in the new records. This can lead to much larger datasets after expansion. It may be necessary to drop unneeded variables before using stsplit.

Example 2: Splitting on age and time-in-study

To use stsplit to expand the records on two time scales simultaneously, such as age and time-in-study, we can first expand on the age scale as described in Example 1, and then on the time-in-study scale, with the command

```
. stsplit timeband, at(0(5)25) after(time=doe)
(767 observations (episodes) created)
. list id _t0 _t ageband timeband fail if id==1 | id==34
```

	id	_t0	_t	ageband	timeband	fail
1.	1	49.615332	50	40	0	.
2.	1	50	54.615332	50	0	.
3.	1	54.615332	59.615332	50	5	.
4.	1	59.615332	60	50	10	.
5.	1	60	61.908282	60	10	0
111.	34	59.843943	60	50	0	.
112.	34	60	64.843943	60	0	.
113.	34	64.843943	67.55373	60	5	3

By splitting the data using two time scales, the data are partitioned into time cells corresponding to a *Lexis diagram* as described, for example, in Clayton and Hills (1993). Also see Keiding (1998) for an overview of Lexis diagrams. Each new observation created by splitting the data records the time that the individual spent in a Lexis cell. We can obtain the time spent in the cell by calculating the difference $\_t - \_t0$. For example, the subject with id = 1 spent .384668 years (50 − 49.615332) in the cell corresponding to age 40 to 49 and study time 0 to 5, and 4.615332 years (54.615332 − 50) in the cell for age 50 to 59 and study time 0 to 5.

Alternatively, we can do these expansions in reverse order. That is, split first on study time and then on age.

Example 3: Explanatory variables that change with time

In the previous examples, time, in the form of age or time-in-study, is the explanatory variable that is to be studied or controlled for, but in some studies there are other explanatory variables that vary with time. The stsplit command can sometimes be used to expand the records so that in each new record such an explanatory variable is constant over time. For example, in the Stanford heart data (see [ST] **stset**), we would like to split the data and generate the explanatory variable posttran, which takes the value 0 before transplantation and 1 thereafter. The follow-up must therefore be divided into time before transplantation and time after.

We first generate for each observation an entry time and an exit time that preserve the correct follow-up time, but in such a way that the time of transplants is the same for all individuals. By summarizing wait, the time to transplant, we obtain its maximum value of 310. By selecting a value greater than this maximum, say 320, we now generate two new variables:

```
. use http://www.stata-press.com/data/r8/stanford, clear
(Heart transplant data)
. generate enter = 320 - wait
. generate exit = 320 + stime
```

Note that we have created a new artificial time scale where all transplants are coded as being performed at time 320. By defining `enter` and `exit` in this manner, we maintain the correct total follow-up time for each patient. We now `stset` and `stsplit` the data:

```
. stset exit, enter(time enter) failure(died) id(id)
                id:  id
     failure event:  died != 0 & died < .
obs. time interval:  (exit[_n-1], exit]
enter on or after:  time enter
exit on or before:  failure

      103  total obs.
        0  exclusions

      103  obs. remaining, representing
      103  subjects
       75  failures in single failure-per-subject data
  34589.1  total analysis time at risk, at risk from t =          0
                             earliest observed entry t =         10
                                 last observed exit t =       2119
. stsplit posttran, at(0,320)
(69 observations (episodes) created)
. replace posttran=0 if transplant==0
(34 real changes made)
. replace posttran=1 if posttran==320
(69 real changes made)
```

We replaced `posttran` in the last command so that it is now a 0/1 indicator variable. We can now `generate` our follow-up time `t1` as the difference between our analysis-time variables, `list` the data, and `stset` the dataset.

```
. generate  t1 =_t - _t0
. list id enter exit _t0 _t posttran if id==16 | id==44
```

	id	enter	exit	_t0	_t	posttran
41.	44	320	360	320	360	0
42.	16	300	320	300	320	0
43.	16	300	363	320	363	1

```
. stset t1, failure(died) id(id)
                id:  id
     failure event:  died != 0 & died < .
obs. time interval:  (t1[_n-1], t1]
exit on or before:  failure

      172  total obs.
        0  exclusions

      172  obs. remaining, representing
      103  subjects
       75  failures in single failure-per-subject data
  31938.1  total analysis time at risk, at risk from t =          0
                             earliest observed entry t =          0
                                 last observed exit t =       1799
```

Using stsplit to split at failure times

stsplit's syntax to split at failure times is, ignoring other options,

$$\text{stsplit} \left[\text{if } exp\right], \text{at(failures)}$$

This form of episode splitting is useful for Cox regression with time-varying covariates. The usefulness of splitting at the failure times is due to a property of the maximum partial likelihood estimator for a Cox regression model: The likelihood is only evaluated at the times at which failures occur in the data, and the computation only depends on the risk pools at those failure times. Changes in covariates between failure times do not affect estimates for a Cox regression model. Thus, to estimate a model with time-varying covariates, all one has to do is define the values of these time-varying covariates at all failure times at which a subject was at risk (e.g., Collett 1994: ch 7). After splitting at failure times, one defines time-varying covariates by referring to the system variable _t (analysis time), or via the *timevar* variable used to stset the data.

After splitting at failure times, all st commands still work fine and will produce the same results as before splitting. Note that to estimate parametric models with time-varying covariates, it does not suffice to specify covariates at failure times. Stata can estimate "piecewise constant" models; the required data manipulation is facilitated by stsplit, {at() | every()}, stegen, and strepl.

Example 4: Splitting on failure times to test the proportional hazards assumption

Collett (1994, 141) presents data on 26 ovarian cancer patients that underwent two different chemotherapy protocols after a surgical intervention. Here are few of the observations:

```
. use http://www.stata-press.com/data/r8/ocancer
. list in 1/6, separator(0)
```

	patient	time	cens	treat	age	rdisea
1.	1	156	1	1	66	2
2.	2	1040	0	1	38	2
3.	3	59	1	1	72	2
4.	4	421	0	2	53	2
5.	5	329	1	1	43	2
6.	6	769	0	2	59	2

The variable treat indicates the chemotherapy protocol administered, age records the age of the patient at the beginning of the treatment, and rdisea records each patient's residual disease after surgery. After stsetting this dataset, we fitted a Cox proportional hazard regression model on age and treat to ascertain the effect of treatment, controlling for age.

```
. stset time, failure(cens) id(patient)
                id:  patient
     failure event:  cens != 0 & cens < .
obs. time interval:  (time[_n-1], time]
 exit on or before:  failure

      26  total obs.
       0  exclusions

      26  obs. remaining, representing
      26  subjects
      12  failures in single failure-per-subject data
   15588  total analysis time at risk, at risk from t =         0
                            earliest observed entry t =         0
                               last observed exit t =      1227
```

```
. stcox age treat, nolog nohr

        failure _d:  cens
  analysis time _t:  time
               id:  patient

Cox regression -- no ties

No. of subjects =          26             Number of obs   =          26
No. of failures =          12
Time at risk    =       15588
                                          LR chi2(2)      =       15.82
Log likelihood  =   -27.073767            Prob > chi2     =      0.0004
```

_t _d	Coef.	Std. Err.	z	P>|z|	[95% Conf. Interval]
age	.1465698	.0458537	3.20	0.001	.0566982 .2364415
treat	-.7959324	.6329411	-1.26	0.209	-2.036474 .4446094

One way of testing the proportional hazards assumption is to include in the model a term for the interaction between age and time-at-risk. This interaction is a continuously varying covariate. This can be easily done by first splitting the data at the failure times and then generating the interaction term.

```
. stsplit, at(failures)
(12 failure times)
(218 observations (episodes) created)

. generate tage = age * _t

. stcox age treat tage, nolog nohr

        failure _d:  cens
  analysis time _t:  time
               id:  patient

Cox regression -- no ties

No. of subjects =          26             Number of obs   =         244
No. of failures =          12
Time at risk    =       15588
                                          LR chi2(3)      =       16.36
Log likelihood  =   -26.806607            Prob > chi2     =      0.0010
```

_t _d	Coef.	Std. Err.	z	P>|z|	[95% Conf. Interval]
age	.2156499	.1126093	1.92	0.055	-.0050602 .43636
treat	-.6635945	.6695492	-0.99	0.322	-1.975887 .6486978
tage	-.0002031	.0002832	-0.72	0.473	-.0007582 .000352

Other time-varying interactions of age and time-at-risk could be generated. For instance,

```
. generate lntage = age * ln(_t)

. generate dage = age * (_t >= 12)
```

While in most analyses in which we include interactions, we also include main effects, if one includes in a Cox regression a multiplicative interaction between analysis time (or any transformation) and some covariate, one should not include the analysis time as a covariate in stcox; The analysis time is constant within each risk set, and hence its effect is not identified.

❏ Technical Note

If our interest really were in just performing this test of the proportional hazards assumption, we would not have had to use stsplit at all. We could have just typed,

```
. stcox age treat, tvc(age)
```

to have estimated a model including $t*$age, and if we wanted to instead include $\ln(t)*$age or age$*t \geq 12$, we could have typed

```
. stcox age treat, tvc(age) texp(ln(_t))
. cstoc age treat, tvc(age) texp(_t>=12)
```

Still, it is worth understanding how stsplit could be used to obtain the same results for instances when stcox's tvc() and texp() options are not rich enough to handle the desired specification.

❏

Assume that we want to control for rdisea as a stratification variable. If the data are already split at all failure times, one can proceed with

```
. stcox age treat tage, strata(rdisea)
```

If the data are not yet split, and memory is scarce, then we could just split the data at the failure times within the respective stratum. That is, with the original data in memory we could

```
. stset time, failure(cens) id(patient)
. stsplit, at(failures) strata(rdisea)
. generate tage = age * _t
. stcox treat age tage, strata(rdisea)
```

This would save memory by reducing the size of the split dataset.

❏ Technical Note

Of course, the above model could also be obtained by typing

```
. stcox treat age, tvc(age) strata(rdisea)
```

without splitting the data.

❏

Example 5: Cox versus conditional logistic regression

Cox regression with the "exact partial" method of handling ties is tightly related to conditional logistic regression. In fact, one can perform Cox regression via clogit as illustrated in the following example using Stata's cancer data. First, let's estimate the Cox model.

```
. use http://www.stata-press.com/data/r8/cancer, clear
(Patient Survival in Drug Trial)
. generate id =_n
```

(Continued on next page)

```
. stset studytime, failure(died) id(id)

                id:  id
     failure event:  died != 0 & died < .
obs. time interval:  (studytime[_n-1], studytime]
 exit on or before:  failure
```

```
    48  total obs.
     0  exclusions
```

```
    48  obs. remaining, representing
    48  subjects
    31  failures in single failure-per-subject data
   744  total analysis time at risk, at risk from t =          0
                            earliest observed entry t =          0
                             last observed exit t =           39
```

```
. stcox age drug, nolog nohr exactp

        failure _d:  died
  analysis time _t:  studytime
                id:  id
```

Cox regression -- exact partial likelihood

No. of subjects =	48	Number of obs	=	48
No. of failures =	31			
Time at risk =	744			
		LR chi2(2)	=	38.13
Log likelihood =	-73.10556	Prob > chi2	=	0.0000

_t _d	Coef.	Std. Err.	z	P>|z|	[95% Conf. Interval]
age	.1169906	.0374955	3.12	0.002	.0435008 .1904805
drug	-1.664873	.3437487	-4.84	0.000	-2.338608 -.9911376

We will now perform the same analysis using clogit. To do this, we first split the data at failure times, specifying the riskset() option so that a risk set identifier is added to each observation. We then fit the conditional logistic regression, using _d as the outcome variable and the risk set identifier as the grouping variable.

```
. stsplit, at(failures) riskset(RS)
(21 failure times)
(534 observations (episodes) created)

. clogit _d age drug, group(RS) nolog
note: multiple positive outcomes within groups encountered.
```

Conditional (fixed-effects) logistic regression	Number of obs	=	573
	LR chi2(2)	=	38.13
	Prob > chi2	=	0.0000
Log likelihood = -73.10556	Pseudo R2	=	0.2069

_d	Coef.	Std. Err.	z	P>|z|	[95% Conf. Interval]
age	.1169906	.0374955	3.12	0.002	.0435008 .1904805
drug	-1.664873	.3437487	-4.84	0.000	-2.338608 -.9911376

Example 6: Joining data split with stsplit

Let's return to the first example. There we split the diet data into age bands using the following commands:

```
. use http://www.stata-press.com/data/r8/diet, clear
(Diet data with dates)
. stset dox, failure(fail) origin(time dob) enter(time doe) scale(365.25) id(id)
 (output omitted )
. stsplit ageband, at(40(10)70)
(418 observations (episodes) created)
```

We can rejoin the data by typing `stjoin`:

```
. stjoin
(option censored(0) assumed)
(0 obs. eliminated)
```

Nothing happened! `stjoin` will combine records that are contiguous and record the same data. In our case, when we split the data, `stsplit` created the new variable `ageband`, and that variable takes on different values across the split observations. Remember to drop the variable that `stsplit` creates:

```
. drop ageband
. stjoin
(option censored(0) assumed)
(418 obs. eliminated)
```

Acknowledgments

`stsplit` and `stjoin` are extensions of `lexis`, by David Clayton, MRC Biostatistical Research Unit, Cambridge, and Michael Hills, London School of Hygiene and Tropical Medicine (retired) (Clayton and Hills 1995). The original `stsplit` and `stjoin` commands are by Jeroen Weesie, Utrecht University, Netherlands (Weesie 1998a, 1998b). The revised `stsplit` command in this release is also by Jeroen Weesie.

Methods and Formulas

`stsplit` and `stjoin` are implemented as ado-files.

References

Clayton, D. G. and M. Hills. 1993. *Statistical Models in Epidemiology*. Oxford: Oxford University Press.

——. 1995. ssa7: Analysis of follow-up studies. *Stata Technical Bulletin* 27: 19–26. Reprinted in *Stata Technical Bulletin Reprints*, vol. 5, pp. 219–227.

Cleves, M. A., W. W. Gould, and R. G. Gutierrez. 2002. *An Introduction to Survival Analysis Using Stata*. College Station, TX: Stata Press.

Collett, D. 1994 *Modelling Survival Data in Medical Research*. London: Chapman and Hall.

Keiding, N. 1998. Lexis diagrams. In *Encyclopedia of Biostatistics*, ed. P. Armitage and T. Colton, 2844–2850. New York: John Wiley & Sons.

Lexis, W. 1875. *Einleitung in die Theorie der Bevölkerungsstatistik*. Strassburg: Trübner.

Mander, A. 1998. gr31: Graphical representation of follow-up by time bands. *Stata Technical Bulletin* 45: 14–17. Reprinted in *Stata Technical Bulletin Reprints*, vol. 8, pp. 50–53.

Morris, J. N., J. W. Marr, and D. G. Clayton. 1977. Diet and heart: a postscript. *British Medical Journal* 19: 1307–1314.

Weesie, J. 1998a. ssa11: Survival analysis with time-varying covariates. *Stata Technical Bulletin* 41: 25–43. Reprinted in *Stata Technical Bulletin Reprints*, vol. 7, pp. 268–292.

——. 1998b. dm62: Joining episodes in multi-record survival time data. *Stata Technical Bulletin* 45: 5–6. Reprinted in *Stata Technical Bulletin Reprints*, vol. 8, pp. 27–28.

Also See

Complementary:	[ST] **stset**
Background:	[ST] **st**, [ST] **survival analysis**

Title

> **stsum** — Summarize survival-time data

Syntax

> stsum [if *exp*] [in *range*] [, by(*varlist*) <u>nosh</u>ow]

> stsum is for use with survival-time data; see [ST] **st**. You must stset your data before using this command.

> by ... : may be used with stsum; see [R] **by**.

Description

stsum presents summary statistics: time at risk, incidence rate, number of subjects, and the 25th, 50th, and 75th percentiles of survival time.

stsum is appropriate for use with single- or multiple-record, single- or multiple-failure, st data.

Options

by(*varlist*) requests separate summaries for each group along with an overall total. Observations are in the same group if they have equal values of the variables in *varlist*. *varlist* may contain any number of variables, each of which may be string or numeric.

noshow prevents stsum from showing the key st variables. This option is rarely used since most people type stset, show or stset, noshow to reset once and for all whether they want to see these variables mentioned at the top of the output of every st command; see [ST] **stset**.

Remarks

Single-failure data

Here is an example of stsum with single-record survival data:

```
. use http://www.stata-press.com/data/r8/page2
. stset, noshow
. stsum
```

	time at risk	incidence rate	no. of subjects	Survival time 25%	50%	75%
total	9118	.0039482	40	205	232	261

```
. stsum, by(group)
```

group	time at risk	incidence rate	no. of subjects	Survival time 25%	50%	75%
1	4095	.0041514	19	190	216	234
2	5023	.0037826	21	232	233	280
total	9118	.0039482	40	205	232	261

stsum works equally well with multiple-record survival data. Here is a summary of the multiple-record Stanford heart-transplant data introduced in [ST] **stset**:

```
. use http://www.stata-press.com/data/r8/stan3, clear
(Heart transplant data)
. stsum
        failure _d:  died
  analysis time _t:  t1
             id:  id
```

	time at risk	incidence rate	no. of subjects	Survival time		
				25%	50%	75%
total	31938.1	.0023483	103	36	100	979

stsum with the by() option may produce results with multiple-record data that, at first, you may think in error.

```
. stsum, by(posttran) noshow
```

posttran	time at risk	incidence rate	no. of subjects	Survival time		
				25%	50%	75%
0	5936	.0050539	103	36	149	340
1	26002.1	.0017306	69	39	96	979
total	31938.1	.0023483	103	36	100	979

Note that, for the time at risk, $5,936 + 26,002.1 = 31,938.1$ but, for the number of subjects, $103 + 69 \neq 103$. Variable posttran is not constant for the subjects in this dataset:

```
. stset, noshow
. stvary posttran
```

	subjects for whom the variable is				
variable	constant	varying	never missing	always missing	sometimes missing
posttran	34	69	103	0	0

In this dataset, subjects have one or two records. All subjects were eligible for heart transplantation. They have one record if they die or are lost due to censoring before transplantation, and they have two if the operation was performed. In that case, the first record records their survival up to transplantation and the second records their subsequent survival. posttran is 0 in the first record and 1 in the second.

Thus, all 103 subjects have records with posttran = 0 and, when stsum reported results for this group, it summarized the pre-transplantation survival. The incidence of death was .005 and median survival time was 149 days.

The posttran = 1 line of stsum's output summarizes the post-transplantation survival: 69 patients underwent transplantation, incidence of death was .002, and median survival time was 96 days. For these data, this is not 96 more days, but 96 days in total. That is, the clock was not reset on transplantation. Thus, without attributing cause, we can describe the differences between the groups as an increased hazard of death at early times followed by a decreased hazard later.

Multiple-failure data

If you simply type `stsum` with multiple-failure data, be aware that the reported survival time is the survival time to the first failure under the assumption that the hazard function is not indexed by number of failures.

Here we have some multiple-failure data:

```
. use http://www.stata-press.com/data/r8/mfail2, clear
. st
-> stset t, id(id) failure(d) time0(t0) exit(time .) noshow
              id:  id
   failure event:  d != 0 & d < .
obs. time interval:  (t0, t]
 exit on or before:  time .
. stsum
```

	time at risk	incidence rate	no. of subjects	Survival time 25%	50%	75%
total	435444	.0018556	926	201	420	703

To understand this output, let's also obtain output for each failure separately:

```
. stgen nf = nfailures()
. stsum, by(nf)
```

nf	time at risk	incidence rate	no. of subjects	Survival time 25%	50%	75%
0	263746	.0020057	926	196	399	604
1	121890	.0018131	529	252	503	816
2	38807	.0014946	221	415	687	.
3	11001	0	58	.	.	.
total	435444	.0018556	926	201	420	703

The `stgen` command added, for each subject, a variable containing the number of previous failures. For a subject, up to and including the first failure, `nf` is 0. Then `nf` is 1 up to and including the second failure, and then it is 2, and so on; see [ST] **stgen**.

You should have no difficulty interpreting the detailed output. The first line, corresponding to $nf = 0$, states that among those who had experienced no failures yet, the incidence rate for (first) failure is .0020. The distribution of the time to (first) failure is as shown.

Similarly, the second line, corresponding to $nf = 1$, is for those who have already experienced one failure. The incidence rate for (second) failures is .0018 and the distribution of time of (second) failures is as shown.

When we simply typed `stsum`, we obtained the same information shown as the total line of the more detailed output. The total incidence rate is easy to interpret, but what is the "total" survival time distribution? Answer: it is an estimate of the distribution of the time to first failure under the assumption that the hazard function $h(t)$ is the same across failures—that the second failure is no different from the first failure. This is an odd definition of same because the clock t is not reset in $h(t)$. What is the hazard of a failure—any failure—at time t? Answer: $h(t)$.

Another definition of the same would have it that the hazard of a failure is given by $h(\tau)$, where τ is the time since last failure—that the process repeats. These definitions are different unless $h()$ is a constant function of t (τ).

So let's examine these multiple-failure data under the process-replication idea. The key variables in these st data are id, t0, t, and d:

```
. st
-> stset t, id(id) failure(d) time0(t0) exit(time .) noshow
                id:  id
     failure event:  d != 0 & d < .
obs. time interval:  (t0, t]
 exit on or before:  time .
```

Our goal is, for each subject, to reset t0 and t to 0 after every failure event. We are going to have to trick Stata, or at least trick stset. stset will not let us set data where the same subject has multiple records summarizing the overlapping periods. So, the trick is create a new id variable that is different for every id−nf combination (remember, nf is the variable we previously created that records the number of prior failures). Then each of the "new" subjects can have their clock start at time 0:

```
. egen newid = group(id nf)

. sort newid t

. by newid: replace t = t - t0[1]
(808 real changes made)

. by newid: gen newt0 = t0 - t0[1]

. stset t, failure(d) id(newid) time0(newt0)
                id:  newid
     failure event:  d != 0 & d < .
obs. time interval:  (newt0, t]
 exit on or before:  failure

    1734  total obs.
       0  exclusions

    1734  obs. remaining, representing
    1734  subjects
     808  failures in single failure-per-subject data
  435444  total analysis time at risk, at risk from t =          0
                           earliest observed entry t =          0
                             last observed exit t =          797
```

Note that stset no longer thinks we have multiple-failure data. Whereas, with id, subjects had multiple failures, newid gives a unique identity to each id−nf combination. Each "new" subject has at most one failure.

```
. stsum, by(nf)

        failure _d:  d
   analysis time _t:  t
                id:  newid
```

		incidence	no. of		Survival time	
nf	time at risk	rate	subjects	25%	50%	75%
0	263746	.0020057	926	196	399	604
1	121890	.0018131	529	194	384	580
2	38807	.0014946	221	210	444	562
3	11001	0	58	.	.	.
total	435444	.0018556	1734	201	404	602

Compare this table with the one we previously obtained. The incidence rates are the same but the survival times differ because now we measure the times from one failure to the next and previously we measured the time from a fixed point. The time between events in these data appears to be independent of event number.

□ Technical Note

The method shown for converting multiple-failure data to replicated-process single-event failure data is completely general. The generic outline of the conversion process is

```
. stgen nf = nfailures()
. egen newid = group(id nf)
. sort newid t
. by newid: replace t = t - t0[1]
. by newid: gen newt0 = t0 - t0[1]
. stset t, failure(d) id(newid) t0(newt0)
```

where *id*, *t*, *t0*, and *d* are the names of your key survival-time variables.

Once you have done this to your data, you need exercise only one caution. If, in estimating models using stcox, stereg, etc., you wish to obtain robust estimates of variance, you should include the option cluster(*id*).

When you specify the robust option, stcox, stereg, etc., assume that you mean robust cluster(*stset_id_variable*) which, in this case, will be newid. The data, however, are really more clustered than that. Two "subjects" with different newid values may, in fact, be the same real subject. cluster(*id*) is what is appropriate.

□

Saved Results

stsum saves in r():

Scalars

r(p25)	25th percentile	r(risk)	time at risk
r(p50)	50th percentile	r(ir)	incidence rate
r(p75)	75th percentile	r(N_sub)	number of subjects

Methods and Formulas

stsum is implemented as an ado-file.

The 25th, 50th, and 75th percentiles of survival times are obtained from $S(t)$, the Kaplan–Meier product-limit estimate of the survivor function. The 25th percentile, for instance, is obtained as the maximum value of t such that $S(t) \leq .75$.

Also See

Complementary:	[ST] **stdes**, [ST] **stir**, [ST] **sts**, [ST] **stgen**, [ST] **stset**, [ST] **stvary**
Background:	[ST] **st**, [ST] **survival analysis**

Title

sttocc — Convert survival-time data to case–control data

Syntax

sttocc [*varlist*] [, <u>m</u>atch(*matchvarlist*) <u>n</u>umber(*#*) <u>gen</u>erate(*genvarlist*) <u>nodot</u>s]

sttocc is for use with survival-time data; see [ST] **st**. You must stset your data before using this command.

Description

sttocc (survival time to case–control) generates a nested case–control study dataset from a cohort study dataset by sampling controls from the risk sets. For each case, the controls are chosen randomly from those members of the cohort who are at risk at the failure time of the case. Said differently, the resulting case–control sample is matched with respect to analysis time, the time scale used to compute risk sets. The following variables are added to the dataset:

_case	coded 0 for controls, 1 for cases
_set	case–control id; matches which cases and controls belong together
_time	analysis time of the case's failure

The names of these three variables can be changed by specifying the generate() option. *varlist* defines variables which, in addition to those used in the creation of the case–control study, will be retained in the final dataset. If *varlist* is not specified, all variables are carried over into the resulting dataset.

When the resulting dataset is analyzed as a matched case–control study, odds ratios will estimate corresponding rate-ratio parameters in the proportional hazards model for the cohort study.

Randomness in the matching is obtained using Stata's uniform() function. To ensure that the sample truly is random, you should set the random-number seed; see [R] **generate**.

Options

match(*matchvarlist*) specifies additional categorical variables for matching controls to cases. When match() is not specified, cases and controls are matched with respect to time only. If match(*matchvarlist*) is specified, the cases will also be matched by *matchvarlist*.

number(*#*) specifies the number of controls to draw for each case. The default is 1, even though this is not a very sensible choice.

generate(*genvarlist*) specifies variable names for the three new variables, _case, _set, and _time.

nodots requests that dots not be placed on the screen at the beginning of each case–control group selection. By default, dots are displayed to provide entertainment.

Remarks

What follows is paraphrased from Clayton and Hills (1997). Any errors are ours.

Nested case–control studies are an attractive alternative to full Cox regression analysis, particularly when time-varying explanatory variables are involved. They are also attractive when some explanatory variables involve laborious coding. For example, you can create a file with a subset of variables for all subjects in the cohort, generate a nested case–control study, and go on to code the remaining data only for those subjects selected.

In the same way as for Cox regression, the results of the analysis are critically dependent on the choice of analysis time (time scale). The choice of analysis time may be calendar time—so that controls would be chosen from subjects still being followed on the date that the case fails—but other time scales, such as age or time-in-study, may be more appropriate in some studies. Remember that the analysis time set in selecting controls is implicitly included in the model in subsequent analysis.

match() requires that controls also be matched to the case with respect to additional categorical variables such as sex. This produces an analysis closely mirroring stratified Cox regression. If we wanted to match on calendar time and 5-year age bands, we could first type 'stsplit ageband ...' to create the age bands and then specify match(ageband) on the sttocc command. Analyzing the resulting data as a matched case–control study would estimate rate ratios in the underlying cohort which are controlled for calendar time (very finely) and age (less finely). Such analysis could be carried out by Mantel–Haenszel (odds ratio) calculations, for example using mhodds, or by conditional logistic regression using clogit.

When ties occur between entry times, censoring times, and failure times, the following convention is adopted:

$$\text{Entry time} < \text{Failure time} < \text{Censoring time}$$

Thus, censored subjects and subjects entering at the failure time of the case are included in the risk set and are available for selection as controls. Tied failure times are broken at random.

Example: Creating a nested case–control study

Using the diet data introduced in [ST] **stsplit**, we shall illustrate the use of sttocc letting age be analysis time. Hence, controls are chosen from subjects still being followed at the age at which the case fails.

```
. use http://www.stata-press.com/data/r8/diet, clear

. stset dox, failure(fail) enter(time doe) id(id) origin(time dob) scale(365.25)

                id:  id
     failure event:  fail != 0 & fail < .
obs. time interval:  (dox[_n-1], dox]
 enter on or after:  time doe
 exit on or before:  failure
     t for analysis:  (time-origin)/365.25
            origin:  time dob

      337  total obs.
        0  exclusions

      337  obs. remaining, representing
      337  subjects
       80  failures in single failure-per-subject data
 4603.669  total analysis time at risk, at risk from t =         0
                            earliest observed entry t =   30.07529
                              last observed exit t =   69.99863
```

```
. set seed 9123456

. sttocc, match(job) n(5) nodots
              failure _d:  fail
        analysis time _t:  (dox-origin)/365.25
                  origin:  time dob
          enter on or after:  time doe
                      id:  id
            matching for:  job

There were 3 tied times involving failure(s)
 - failures assumed to precede censorings,
 - tied failure times split at random

There are 80 cases
Sampling 5 controls for each case
```

The above two commands create a new dataset in which there are 5 controls per case, matched on job, with the age of the subjects when the case failed recorded in the variable _time. The case indicator is given in _case and the matched set number in _set. Because we did not specify the optional *varlist*, all variables are carried over into the new dataset.

We can verify that the controls were correctly selected:

```
. generate ageentry=(doe-dob)/365.25

. generate ageexit=(dox-dob)/365.25

. sort _set _case id

. list _set id _case _time ageentry ageexit, sepby(_set)
```

	_set	id	_case	_time	ageentry	ageexit
1.	1	65	0	42.57358	40.11225	56.82409
2.	1	66	0	42.57358	40.09309	56.9692
3.	1	74	0	42.57358	37.09788	53.39083
4.	1	83	0	42.57358	30.07529	46.20123
5.	1	86	0	42.57358	38.14921	54.10815
6.	1	90	1	42.57358	31.4141	42.57358
7.	2	235	0	47.8987	44.58043	51.70431
8.	2	250	0	47.8987	43.9562	62.91581
9.	2	292	0	47.8987	46.24504	62.28611
10.	2	313	0	47.8987	41.50582	57.05133
11.	2	334	0	47.8987	47.32923	62.70773
12.	2	196	1	47.8987	45.46475	47.8987
13.	3	7	0	47.964408	45.13895	63.18138
14.	3	58	0	47.964408	45.81793	57.53046
				(output omitted)		
479.	80	171	0	68.596851	61.31417	69.99863
480.	80	108	1	68.596851	55.72074	68.59686

The controls do indeed belong to the appropriate risk set. Note that the controls in each set enter at an age that is less than the age of the case at failure, and exit at an age that is greater than the age of the case at failure. To estimate the effect of high energy, use clogit, just as you would for any matched case–control study:

```
. clogit _case hienergy, group(_set) or
Iteration 0:   log likelihood = -143.32699
Iteration 1:   log likelihood = -143.28861
Iteration 2:   log likelihood = -143.28861
Conditional (fixed-effects) logistic regression    Number of obs   =      480
                                                   LR chi2(1)      =     0.10
                                                   Prob > chi2     =   0.7467
Log likelihood = -143.28861                        Pseudo R2       =   0.0004
```

_case	Odds Ratio	Std. Err.	z	P>|z|	[95% Conf. Interval]	
hienergy	.9247363	.2241581	-0.32	0.747	.5750225	1.487137

Methods and Formulas

sttocc is implemented as an ado-file.

Acknowledgments

The original version of sttocc was written by David Clayton, MRC Biostatistical Research Unit, Cambridge, and Michael Hills, London School of Hygiene and Tropical Medicine (retired).

References

Clayton, D. G. and M. Hills. 1993. *Statistical Models in Epidemiology*. Oxford: Oxford University Press.

——. 1995. ssa7: Analysis of follow-up studies. *Stata Technical Bulletin* 27: 19–26. Reprinted in *Stata Technical Bulletin Reprints*, vol. 5, pp. 219–227.

——. 1997. ssa10: Analysis of follow-up studies with Stata 5.0. *Stata Technical Bulletin* 40: 27–39. Reprinted in *Stata Technical Bulletin Reprints*, vol. 7, pp. 253–268.

Coviello, V. 2001. sbe41: Ordinary case–cohort design and analysis. *Stata Technical Bulletin* 59: 12–18. Reprinted in *Stata Technical Bulletin Reprints*, vol. 10, pp. 121–129.

Langholz, B. and D. C. Thomas. 1990. Nested case–control and case–cohort methods of sampling from a cohort: a critical comparison. *American Journal of Epidemiology* 131: 169–176.

Also See

Complementary:	[ST] **stbase**, [ST] **stdes**, [ST] **stsplit**
Background:	[ST] **st**, [ST] **survival analysis**

Title

> **sttoct** — Convert survival-time data to count-time data

Syntax

> sttoct *newfailvar newcensvar* [*newentvar*] [, by(*varlist*) replace <u>nosh</u>ow]

sttoct is for use with survival-time data; see [ST] **st**. You must stset your data before using this command.

Description

sttoct converts survival-time (st) data to count-time (ct) data; see [ST] **ct**.

There is, currently, absolutely no reason you would want to do this.

Options

by(*varlist*) specifies that counts are to reflect counts by group where the groups are defined by observations with equal values of *varlist*.

replace specifies that it is OK to proceed with the transformation even though the current dataset has not been saved on disk.

noshow prevents sttoct from showing the key st variables. This option is rarely used since most people type stset, show or stset, noshow to reset once and for all whether they want to see these variables mentioned at the top of every st command; see [ST] **stset**.

Remarks

sttoct is a never-used command and is included for completeness. The definition of ct data is found in [ST] **ct**. In the current version of Stata, all you can do with ct data is convert it to st data (which thus provides access to Stata's survival-analysis capabilities to those with ct data), so there is little point in converting st to ct data.

The converted dataset will contain

varlist	from by(*varlist*) if specified
t	the exit time variable previously stset
newfailvar	number of failures at *t*
newcensvar	number of censored at *t* (after failures)
newentvar	if specified, number of entries at *t* (after censorings)

The resulting dataset will be ctset automatically.

There are two forms of the sttoct command:

1. sttoct *failvar censvar*, ...

2. sttoct *failvar censvar entvar*, ...

That is, whether *entvar* is specified makes a difference.

Case 1: entvar not specified

This is possible only if

a. the risk is not recurring;

b. the original st data is single-record data or, if multiple record, all subjects enter at time 0 and have no gaps thereafter; and

c. if by(*varlist*) is specified, subjects do not have changing values of the variables in *varlist* over their histories.

If you do not specify *entvar*, cttost verifies that (a), (b), and (c) are true. If the assumptions are true, cttost converts your data and counts each subject only once. That is, in multiple-record data, all thrashing (censoring followed by immediate re-enter with different covariates) is removed.

Case 2: entvar specified

Any kind of survival-time data can be converted to count-data with an entry variable. You can convert your data in this way whether assumptions (a), (b), and (c) are true or not.

When you specify a third variable, thrashing is not removed even if it could be (even if assumptions (a), (b), and (c) are true).

Methods and Formulas

sttoct is implemented as an ado-file.

Also See

Complementary:	[ST] **ct**, [ST] **st_is**, [ST] **sttocc**
Background:	[ST] **st**, [ST] **survival analysis**

Title

> **stvary** — Report which variables vary over time

Syntax

stvary [*varlist*] [if *exp*] [in *range*] [, <u>nosh</u>ow]

stvary is for use with survival-time data; see [ST] **st**. You must stset your data before using this command.

by ... : may be used with stvary; see [R] **by**.

Description

stvary is for use with multiple-record datasets—datasets for which id() has been stset. It reports whether values of variables within subject vary over time and on their pattern of missing values. While stvary is intended for use with multiple-record st data, it may be used with single-record data as well, but this produces little useful information.

stvary ignores weights even if you have set them. stvary is intended to provide a summary of the variables in the computer or data-based sense of the word.

Options

noshow prevents stvary from showing the key st variables. This option is rarely used since most people type stset, show or stset, noshow to reset once and for all whether they want to see these variables mentioned at the top of the output of every st command; see [ST] **stset**.

Remarks

Consider a multiple-record dataset. A subject's gender, presumably, does not change. His or her age very well might. stvary allows you to verify that values vary in the way that you expect:

```
. use http://www.stata-press.com/data/r8/stan3
(Heart transplant data)

. stvary

       failure _d:  died
 analysis time _t:  t1
             id:  id

            subjects for whom the variable is
                                      never    always  sometimes
   variable | constant   varying    missing   missing   missing
 -----------+-----------------------------------------------------
       year |   103         0         103        0         0
        age |   103         0         103        0         0
      stime |   103         0         103        0         0
    surgery |   103         0         103        0         0
  transplant|   103         0         103        0         0
       wait |   103         0         103        0         0
    posttran |    34        69         103        0         0
```

That 103 values for year are "constant" does not mean year itself is a constant—it means merely that, for each subject, the value of year does not change across the records. Whether the values of year vary across subjects is still an open question.

Now look at the bottom of the table: posttran is constant over time for 34 subjects and varies for the remaining 69.

Below we have another dataset and we will examine just two of the variables:

```
. use http://www.stata-press.com/data/r8/stvaryex
. stvary sex drug
```

	subjects for whom the variable is				
variable	constant	varying	never missing	always missing	sometimes missing
sex	119	1	119	3	1
drug	121	2	123	0	0

Clearly, there are errors in the variable sex; for 119 of the subjects, sex does not change over time, but for one, it does. In addition, we see that we do not know the sex of 3 of the patients, but for another, we sometimes know it and sometimes do not. The latter must be a simple data construction error. As for drug, we see that for two of our patients, the drug administered varied over time. Perhaps this is an error, or perhaps those two patients were treated differently from all the rest.

Saved Results

stvary saves in r():

Scalars

r(cons)	number of subjects for whom variable is constant when not missing
r(varies)	number of subjects for whom nonmissing values vary
r(never)	number of subjects for whom variable is never missing
r(always)	number of subjects for whom variable is always missing
r(miss)	number of subjects for whom variable is sometimes missing

Methods and Formulas

stvary is implemented as an ado-file.

References

Cleves, M. A., W. W. Gould, and R. G. Gutierrez. 2002. *An Introduction to Survival Analysis Using Stata.* College Station, TX: Stata Press.

Also See

Complementary:	[ST] **stdes**, [ST] **stfill**, [ST] **stset**
Background:	[ST] **st**, [ST] **survival analysis**

Subject and author index

This is the subject and author index for the *Stata Survival Analysis & Epidemiological Tables Reference Manual*. Readers interested in topics other than survival analysis and graphics should see the combined subject index at the end of Volume 4 of the *Stata Base Reference Manual*, which indexes the *Stata Base Reference Manual*, the *Stata User's Guide*, the *Stata Cluster Analysis Reference Manual*, the *Stata Cross-Sectional Time-Series Reference Manual*, the *Stata Programming Reference Manual*, the *Stata Survey Data Reference Manual*, the *Stata Time-Series Reference Manual*, and this manual.

Readers interested in non-survival graphics topics should see the index at the end of the *Stata Graphics Reference Manual*.

Semicolons set off the most important entries from the rest. Sometimes no entry will be set off with semicolons; this means all entries are equally important.

A

Aalen–Nelson cumulative hazard, *see* Nelson–Aalen cumulative hazard
accelerated failure-time model, [ST] **streg**
actuarial tables, [ST] **ltable**
adjusted Kaplan–Meier survivor function, [ST] **sts**
AIC, [ST] **streg**
attributable proportion, [ST] **epitab**

B

baseline hazard and survivor functions, [ST] **stcox**, [ST] **stphplot**

C

case–cohort data, [ST] **sttocc**
case–control data, [ST] **epitab**, [ST] **sttocc**
categorical data, [ST] **epitab**
cc and cci commands, [ST] **epitab**
chi-squared,
    test of independence, [ST] **epitab**
cluster sampling, [ST] **stcox**, [ST] **stphplot**, [ST] **streg**
confidence intervals,
    for odds and risk ratios, [ST] **epitab**, [ST] **stci**
contingency tables, [ST] **epitab**
Cornfield confidence intervals, [ST] **epitab**
count-time data, [ST] **ct**; [ST] **ctset**, [ST] **cttost**, [ST] **sttoct**
Cox proportional hazards model, [ST] **stcox**
    test of assumption, [ST] **stcox**, [ST] **stphplot**
Cox–Snell residuals, [ST] **stcox**, [ST] **streg**
crude estimates, [ST] **epitab**
cs and csi commands, [ST] **epitab**
ct, [ST] **ct**, [ST] **ctset**, [ST] **cttost**, [ST] **sttoct**

ctset command, [ST] **ctset**
cttost command, [ST] **cttost**
cumulative hazard function, [ST] **sts**, [ST] **sts generate**, [ST] **sts graph**, [ST] **sts list**
    graph of, [ST] **streg**
cumulative incidence data, [ST] **epitab**

D

data,
    case–cohort, *see* case–cohort data
    case–control, *see* case–control data
    categorical, *see* categorical data
    count-time, *see* count-time data
deviance residual, [ST] **stcox**, [ST] **streg**
discrete survival data, [ST] **discrete**

E

epidemiological tables, [ST] **epitab**
equality tests,
    survivor functions, [ST] **sts test**
etiologic fraction, [ST] **epitab**
exact test, Fisher's, [ST] **epitab**
excess fraction, [ST] **epitab**
expand command, [ST] **epitab**, [ST] **stset**
exponential distribution, [ST] **streg**
exponential survival regression, [ST] **streg**

F

failure tables, [ST] **ltable**
failure-time models, [ST] **stcox**, [ST] **stphplot**, [ST] **streg**
Fisher's exact test, [ST] **epitab**
fourfold tables, [ST] **epitab**
frailty models, [ST] **stcox**, [ST] **streg**; [ST] **discrete**

G

generalized gamma survival regression, [ST] **streg**
Gompertz survival regression, [ST] **streg**
graphs, *also see Graphics Reference Manual*
    adjusted Kaplan–Meier survival curves, [ST] **sts**
    baseline hazard and survival, [ST] **stcox**, [ST] **sts**
    cumulative hazard function, [ST] **streg**
    hazard function, [ST] **streg**
    Kaplan–Meier survival curves, [ST] **sts**
    log-log curves, [ST] **stphplot**
    survivor function, [ST] **streg**, [ST] **sts graph**
Greenwood confidence intervals, [ST] **sts**; [ST] **ltable**

H

hazard function, graph of, [ST] **streg**
heterogeneity tests, [ST] **epitab**
homogeneity tests, [ST] **epitab**